Dipti Srinivasan and Lakhmi C. Jain (Eds.)

Innovations in Multi-Agent Systems and Applications – 1

# Studies in Computational Intelligence, Volume 310

**Editor-in-Chief**
Prof. Janusz Kacprzyk
Systems Research Institute
Polish Academy of Sciences
ul. Newelska 6
01-447 Warsaw
Poland
*E-mail:* kacprzyk@ibspan.waw.pl

Further volumes of this series can be found on our homepage:
springer.com

Vol. 287. Dan Schonfeld, Caifeng Shan, Dacheng Tao, and
Liang Wang (Eds.)
*Video Search and Mining,* 2010
ISBN 978-3-642-12899-8

Vol. 288. I-Hsien Ting, Hui-Ju Wu, Tien-Hwa Ho (Eds.)
*Mining and Analyzing Social Networks,* 2010
ISBN 978-3-642-13421-0

Vol. 289. Anne Håkansson, Ronald Hartung, and
Ngoc Thanh Nguyen (Eds.)
*Agent and Multi-agent Technology for Internet and Enterprise
Systems,* 2010
ISBN 978-3-642-13525-5

Vol. 290. Weiliang Xu and John Bronlund
*Mastication Robots,* 2010
ISBN 978-3-540-93902-3

Vol. 291. Shimon Whiteson
*Adaptive Representations for Reinforcement Learning,* 2010
ISBN 978-3-642-13931-4

Vol. 292. Fabrice Guillet, Gilbert Ritschard,
Henri Briand, Djamel A. Zighed (Eds.)
*Advances in Knowledge Discovery and Management,* 2010
ISBN 978-3-642-00579-4

Vol. 293. Anthony Brabazon, Michael O'Neill, and
Dietmar Maringer (Eds.)
*Natural Computing in Computational Finance,* 2010
ISBN 978-3-642-13949-9

Vol. 294. Manuel F.M. Barros, Jorge M.C. Guilherme, and
Nuno C.G. Horta
*Analog Circuits and Systems Optimization based on
Evolutionary Computation Techniques,* 2010
ISBN 978-3-642-12345-0

Vol. 295. Roger Lee (Ed.)
*Software Engineering, Artificial Intelligence, Networking and
Parallel/Distributed Computing,* 2010
ISBN 978-3-642-13264-3

Vol. 296. Roger Lee (Ed.)
*Software Engineering Research, Management and Applications,*
2010
ISBN 978-3-642-13272-8

Vol. 297. Tania Tronco (Ed.)
*New Network Architectures,* 2010
ISBN 978-3-642-13246-9

Vol. 298. Adam Wierzbicki
*Trust and Fairness in Open, Distributed Systems,* 2010
ISBN 978-3-642-13450-0

Vol. 299. Vassil Sgurev, Mincho Hadjiski, and
Janusz Kacprzyk (Eds.)
*Intelligent Systems: From Theory to Practice,* 2010
ISBN 978-3-642-13427-2

Vol. 300. Baoding Liu (Ed.)
*Uncertainty Theory,* 2010
ISBN 978-3-642-13958-1

Vol. 301. Giuliano Armano, Marco de Gemmis,
Giovanni Semeraro, and Eloisa Vargiu (Eds.)
*Intelligent Information Access,* 2010
ISBN 978-3-642-13999-4

Vol. 302. Bijaya Ketan Panigrahi, Ajith Abraham,
and Swagatam Das (Eds.)
*Computational Intelligence in Power Engineering,* 2010
ISBN 978-3-642-14012-9

Vol. 303. Joachim Diederich, Cengiz Gunay, and
James M. Hogan
*Recruitment Learning,* 2010
ISBN 978-3-642-14027-3

Vol. 304. Anthony Finn and Lakhmi C. Jain (Eds.)
*Innovations in Defence Support Systems,* 2010
ISBN 978-3-642-14083-9

Vol. 305. Stefania Montani and Lakhmi C. Jain (Eds.)
*Successful Case-based Reasoning Applications,* 2010
ISBN 978-3-642-14077-8

Vol. 306. Tru Hoang Cao
*Conceptual Graphs and Fuzzy Logic,* 2010
ISBN 978-3-642-14086-0

Vol. 307. Anupam Shukla, Ritu Tiwari, and Rahul Kala
*Towards Hybrid and Adaptive Computing,* 2010
ISBN 978-3-642-14343-4

Vol. 308. Roger Nkambou, Jacqueline Bourdeau, and
Riichiro Mizoguchi (Eds.)
*Advances in Intelligent Tutoring Systems,* 2010
ISBN 978-3-642-14362-5

Vol. 309. Isabelle Bichindaritz, Lakhmi C. Jain, Sachin Vaidya,
and Ashlesha Jain (Eds.)
*Computational Intelligence in Healthcare 4,* 2010
ISBN 978-3-642-14463-9

Vol. 310. Dipti Srinivasan and Lakhmi C. Jain (Eds.)
*Innovations in Multi-Agent Systems and Applications – 1,* 2010
ISBN 978-3-642-14434-9

Dipti Srinivasan and Lakhmi C. Jain (Eds.)

# Innovations in Multi-Agent Systems and Applications – 1

 Springer

Prof. Dr. Dipti Srinivasan
Department of Electrical & Computer Engineering
National University of Singapore
10 Kent Ridge Crescent,
Singapore 119260

Prof. Lakhmi C. Jain
School of Electrical and Information Engineering
University of South Australia
Adelaide
Mawson Lakes Campus
South Australia
Australia
E-mail: lakhmi.jain@unisa.edu.au

ISBN 978-3-642-14434-9                    e-ISBN 978-3-642-14435-6

DOI 10.1007/978-3-642-14435-6

Studies in Computational Intelligence          ISSN 1860-949X

Library of Congress Control Number: 2010932015

*Typeset & Cover Design:* Scientific Publishing Services Pvt. Ltd., Chennai, India.

Printed on acid-free paper

9 8 7 6 5 4 3 2 1

springer.com

# Preface

In today's world, the increasing requirement for emulating the behavior of real-world applications for achieving effective management and control has necessitated the usage of advanced computational techniques. Computational intelligence-based techniques that combine a variety of problem solvers are becoming increasingly pervasive. The ability of these methods to adapt to the dynamically changing environment and learn in an online manner has increased their usefulness in simulating intelligent behaviors as observed in humans. These intelligent systems are able to handle the stochastic and uncertain nature of the real-world problems. Application domains requiring interaction of people or organizations with different, even possibly conflicting goals and proprietary information handling are growing exponentially. To efficiently handle these types of complex interactions, distributed problem solving systems like multiagent systems have become a necessity. The rapid advancements in network communication technologies have provided the platform for successful implementation of such intelligent agent-based problem solvers.

An agent can be viewed as a self-contained, concurrently executing thread of control that encapsulates some state and communicates with its environment, and possibly other agents via message passing. Agent-based systems offer advantages when independently developed components must interoperate in a heterogenous environment. Such agent-based systems are increasingly being applied in a wide range of areas including telecommunications, Business process modeling, computer games, distributed system control and robot systems.

Multi-agent systems is an area of distributed artificial intelligence that emphasizes the joint behaviors of agents with some degree of autonomy and the complexities arising from their interactions. Multi-agent systems allow the subproblems of a constraint satisfaction problem to be subcontracted to different problem solving agents with their own interest and goals. This increases the speed, creates parallelism and reduces the risk of system collapse on a single point of failure. Different multi-agent architectures, that are tailor-made for a specific application is possible. They are able to synergistically combine the various computational intelligent techniques for attaining a superior performance. This gives an opportunity for bringing the advantages of various techniques into a single framework. It also provides the freedom to model the behavior of the system to be as competitive or coordinating, each having its own advantages and disadvantages.

This book provides an overview of multi-agent systems and several applications that have been developed for real-world problems.

We wish to express our appreciation to the authors and reviewers for their contributions. We acknowledge the excellent editorial assistance by the Springer-Verlag.

Dipti Srinivasan, Singapore
Lakhmi C. Jain, Australia

# Editors

Dr. Dipti Srinivasan obtained her M.Eng. and Ph.D. degrees in Electrical Engineering from the National University of Singapore (NUS) in 1991 and 1994 respectively. She worked at the University of California at Berkeley's Computer Science Division as a postdoctoral researcher from 1994 to 1995. In June 1995, she joined the faculty of the Electrical & Computer Engineering department at the National University of Singapore, where she is an Associate Professor. From 1998-1999 she was a Visiting Faculty in the Department of Electrical & Computer Engineering at the Indian Institute of Science, Bangalore, India.

Her main areas of interest are neural networks, evolutionary computation, intelligent multi-agent systems, and application of computational intelligence techniques to engineering optimization, planning and control problems. Her research has focused on the development of hybrid neural network architectures, learning methods and their practical applications for large complex engineered systems, such as the electric power system and urban transportation systems. These systems are examined in various projects by applying multidisciplinary methods that are able to cope with the problems of imprecision, learning, uncertainty and optimization, when concrete models are constructed.

Dipti Srinivasan is a senior member of IEEE, and a member of IES, Singapore. She has published about 170 technical papers in international journals and conferences. She is an active member of IEEE, and has been contributing at Society level, Section level and Chapter level. She is currently serving as the Vice-Chair of IEEE Women in Computational Intelligence (WCI) committee. She is the current chair of the IEEE Power Engineering Chapter, Singapore, member of the IEEE Computational Intelligence Chapter, and a member of the Intelligent Transport System technical committee.

Dipti Srinivasan is an Associate Editor of IEEE Transactions on Intelligent Transportation Systems, Associate Editor of IEEE Transactions on Neural Networks, Area Editor of International Journal of Uncertainty, Fuzziness and Knowledge-based Systems, and Managing Editor of Neurocomputing journal.

Professor Lakhmi C. Jain is a Director/Founder of the Knowledge-Based Intelligent Engineering Systems (KES) Centre, located in the University of South Australia. He is a fellow of the Institution of Engineers Australia.

His interests focus on the artificial intelligence paradigms and their applications in complex systems, art-science fusion, e-education, e-healthcare, unmanned air vehicles and intelligent agents.

# Table of Contents

# Chapter 1

# An Introduction to Multi-Agent Systems

P.G. Balaji and D. Srinivasan

Department of Electrical and Computer Engineering
National university of Singapore
g0501086@nus.edu.sg, dipti@nus.edu.sg

**Summary.** Multi-agent systems is a subfield of Distributed Artificial Intelligence that has experienced rapid growth because of the flexibility and the intelligence available solve distributed problems. In this chapter, a brief survey of multi-agent systems has been presented. These encompass different attributes such as architecture, communication, coordination strategies, decision making and learning abilities. The goal of this chapter is to provide a quick reference to assist in the design of multi-agent systems and to highlight the merit and demerits of the existing methods.

**Keywords:** Multi-agent systems, Agent architecture, Coordination strategies and MAS communication.

## 1 Distributed Artificial Intelligence (DAI)

Distributed artificial intelligence (DAI) is a subfield of Artificial Intelligence [1] that has gained considerable importance due to its ability to solve complex real-world problems. The primary focus of research in the field of distributed artificial intelligence has included three different areas. These are parallel AI, Distributed problem solving(DPS) and Multi-agent systems (MAS). Parallel AI primarily refers to methodologies used to facilitate classical AI [2-8] techniques when applied to distributed hardware architectures like multiprocessor or cluster based computing. The main aim of parallel AI is to increase the speed of operation and to work on parallel threads in order to arrive at a global solution for a particular problem. Distributed problem solving is similar to parallel AI and considers how a problem can be solved by sharing the resources and knowledge between large number of cooperating modules known as Computing entity. In distributed problem solving, communication between computing entities, quantity of information shared are predetermined and embedded in design of computing entity. Distributed problem solving is rigid due to the embedded strategies and consequently offers little or no flexibility.

In contrast to distributed problem solving, Multi-agent systems (MAS) [9-11] deal with the behaviour of the computing entities available to solve a given problem. In a multi-agent system each computing entity is referred to as an agent. MAS can be defined as a network of individual agents that share knowledge and communicate

D. Srinivasan & L.C. Jain (Eds.): Innovations in MASs and Applications – 1, SCI 310, pp. 1–27.
springerlink.com                © Springer-Verlag Berlin Heidelberg 2010

with each other in order to solve a problem that is beyond the scope of a single agent. It is imperative to understand the characteristics of the individual agent or computing entity to distinguish a simple distributed system and multi-agent system.

The chapter is organized into nine sections. Section 2 gives a brief overview of an agent and its properties. The characteristics of multi-agent system is given in section 3. Section 4 shows the general classification of MAS based on their organization and structure. Section 5 gives details of various mechanisms used in the communication of information between agents. Section 6 gives details of the decision making strategies used in MAS and is  followed by the coordination principles in section 7. Section 8 gives an insight into the learning process in multi-agent systems,. The advantages and their disadvantages are highlighted. These are followed by section 9 which contains some of  the concluding remarks.

## 2  Agent

Researchers in the field of artificial intelligence have so far failed to agree on a consensus definition of the word "Agent". The first and foremost reason for this is due to the universality of the word Agent. It cannot be owned by a single community. Secondly, the agents can be present in many physical forms which vary from robots to computer networks. Thirdly, the application domain of the agent is vastly varied and it is impossible to generalize. Researchers have used terms like softbots (software agents), knowbots (Knowledge agents), taskbots (task-based agents) based on the application domain where the agents were employed [12]. The most agreed definition of agent was that of Russell and Norvig. They define an agent as a flexible autonomous entity capable of perceiving the environment through the sensors connected to it. These act on the environment through actuators. The definition provided does not cover the entire range of characteristics that an agent should possess. It can be distinguished from expert systems and distributed controllers. Some important traits that differentiate an agent from simple controllers are as follows.

*Situatedness:* This refers to the interaction of an agent with the environment through the use of sensors and the resultant actions of the actuators. Environment in which an agent is present is an integral part of its design. All of the inputs are received directly as a consequence of the agents interactions with its environment. The agent's directly act upon the environment through the actuators and do not serve merely as a meta level advisor. This attribute makes differentiates it from expert systems in which the decision making node or entity suggests for changes through a middle agent and does not directly influence the environment.

*Autonomy:* This can be defined as the ability of an agent to choose its actions independently without external intervention by other agents in the network (case of multi-agent systems) or human interference. These attribute protects the internal states of agent from external influence. It isolates the agent from instability caused by external disturbances.

*Inferential capability:* The ability of an agent to work on abstract goal specifications such as deducing an observation by generalizing the information. This could be done by  utilizing relevant contents of available information.

*Responsiveness:* The ability to perceive the condition of environment and respond to it in a timely fashion to take account of any changes in the environment. This latter property is of critical importance in real-time applications.

*Pro-activeness:* Agent must exhibit a good response to opportunistic behaviour. This is in order to enhance actions that are goal-directed rather than just being responsive to a specific change in environment. It must have the ability to adapt to any changes in the dynamic environment.

*Social behaviour:* Even though the agent's decision must be free from external intervention, it must still be able to interact with the external sources when the need arises to achieve a specific goal. It must also be able to share this knowledge and help other agents (MAS) solve a specific problem. That is agents must be able to learn from the experience of other communicating entities which may be human, other agents in the network or statistical controllers.

Some other properties that are associated with the agents include mobility, temporal continuity, collaborative behaviour etc. Based on whether a computing entity is able to satisfy all or a few of the above properties , agents could be further specified as exhibiting either weak or a strong agency.

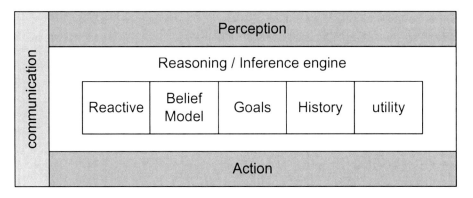

**Fig. 1.** Typical building blocks of an autonomous agent

It is however extremely difficult to characterize agents based only on these properties. It must also be based on the complexity involved in the design, the function which is to be performed and rationality which is exhibited.

## 3  Multi-Agent Systems

A Multi-Agent System (MAS) is an extension of the agent technology where a group of loosely connected autonomous agents act in an environment to achieve a common goal. This is done either by cooperating or competing, sharing or not sharing knowledge with each other.

Multi-agent systems have been widely adopted in many application domains because of the beneficial advantages offered. Some of the benefits available by using MAS technology in large systems [13] are

1. An increase in the speed and efficiency of the operation due to parallel computation and asynchronous operation
2. A graceful degradation of the system when one or more of the agent fail. It thereby increases the reliability and robustness of the system
3. Scalability and flexibility- Agents can be added as and when necessary
4. Reduced cost- This is because individual agents cost much less than a centralized architecture
5. Reusability-Agents have a modular structure and they can be easily replaced in other systems or be upgraded more easily than a monolithic system

Though multi-agent systems have features that are more beneficial than single agent systems, they also present some critical challenges. Some of the challenges are highlighted in the following section.

*Environment:* In a multi-agent system, the action of an agent not only modifies its own environment but also that of its neighbours. This necessitates that each agent must predict the action of the other agents in order to decide the optimal action that would be goal directed. This type of concurrent learning could result in non-stable behaviour and can possibly cause chaos. The problem is further complicated, if the environment is dynamic. Then each agent needs to differentiate between the effects caused due to other agent actions and variations in environment itself.

*Perception:* In a distributed multi-agent system, the agents are scattered all over the environment. Each agent has a limited sensing capability because of the limited range and coverage of the sensors connected to it. This limits the view available to each of the agents in the environment. Therefore decisions based on the partial observations made by each of the agents could be sub-optimal and achieving a global solution by this means becomes intractable.

*Abstraction:* In agent system, it is assumed that an agent knows its entire action space and mapping of the state space to action space could be done by experience. In MAS, every agent does not experience all of the states. To create a map, it must be able to learn from the experience of other agents with similar capabilities or decision making powers. In the case of cooperating agents with similar goals, this can be done easily by creating communication between the agents. In case of competing agents it is not possible to share the information as each of the agents tries to increase its own chance of winning. It is therefore essential to quantify how much of the local information and the capabilities of other agent must be known to create an improved modelling of the environment.

*Conflict resolution:* Conflicts stem from the lack of global view available to each of the agents. An action selected by an agent to modify a specific internal state may be bad for another agent. Under these circumstances, information on the constraints, action preferences and goal priorities of agents must be shared between to improve cooperation. A major problem is knowing when to communicate this information and to which of the agents.

*Inference:* A single agent system inference could be easily drawn by mapping the State Space to the Action Space based on trial and error methods. However in MAS, this is difficult as the environment is being modified by multiple agents that may or may not be interacting with each other. Further, the MAS might consists of heterogeneous agents, that is agents having different goals and capabilities. These may be not cooperating and competing with each other. Identifying a suitable inference mechanism in accordance of the capabilities of each agent is crucial in achieving global optimal solution.

It is not necessary to use multi-agent systems for all applications. Some specific application domains which may require interaction with different people or organizations having conflicting or common goals can be able to utilize the advantages presented by MAS in its design.

## 4   Classification of Multi Agent System

The classification of MAS is a difficult task as it can be done based on several different attributes such as Architecture [14], Learning [15][16][17], Communication [14], Coordination [18]. A general classification encompassing most of these features is shown in figure 2.

### 4.1   Internal Architecture

Based on the internal architecture of the particular individual agents forming the multi-agent system, it may be classified as two types:

1. Homogeneous structure
2. Heterogeneous structure

*a) Homogeneous Structure*
In a homogeneous architecture, all agents forming the multi-agent system have the same internal architecture. Internal architecture refers to the Local Goals, Sensor Capabilities, Internal states, Inference Mechanism and Possible Actions [19]. The difference between the agents is its physical location and the part of the environment where the action is done. Each agent receives an input from different parts of the environment. There may be overlap in the sensor inputs received. In a typical distributed environment, overlap of sensory inputs is rarely present [20].

*b) Heterogeneous Structure*
In a heterogeneous architecture, the agents may differ in ability, structure and functionality [21]. Based on the dynamics of the environment and the location of the particular agent, the actions chosen by agent might differ from the agent located in a different part but it will have the same functionality. Heterogeneous architecture helps to make modelling applications much closer to real-world [22].Each agent can have different local goals that may contradict the objective of other agents. A typical example of this can be seen in the Predator-Prey game [23]. Here both the prey and the predator can be modelled as agents. The objectives of the two agents are likely to be in direct contradiction one to the other.

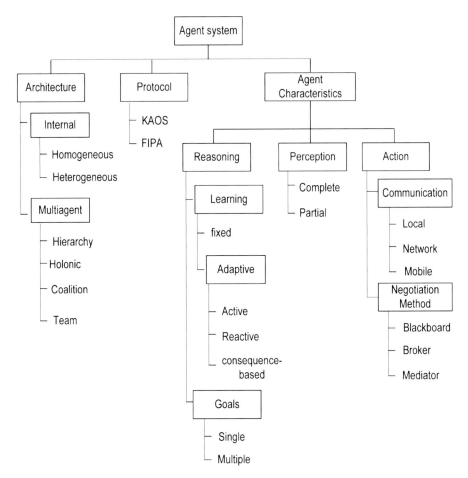

**Fig. 2.** Classification of a multi agent system based on the use of different attributes

### 4.2 Overall Agent Organization

*a) Hierarchical Organization*

Hierarchical Organization [24] is one of the earliest organizational design in multi-agent systems. Hierarchical architecture has been applied to a large number of distributed problems. In the hierarchical agent architecture, the agents are arranged in a typical tree like structure. The agents at different levels on the tree structure have different levels of autonomy. The data from the lower levels of hierarchy typically flow upwards to agents with a higher hierarchy. The Control Signal or Supervisory Signals flow from higher to a lower hierarchy [25]. Figure.3 shows a typical Three Hierarchical Multi-Agent Architecture. The flow of control signals is from a higher to lower priority agents.

According to the distribution of the control between the agents, hierarchical architecture can be further classified as being a simple and uniform hierarchy.

*Simple Hierarchy:* In a simple hierarchy [26], the decision making authority is bestowed using a single agent of highest level of the hierarchy. The problem with a simple hierarchy is that a single point failure of the agent in the highest hierarchy may cause the entire system to fail.

*Uniform Hierarchy:* In a uniform hierarchy, the authority is distributed among the various agents in order to increase the efficiency, fault tolerance having a graceful degradation in case of single and multi point failures. Decisions are made by agents having the appropriate information. These decisions are sent up the hierarchy only where there is a conflict of interest between agents in different hierarchy.

Reference [25], provides an example of a uniform hierarchical multi-agent system applied to a urban traffic signal control problem. The objective is to provide a distributed control of traffic signal timings. This is to reduce the total delay time experienced by vehicles in a road network. A Three Level Hierarchical Multi-Agent System where each intersection is modelled as an agent forming the agents at lowest hierarchy followed by zonal agents which supervise a group of lower level agents and finally a single apex supervisor agent at the top of hierarchy. The agent in the lower level of the hierarchy decides on the optimal green time. This is based on the local information collected at each of the intersections. The agents at the higher level of the hierarchy modify decision of the lower hierarchical agents. From time to time there may be a conflict of interest or the overall delay experienced at a group of intersections increases due to a selected action. Here, the overall control is uniformly distributed among the agents. A disadvantage is that the amount and the type of information which must be transmitted to agents at higher hierarchy. This is a non-trivial problem which increases as the network size increases.

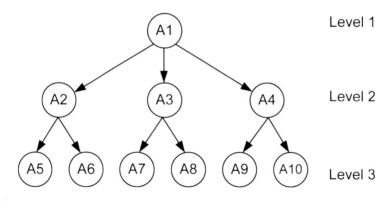

**Fig. 3.** A Hierarchical Agent Architecture

*b) Holonic Agent Organization*
A 'Holon' is a stable and coherent similar or fractal structure that consists of several 'holons' as its sub-structure and is itself a part of a larger framework. The concept of a holon was proposed by Arthur Koestler [27] to explain the social behaviour of biological species. However, the hierarchical structure of the holon and its interactions have been used to model a large organizational behaviours in manufacturing and business domains [28][29].

In a holonic multi-agent system, an agent that appears as a single entity to may be composed of many sub-agents bound together by commitments. The sub-agents are not bound by a hard constraints or by pre-defined rule but through commitments. These refer to the relationships agreed to by all of the participating agents inside the holon.

Each holon appoints or selects a Head Agent that can communicate with the environment or with other agents located in the environment. The selection of the head agent is usually based on the resource availability, communication capability and the internal architecture of each agent. In a homogeneous multi-agent system, the selection can be random and a rotation policy could be employed similar to that used with distributed wireless sensor networks. In the heterogeneous architecture, the selection is based on the capability of the agent. The holons formed may group further in accordance to benefits foreseen in forming a coherent structure. They form Superholons. Figure 4. shows a Superholon formed by grouping two holons. Agents A1 and A4 are the heads of the holons and communicate with agent A7. This is the head of the superholon. The architecture appears to be similar to that of hierarchical organization. However in holonic architecture, cross tree interactions and overlapping or agents forming part of two different holons are allowed.

In recent times, [30] had proved the superiority of the holonic multi-agent organization and how the autonomy of the agents increases when in a holonic group. The abstraction of the internal working of holons provides an increased degree of freedom when selecting the behaviour. A major disadvantage[30-31] is the lack of a model or of a knowledge of the internal architecture of the holons. This makes it difficult for other agents to predict the resulting actions of the holons.

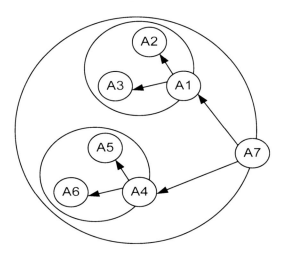

**Fig. 4.** An example of Superholon with Nested Holons resembling the Hierarchical MAS

*c) Coalitions*

In coalition architecture, a group of agents come together for a short time to increase the utility or performance of the individual agents in a group. The coalition ceases to exist when the performance goal is achieved. Figure 5. shows a typical coalition multiagent system. The agents forming the coalition may have either a flat or a hierarchical architecture. Even when using a flat architecture, it is possible to have a leading agent to act as a representative of the coalition group. The overlap of agents among coalition groups is allowed as this increases the common knowledge within the coalition group. It helps to build a belief model. However the use of overlap increases the complexity of computation of the negotiation strategy. Coalition is difficult to maintain in a dynamic environment due to the shift in the performance of group. It may be necessary to regroup agents in order to maximize system performance.

Theoretically, forming a single group consisting of all the agents in the environment will maximize the performance of the system. This is because each agent has access to all of the information and resources necessary to calculate the condition for optimal action.It is impractical to form such a coalition due to restraints on the communication and resources.

The number of coalition groups created must be minimized in order to reduce the cost associated with creating and dissolving a coalition group. The group formation may be pre-defined based on a threshold set for performance measure or alternatively could be evolved online.

In reference [32], a coalition multi-agent architecture for urban traffic signal control was mentioned. Each intersection was modelled as an agent with capability to decide the optimal green time required for that intersection. A distributed neuro-fuzzy inference engine was used to compute the level of cooperation required and the agents which must be grouped together.

The coalition groups reorganize and regroup dynamically with respect to the changing traffic input pattern. The disadvantage is the increased computational complexity involved in creating ensembles or coalition groups. The coalition MAS may have a better short term performance than the other agent architectures [33].

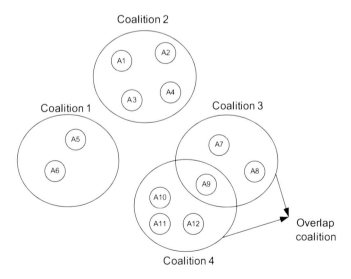

**Fig. 5.** Coalition multi-agent architecture using overlapping groups

*d) Teams*

Team MAS architecture [34] is similar to coalition architecture in design except that the agents in a team work together to increase the overall performance of the group. Rather than each working as individual agents. The interactions of the agents within a team can be quite arbitrary and the goals or the roles assigned to each of the agents can vary with time based on improvements resulting from the team performance. Reference [35] , deals with a team based multi-agent architecture having a  partially observable environment. In other words, teams that cannot communicate with each other has been proposed for the Arthur's bar problem. Each team decides on whether to attend a bar by means of predictions based on the previous behavioural pattern and the crowd level experienced which is the reward or the utility received associated with the specific period of time. Based on the observations made in [35], it can be concluded that a large team size is not beneficial under all conditions. Consequently some compromise must be made between the amount of information, number of agents in the team and the learning capabilities of the agents.

   Large teams offer a better visibility of the environment and larger amount of relevant information. However, learning or incorporating the experiences of individual agents into a single framework team is affected. A smaller team size offers faster learning possibilities but result in sub-optimal performance due to a limited view of  the environment. Tradeoffs between learning and performance need to be made in the selection of the optimal team size. This increases the computational cost much greater than that experienced in coalition multi-agent system architecture. Figure 6. shows a typical team based on architecture with partial view. The team 1 and 3 can see each other but not teams 2 ,4 and vice versa. The internal behaviour of the agents and their roles are arbitrary and vary with teams even in  homogeneous agent structure.

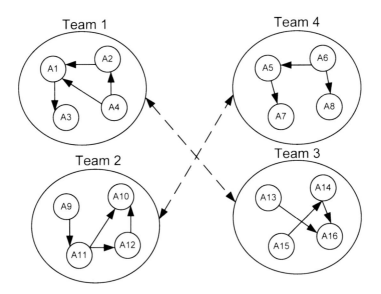

**Fig. 6.** Team based multi-agent architecture with a partial view of the other teams

Variations and constraints on aspects of the four agent architecture mentioned before can produce other architectures such as federations, societies and congregations. Most of these architectures are inspired by behavioural patterns in governments, institutions and large industrial organizations. A detailed description of these architectures, their formation and characteristics may be found in [34].

# 5  Communication in Multi-Agent System

Communication is one of the crucial components in multi-agent systems that needs careful consideration. Unnecessary or redundant intra-agent communication can increase the cost and cause instability. Communication in a multi-agent system can be classified as two types. This is based on the architecture of the agent system and the type of information which is to be communicated between the agents. In [14], the various issues arising in MAS system with homogeneous and heterogeneous architecture has been considered and explained by using a predator/prey and by the use of robotic soccer games. Based on the information communication between the agents [36], MAS can classified as local communication or message passing and network communication or Blackboard. Mobile communication can be categorized into class of local communication.

## 5.1  Local Communication

Local communication has no place to store the information and there is no intermediate communication media present to act as a facilitator. The term message passing is used to emphasize the direct communication between the agents. Figure 7. shows the structure of the message passing communication between agents. In this type of communication, the information flow is bidirectional. It creates a distributed architecture and it reduces the bottleneck caused by failure of central agents. This type of communication has been used in [25] [37] [38].

## 5.2  Blackboards

Another way of exchanging information between agents is through Blackboards [39]. Agent-based blackboards, like federation systems, use grouping to manage the interactions between agents. There are significant differences between the federation agent architecture and the blackboard communication.

In blackboard communication, a group of agents share a data repository which is provided for efficient storage and retrieval of data actively shared between the agents. The repository can hold both the design data as well as the control knowledge that can be accessed by the agents. The type of data that can be accessed by an agent can be controlled through the use of a control shell. This acts as a network interface that notifies the agent when relevant data is available in the repository. The control shell can be programmed to establish different types of coordination among the agents. Neither the agent groups nor the individual agents in the group need to be physically located near the blackboards. It is possible to establish communication between various groups by remote interface communication. The major issue is due to the failure of blackboards. This could render the group of agents useless depending on the specific blackboard. However, it is possible to establish some redundancy and share resources between various blackboards. Figure 8a. shows a single blackboard with the group of agents associated with it. Figure 8b. shows blackboard communication between two different agent groups and also the facilitator agents present in each group.

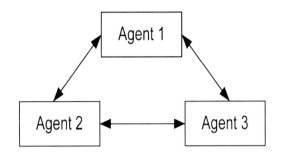

**Fig. 7.** Message Passing Communication between agents

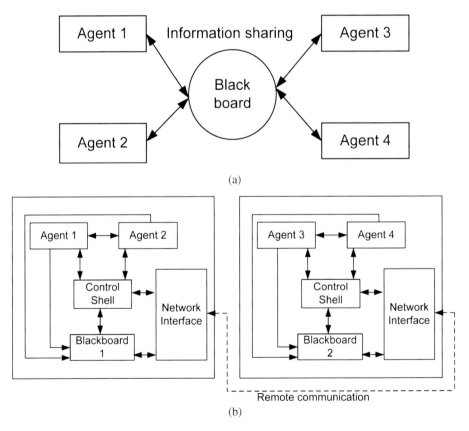

(a)

(b)

**Fig. 8.** (a) Blackboard type communication between agents. (b) Blackboard communication using remote communication between agent groups.

## 5.3  Agent Communication Language

An increase in the number of agents and the heterogeneity of the group necessitates a common framework to help in proper interaction and information sharing. This common framework is provided by the agent communication languages (ACL). The elements that are of prime importance in the design of ACL were highlighted in [40]. They are the availability of the following.

- A common language and interaction format (Protocol) that can be understood by all of the participating agents.
- A shared Ontology where the message communicated has the same meaning in all contexts and follows agent independent semantics.

There are two popular approaches in the design of an agent communication language. They are Procedural approach and Declarative approach. In Procedural approach, the communication between the agents is modelled as a sharing of the procedural directives. Procedural directives shared could be a part of how the specific agents does a specific task or the entire working of the agent itself. Scripting languages are commonly used in the procedural approach. Some of the most common scripting languages employed are JAVA, TCL, Applescript and Telescript. The major disadvantage of the procedural approach is the necessity of providing information on the recipient agent which in most cases is not known or only partially known. In case of making a wrong model assumption, the procedural approach may have a destructive effect on the performance of the agents. The second major concern is the merging of shared procedural scripts into a single large executable relevant script for the agent. Owing to these disadvantages, the procedural approach is not the preferred method for designing agent communication language.

In the declarative approach, the agent communication language is designed and based on the sharing of the declarative statements that specifies definitions, assumptions, assertions, axioms etc. For the proper design of an ACL using a declarative approach, the declarative statements must be sufficiently expressive to encompass the use of a wide-variety of information. This would increase the scope of the agent system and also avoid the necessity of using specialized methods to pass certain functions. The declarative statements must be short and precise as to increase in the length affects the cost of communication and also the probability of information corruption. The declarative statements also needs to be simple enough to avoid the use of a high level language. This means that the use of the language is not required to interpret the message passed. To meet all of the above requirements of the declarative approach based ACL, the ARPA knowledge sharing effort has devised an agent communication language to satisfy all requirements.

The ACL designed consists of three parts [41]: A Vocabulary part, "Inner language" and "Outer language". The Inner language is responsible for the translation of the communication information into a logical form that is understood by all agents. There is still no consensus on a single language and many inner language representations like KIF (Knowledge Interchange Format)[42], KRSL, LOOM are available. The linguistic representation created by these inner languages are concise, unambiguous and context-dependent. The receivers must derive from them the original logical form. For each linguistic representation, ACL maintains a large vocabulary repository. A good ACL maintains this repository open-ended so that modifications and additions can be made to include increased functionality. The repository must also maintain multiple ontology's and its usage will depends on the application domain.

Knowledge Interchange Format [43] is one of the best known inner languages and it is an extension of the First-Order Predicate Calculus (FOPC). Some of the information that can be encoded using KIF are simple data, constraints, negations, disjunctions, rules, meta-level information that aids in the final decision process. It is not possible to use just the KIF for information exchange as much implicit information needs to be embedded. This is so that the receiving agent can interpret it with a minimal knowledge of the sender's structure. This is difficult to achieve as the packet size grows with the increase in embedded information. To overcome this bottleneck, a high level language that utilizes the inner language as its backbone were introduced. These high-level languages make the information exchange independent of the content syntax and ontology. One well known Outer language that satisfies this category is the KQML (Knowledge Query and Manipulation Language) [44]. A typical information exchange between two agents utilizing the KQML and KIF agent communication language is as follows.

*(ask :Content (geolocation lax(?long ?lat))*

*: language KIF*

*:ontology STD_GEO*

*: from location_agent*

*: to  location_server*

*: label Query- "Query identifier")*

*(tell : content "geolocation(lax, [55.33,45.56])"*

*: language standard_prolog*

*: ontology STD_GEO)*

The KQML is conceived as both message format and message handling protocol to facilitate smooth communication between agents. From the above example provided, it can be seen that KQML consists of three layers (Figure 9): A communication layer which indicates the origin and destination agent information and query label or identifier, a message layer that specifies the function to be performed (eg: In the example provided, the first agent asks for the geographic location and the second agent replies to the query), and a content layer to provide the necessary details to perform the specific query.

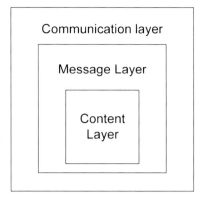

**Fig. 9.** KQML - Layered language structure

In KQML, the communication layer is at a low level and packet oriented. A stream oriented approach is yet to be developed. The communication streams could be built on TCP/IP, RDP, UDP or any other packet communication media. The content layer specifies the language to be employed by the agent. It should be noted that agents can use different languages to communicate with each other and interpretation can be performed locally by higher level languages.

## 6  Decision Making in Multi-Agent System

Multi-agent decision making is different from a simple single agent decision system. The uncertainty associated with the effects of a specific action on the environment and the dynamic variation in the environment as a result of the action of other agents makes multi-agent decision making a difficult task. Usually the decision making in MAS is considered as a methodology to find a joint action or the equilibrium point which maximizes the reward received by every agent participating in decision making process. The decision making in MAS can be typically modelled as a game theoretic method. Strategic game is the most simplest form of decision making process. Here every agent chooses its actions at the beginning of the game and the simultaneous execution of the chosen action by all agents.

A strategic game [45][46] consists of

- a set of players - in multi-agent scenario, the agents are assumed to be the players
- For each player, there is a set of actions
- For each player, the preferences over a set of action profiles

There is a payoff associated with each of the combination of action values for the participating players. The payoff function is assumed to be predefined and known in the case of a simple strategic game. It is also assumed that the actions of all agents are observable and is a common knowledge available to all agents. A solution to a specific game is the prediction of the outcome of the game making the assumption that all participating agents are rational.

The prisoner's dilemma is a best case for demonstrating the application of game theory in decision making involving multiple agents. The prisoner's dilemma problem can be states as

*Two suspects involved in the same crime are interrogated independently. If both the prisoner's confess to the crime, each of them will end up spending three years in prison. If only one of the prisoner confesses to the crime, the confessor is free while the other person will spend four years in prison. If they both do not confess to the crime, each will spend a year in prison.*

This scenario can be represented as a strategic game.

*Players:* Two suspects involved in the crime

*Actions:* Each agent's set of actions is {*Not confess, confess*}

*Preferences:* Ordering of the action profile for agent 1, from best to worst case scenario, is {*confess, Not confess*}, {*Not Confess, Not confess*}, {*Confess, Confess*} and {*Not confess, Confess*}. Similar ordering could be performed by agent 2.

A payoff matrix that represents the particular preferences of the agents needs to be created. Simple payoff matrix can be $u1\{Confess, Not\ confess\} = 3$, $u1\{Not\ confess, Not\ confess\} = 2$. $u1\{Confess, Confess\} = 1$, $u1\{Not\ confess, confess\} = 0$. Similarly the utility or payoff for agent 2 can be represented as $u2\{Not\ confess, confess\} = 3$, $u2\{Not\ confess, Not\ confess\} = 2$, $u2\{confess, Not\ confess\} = 0$ and $u2\{confess, confess\} = 1$.

The reward or payoff received by each agent for choosing a specific joint action can be represented in a matrix format called as payoff matrix table. The problem depicts a scenario where the agents can gain if they cooperate with each other but there is also a possibility to be free if a confession is made. The particular problem can be represented as a payoff matrix as shown in Figure 10. In this case it can be seen that the solution "*Not confess*" is strictly dominated. By strictly dominated solution, it means that a specific action of an agent always increases the payoff of the agent irrespective of the other agents actions.

|           |             | Agent 2 | |
|-----------|-------------|-------------|-------------|
|           |             | *Not Confess* | *Confess* |
| **Agent 1** | *Not confess* | 2,2 | 0,3 |
|           | *Confess* | 3,0 | 1,1 |

**Fig. 10.** Payoff matrix in the Prisoner's Dilemma Problem

However, there can be variations to the prisoner's dilemma problem by introducing an altruistic preference while still calculating the payoff of the actions. Under this circumstance, there is no action strictly dominated by the other.

## 6.1  Nash Equilibrium

To obtain the best solution based on the constructed payoff matrix, the most common method employed is the Nash Equilibrium. Nash Equilibrium [47] can be stated as follows

*A Nash Equilibrium is an action profile a\* with the property that no player i can do better by choosing an action different from a\* of i, given that every other player adheres to a\* of j.*

In the most idealistic conditions, where the components of the game are drawn randomly from a collection of populations or agents, a Nash equilibrium corresponds to a steady state value. In a strategic game, there always exists a Nash equilibrium but it is not necessarily a unique solution. Examining the payoff matrix in Figure. 11 shows that {*confess, confess*} is the Nash equilibrium for the particular problem. The action pair {*confess, confess*} is a Nash equilibrium because given that agent 2 chooses to confess, agent 1 is better off choosing confess than Non confess. By a similar argument with respect to agent 2 it can be concluded that {*confess, confess*} is a Nash Equilibrium. In particular, the incentive to have a  free ride on confession eliminates any possibility of selecting mutually desirable outcome of the type {*Not Confess, Not Confess*}. If the payoff matrix could be modified to add value based on the trust or reward to create altruistic behaviour and feeling of indignation, then the subtle balance that exists shifts and the problem would have a multiple number of Nash equilibrium points as shown in Figure 11.

|  | | Agent 2 | |
|---|---|---|---|
|  |  | *Not Confess* | *Confess* |
|  | *Not confess* | 2,2 | -2,-1 |
| *Agent 1* | | | |
|  | *Confess* | -1,-2 | 1,1 |

**Fig. 11.** Modified Payoff matrix in the Prisoner's Dilemma Problem

In this particular case, there are no dominated solution and multiple Nash equilibrium would exist. To obtain a solution for the type of problem the coordination between the agents is an essential requirement.

## 6.2  The Iterated Elimination Method

The solution to the Prisoner's dilemma problem can also be obtained by using the iterated elimination method [48]. In this method, the strongly dominated actions are iteratively eliminated until no more actions are strictly dominated. The iterated elimination method assumes that all agents are rational and it would not choose a strictly dominated solution. This method is weaker than the Nash equilibrium as it finds the solution by means of a algorithm. Iterated elimination method fails when there are no strictly dominated actions available in the solution space. This limits the applicability of the method in multi-agent scenario where mostly weakly-dominated actions are encountered.

## 7  Coordination in Multi-Agent System

Coordination is the central issue in the design of multi-agent systems. Agents are seldom stand-alone systems and usually involve more than one agent working in parallel to achieve a common goal. When multiple agents are employed to achieve a goal, there is a necessity to coordinate or synchronize the actions to ensure the stability of the system. Coordination between agents increases the chances of attaining a optimal global solution. In [49] major reasons necessitating coordination between the agents were highlighted. The requirements are

- To prevent chaos and anarchy
- To meet global constraints
- To utilize distributed resources, expertise and information
- To prevent conflicts between agents
- To improve the overall efficiency of the system

Coordination can be achieved by applying constraints on the joint action choices of each agent or by utilizing the information collated from neighbouring agents. These are used to compute the equilibrium action point that could effectively enhance the utility of all the participating agents. Applying constraints on the joint actions requires an extensive knowledge of the application domain. This may not be readily available. It necessitates the selection of the proper action taken by each agent. It is based on the equilibrium action computed. However, the payoff matrix necessary to compute the utility value of all action choices might be difficult to determine. The dimension of the payoff matrix grows exponentially with the increasing the number of agents and the available action choices. This may create a bottleneck when computing the optimal solution.

The problem of this dimensional explosion can be solved by dividing the game into a number of sub-games that can be more effectively solved. A simple mechanism which can  reduce the number of action choices is to apply constraints or assign roles to each agent. Once a specific role is assigned, the number of permitted action choices is reduced and are made more computationally feasible. This approach is of particular importance in a distributed coordination mechanism. However, in centralized coordination techniques this is not a major concern as it is possible to construct belief models for all agents. The payoff matrix can be computed centrally and provided to all of the agents as shared resource. The centralized coordination is adopted from the basic client/server model of coordination. Most of the centralized coordination techniques uses blackboards as a way in which to exchange information. Master agent schedules of all the connected agents are required to read and write information from and to the central information repository. Some of the commonly adopted client/server models are KASBAH[50] and MAGMA[51]. The model uses a global blackboard to achieve the required coordination. Disadvantage in using the centralized coordination is that of system disintegration resulting from a single point failure of the repository or of the mediating agent. Further use of the centralized coordination technique is contradictory to the basic assumption of DAI[52][49].

## 7.1  Coordination through Protocol

A classic coordination technique among agents in a distributed architecture is through the communication protocol. Protocol is usually in high level language which specifies the method of coordination between the agents and is a series of task and resource allocation methods. The most widely used protocol is the Contract Net Protocol [53] which facilitates the use of distributed control of cooperative task execution. The protocol specifies what information is to be communicated between the agents and the format of the information of dissemination is handled by the protocol. A low-level communication language such as KIF that can handle the communication streams is assumed to be available. The protocol engages in negotiation between the agents to arrive at an appropriate solution. The negotiation process must adhere to the following characteristics

1. Negotiation is a local process between agents and it involves no central control
2. Two way communication is available between all participating agents exists
3. Each agent makes its evaluation based on its own perception of the environment
4. The final agreement is made through a mutual selection of the action plan

Each agent assumes the role of *Manager* and *Contractor* as necessary. The manager essentially serves to break a larger problem into smaller sub-problems and finds contractors that can perform these functions effectively. A contractor can become a manager and decompose the sub-problem so as to reduce the computational cost and increase efficiency. The manager contracts with a contractor through a process of bidding. In the bidding process, the manager specifies the type of resource required and a description of the problem to be solved. Agents that are free or idle and have the resources required to perform the operation submits a bid indicating their capabilities. The manager agent then evaluates the received bids, chooses an appropriate contractor agent and awards the contract. In case of non-availability of any suitable contracting agent, the manager agent waits for a pre-specified period before rebroadcasting the contract to all agents. The contracting agent may negotiate with the manager agent seeking an access to a particular resource as a condition before accepting the contract.

The FIPA model [54] is the best example of an agent platform that utilizes the contract net protocol to achieve coordination in between the agents. FIPA - Foundation for Intelligent Physical Agents is a model developed to standardize agent technology. The FIPA has its own ACL (Agent Communication Language) that serves as the backbone for the high-level contract net protocol.

Disadvantage of the protocol based coordination is the assumption of the existence of an cooperative agent. The negotiation strategy is passive and does not involve any punitive measures which attempts to force an agent to adopt a specific strategy. Usually a common strategy is achieved through iterative communications where the negotiation parameters are modified progressively to achieve equilibrium. This makes the contract net protocol to be communication intensive.

## 7.2  Coordination via Graphs

Coordination graphs were introduced in [55] to serve as a framework to solve large scale distributed coordination problems. In coordination graphs, each problem is sub-divided into smaller problems that are easier to solve. The main assumption with coordination graphs is that the payoffs can be expressed as a linear combination of the local payoffs of the sub-game. Based on this assumption, algorithm such as variable elimination method can compute the optimal joint actions by iteratively eliminating agents and creating new conditional functions that compute the maximal value the agent can achieve given the action of other agents on which it depends. The joint action choice is only known after the completion of the entire computation process, which scales with the increase in agents and available action choices and is of concern in time critical processes. An alternate method using max-plus which reduces the computation time required was used in [56]. This was to achieve coordination in multi-agent system when applied to urban traffic signal control.

## 7.3  Coordination through Belief Models

In scenarios where time is of critical importance, coordination through protocols fail to succeed when an agent with a specific resource to solve the sub-problem reject the bid. In such scenarios, agents with an internal belief model of the neighbouring agents could solve the problem. The internal belief model could be either evolved by observing the variation in the dynamics of the environment or developed based on heuristic knowledge and domain expertise. When the internal model is evolved, the agent has to be intelligent enough to differentiate between the change in environment due to other agent actions and natural variations occurring in the environment. In [20] [38], a heuristics based belief model has been employed to create coordination between agents and to effectively change the green time. In [57], evolutionary methods combined with neural networks have been employed to dynamically compute the level of cooperation required between the agents. This is based on the internal state model of the agents. The internal state model was updated using reinforcement learning methods.

A disadvantage using the coordination based on belief model for the agents is an incorrect model could cause chaos due to the actions selected.

# 8  Learning in a Multi-Agent System

The learning of an agent can be defined as building or modifying the belief structure based on the knowledge base, input information available and the consequences or actions needed to achieve the local goal [58]. Based on the above definition, agent learning can be classified into three types.

1. Active learning
2. Reactive learning
3. Learning based on consequence

In active and reactive learning, the update of the belief part of the agent is given preference over an optimal action selection strategy as a better belief model could increase the probability of the selection of an appropriate action.

## 8.1  Active Learning

Active learning can be described as a process of analysing the observations to create a belief or internal model of the corresponding situated agent's environment. The active learning process can be performed by using a deductive, inductive or probabilistic reasoning approach.

In the deductive learning approach, the agent draws a deductive inference to explain a particular instance or state-action sequence using its knowledge base. Since the result learned is implied or deduced from the original knowledge base which already exists, the information learnt by each agent is not a new but useful inference. The local goal of each agent could form a part of the knowledge base. In the deductive learning approach, the uncertainty or the inconsistency associated with the agent knowledge is usually disregarded. This makes it not suitable for real-time applications.

In an inductive learning approach, the agent learns from observations of state-action pair. These viewed as the instantiation of some underlying general rules or theories without the aid of a teacher or a reference model. Inductive learning is effective when the environment can be presented in terms of some generalized statements. Well known inductive learning approaches utilize the correlation between the observations and the final action space to create the internal state model of the agent. The functionality of inductive learning may be enhanced if the knowledge base is used as a supplement to infer the state model. The inductive learning approach suffers at the beginning of operation as statistically significant data pertaining to the agent may not be available.

The probabilistic learning approach is based on the assumption that the agent knowledge base or the belief model can be represented as probabilities of occurrence of events. The agent's observation of the environment is used to predict the internal state of the agent. One of the best examples of probabilistic learning is that of the Bayesian theorem. According to the Bayesian theorem, the posterior probability of an event can be determined by the prior probability of that event and the likelihood of its occurrence. The likelihood probability can be calculated based on observations of the samples collected from the environment and prior probability can be updated using the posterior probability calculated in the previous time step of the learning process. In a multi-agent scenario where the action of one agent influences the state of other agent, the application of using the probabilistic learning approach is difficult. This stems from the major knowledge requirement of the joint probability of actions and state space of different agents. With an increase in the number of agents, it is difficult in practice to calculate and infeasible computationally. The other limitation is the limited number of the sample observations available to estimate the correct trajectory.

## 8.2  Reactive Learning

The process of updating a belief without having the actual knowledge of what needs to be learnt or observed is called as Reactive Learning. This method is particularly useful when the underlying model of the agent or the environment is not known clearly and are designated as black box. Reactive learning can be seen in agents which utilize connectionist systems such as neural networks. Neural networks depend on the mechanism which maps the inputs to output data samples using inter-connected computational layers. Learning is done by the adjustment of the synaptic weights between the layers. In [59], reactive multi-agent based feed forward neural networks have been used and its application to the identification of non-linear dynamic system have been demonstrated.  In [60] many other reactive learning methods such as accidental learning, go-with-the-flow, channel multiplexing and a shopping around approach have been discussed. Most of these methods are rarely employed in a real application environment as they depend on the application domain.

## 8.3  Learning Based on Consequences

Learning methods presented in the previous sections were concerned with understanding  the environment based on the belief model update and analysis of patterns in sample observations. This section will deal with the learning methods based on the evaluation of the goodness of selected action. This may be performed by reinforcement learning methods.

Reinforcement learning is a way of programming the agents using reward and punishment scalar signals without specifying how the task is to be achieved. In reinforcement learning, the behaviour of the agent is learnt through trial and error interaction with the dynamic environment without an external teacher or supervisor that knows the right solution. Conventionally, reinforcement learning methods are used when the action space is small and discrete. Recent developments in reinforcement learning techniques have made it possible to use the methods in continuous and large state-action space scenarios too. Examples of applications using reinforcement learning techniques in reactive agents are given in [61] [62].

In reinforcement learning [63], the agent attempts to maximize the discounted scalar reward received from the environment over a finite period of time. To represent this, an agent is represented as a Markov Decision Process.

- A discrete number of states  $s \in S$
- A discrete set of actions  $a \in A$
- State transition probability  $p(s' | s, a)$
- Reward function  $R : SXA \rightarrow \infty$

The reward function can be written as  $R = \sum_{t=0}^{\infty} \gamma^t R(s_t, a_t)$ . The objective is to maximize this function for a given policy function. A policy is a mapping from the state to the action values. The optimal value of a specific state $s$ can be defined as the maximum discounted future reward which is received by following a specific stationary policy and can be written as

$$V*(s) = \max_{\pi} E\left[ \sum_{t=0}^{\infty} \gamma^t R(s_t, a_t) | s_0 = s, a_t = \pi(s) \right] \qquad (1)$$

The expectation operator averages the transition values. In a similar manner the Q value can be written as

$$Q^*(s,a) = \max_{\pi} E\left[\sum_{t=0}^{\infty} \gamma^t R(s_t,a_t) \mid s_0 = s, a_0 = a\right] \tag{2}$$

The optimal policy can then be determined as argmax of the Q-value. To compute the optimal value function and the Q-value, the Bellman equation (3) and (4) is used. The solution to Bellman equation can be obtained by recursively computing the values using dynamic programming techniques. However, the transition probability values are difficult to obtain. Therefore the solution is obtained iteratively by using the temporal difference error between the value of successive iterations as shown in (5) and (6).

$$V^*(s) = \max_{\pi}\left[R(s,a) + \gamma \sum_{s'} p(s' \mid s,a) V^*(s')\right] \tag{3}$$

$$Q^*(s,a) = R(s,a) + \gamma \sum_{s'} p(s' \mid s,a) V(s') \tag{4}$$

$$V(s) \leftarrow V(s) + \alpha\left[r + \gamma V(s') - V(s)\right] \tag{5}$$

$$Q(s_t,a_t) \leftarrow Q(s,a) + \alpha\left[r + \gamma \max_{a'} Q(s',a') - Q(s,a)\right] \tag{6}$$

The solution to (6) is referred to as the q-learning method. The Q-value computed for each state action pair is stored in Q-map and is used to update the Q-values. Based on the Q-values, the appropriate actions are selected. The major disadvantage is that the exploration and exploitation trade-off must be determined. To build an efficient Q-map, it is essential to compute the Q-values corresponding to all the state-action pair. The convergence is guaranteed if all the state-action pairs have been visited infinite number of times (theoretical).

In single agent RL, the convergence and methodologies are well defined and proven. In a distributed MAS, the reinforcement learning method faces the problem of combinatorial explosion in the state action pairs. Another major concern is the information must be passed between the agents for effective learning. In [64], distributed value function based on RL has been described. A complete survey of reinforcement learning can be found in [65].

## 9  Conclusion

In this chapter, a brief survey of the existing architectures, communication requirements, coordination mechanism, decision making and learning in multi-agent systems have been presented. Rapid advances made in multi-agent system have created a scope for applying it to several applications that require distributed intelligent computing. The communication mechanism, coordination strategies and decision making are still in development stages and may offer a greater scope for further improvement and innovation. The increase in computing power has further increased the scope of MAS employability in real-world applications. These include many applications such as grid computing, robotic soccer, urban traffic signal control.

# References

[1] Jennings, N.R., Sycara, K., Woolridge, M.: A roadmap of agent research and development. Autonomous Agents and Multi-Agent Systems Journal 1(1), 7–38 (1998)

[2] Jain, L.C., Jain, R.K. (eds.): Hybrid Intelligent Engineering Systems. World Scientific Publishing Company, Singapore (1997)

[3] Mumford, C., Jain, L.C. (eds.): Computational intelligence: Collaboration, Fusion and Emergence. Springer, Heidelberg (2009)

[4] Fulcher, J., Jain, L.C. (eds.): Computational Intelligence: A Compendium. Springer, Heidelberg (2008)

[5] Jain, L.C., Sato, M., Virvou, M., Tsihrintzis, G., Balas, V., Abeynayake, C. (eds.): Computational Intelligence Paradigms. Innovative Applications, vol. 1. Springer, Heidelberg (2008)

[6] Jain, L.C., De Wilde, P. (eds.): Practical Applications of Computational Intelligence Techniques. Kluwer Academic Publishers, USA (2001)

[7] Jain, L.C., Martin, N.M. (eds.): Fusion of Neural Networks, Fuzzy Logic and Evolutionary Computing and their Applications. CRC Press, USA (1999)

[8] Tedorescu, H.N., Kandel, A., Jain, L.C. (eds.): Fuzzy and Neuro-fuzzy Systems in Medicine. CRC Press, USA (1998)

[9] Khosla, R., Ichalkaranje, N., Jain, L.C. (eds.): Design of Intelligent Multi-Agent Systems. Springer, Germany (2005)

[10] Jain, L.C., Chen, Z., Ichalkaranje, N. (eds.): Intelligent Agents and Their Applications. Springer, Germany (2002)

[11] Resconi, G., Jain, L.C. (eds.): Intelligent Agents: Theory and Applications. Springer, Germany (2004)

[12] Nwana, H.: Software agents: An overview. Knowledge and Engineering Review 11(3) (1996)

[13] Vlassis, N.: A Concise introduction to multiagent systems and distributed artifical intelligence. Synthesis Lectures On Artificial Intelligence And Machine Learning, 1st edn. (2007)

[14] Stone, P., Veloso, M.: Multiagent systems: A survey from the machine learning perspective. Autonomous Robotics 8(3), 1–56 (2000)

[15] Ren, Z., Anumba, C.J.: Learning in multi-agent systems: a case study of construction claim negotiation. Advanced Engineering Informatics 16(4), 265–275 (2002)

[16] Goldman, C.V.: Learning in multi-agent systems. In: Proceedings of the Thirteenth National Conference on Artificial Intelligence and the Eighth Innovative Applications of Artificial Intelligence Conference, vol. 2, p. 1363 (1996)

[17] Alonso, E., D'Inverno, M., Kudenko, D., Luck, M., Noble, J.: Learning in multi-agent systems. The Knowledge Engineering Review 13(3), 277–284 (2001)

[18] Bergenti, F., Ricci, A.: Three approaches to the coordination of multiagent systems. In: Proceedings of the 2002 ACM Symposium on Applied Computing, March 2002, pp. 367–373 (2002)

[19]  Hsia, T.C., Soderstrand, M.: Development of a micro robot system for playing soccer games. In: Proceedings of the Micro-Robot World Cup Soccer Tournament, pp. 149–152 (1996)

[20]  Balaji, P.G., Srinivasan, D.: Distributed multi-agent type-2 fuzzy architecture for urban traffic signal control. In: IEEE International Conference on Fuzzy Systems, pp. 1624–1632 (2009)

[21]  Parker, L.E.: Heterogeneous multi-robot cooperation. PhD Thesis, Massachusetts Institute of Technology (1994)

[22]  Parker, L.E.: Life-long adaptation in hertergeneous multi-robot teams: response to continual variation in robot performance. Autonomous Robots 8(3) (2000)

[23]  Drezewski, R., Siwik, L.: Co-evolutionary multi-agent system with predator-prey mechanism for multi-objective optimization. In: Beliczynski, B., Dzielinski, A., Iwanowski, M., Ribeiro, B. (eds.) ICANNGA 2007. LNCS, vol. 4431, pp. 67–76. Springer, Heidelberg (2007)

[24]  Damba, A., Watanabe, S.: Hierarchical control in a multiagent system. International Journal of innovative computing, Information & Control 4(2), 3091–3100 (2008)

[25]  Choy, M.C., Srinivasan, D., Cheu, R.L.: Neural Networks for Continuous Online Learning and Control. IEEE Transactions on Neural Networks 17(6), 1511–1531 (2006)

[26]  Balaji, P.G., Sachdeva, G., Srinivasan, D., Tham, C.K.: Multi-agent system based urban traffic management. In: Proceedings of IEEE Congress on Evolutionary Computation, pp. 1740–1747 (2007)

[27]  Koestler, A.: The ghost in the machine. Hutchinson Publication Group, London (1967)

[28]  Leitao, P., Valckenaers, P., Adam, E.: Self-adaptation for robustness and cooperation in holonic multi-agent systems. In: Hameurlain, A., Küng, J., Wagner, R. (eds.) Transactions on Large-Scale Data- and Knowledge-Centered Systems I. LNCS, vol. 5740, pp. 267–288. Springer, Heidelberg (2009)

[29]  Yadgar, O., Kraus, S., Oritz, C.: Scaling up distributed sensor networks: Cooperative large scale mobile agent organizations. In: Distributed Sensor Networks: A Multiagent Perspective, pp. 185–218 (2003)

[30]  Schillo, M., Fischer, K.: A taxanomy of autonomy in multiagent organisation. In: Nickles, M., Rovatsos, M., Weiss, G. (eds.) AUTONOMY 2003. LNCS (LNAI), vol. 2969, pp. 68–82. Springer, Heidelberg (2004)

[31]  Bongearts, L.: Integration of scheduling and control in holonic manufacturing systems. PhD Thesis. Katholieke Universiteit Leuven, Belgium (1998)

[32]  Srinivasan, D., Choy, M.C.: Distributed problem solving using evolutionary learning in multi-agent systems. Studies in Computational Intelligence, vol. 66, pp. 211–227 (2007)

[33]  Van De Vijsel, M., Anderson, J.: Coalition formation in multi-agent systems under real-world conditions. AAAI Workshop – Technical Report, v WS-04-06, pp. 54–60 (2004)

[34]  Horling, B., Lesser, V.: A survey of multi-agent organizational paradigms. Knowledge Engineering Review 19(4), 281–316 (2004)

[35]  Agogino, A.K., Tumer, K.: Team formation in partially observable multi-agent systems. NASA Ames Research Center, NTIS (2004)

[36]  Budianto: An overview and survey on multi agent system, in Seminar Nasional Soft Computing, Intelligent Systems and Information Technology, SIIT 2005 (2005)

[37]  Choy, M.C., Srinivasan, D., Cheu, R.L.: Cooperative, hybrid agent architecture for real-time traffic signal control. IEEE Trans. On Systems, Man and Cybernetics-Part A: Systems and Humans 33(5), 597–607 (2003)

[38] Balaji, P.G., Srinivasan, D., Tham, C.K.: Coordination in distributed multi-agent system using type-2 fuzzy decision systems. In: Proceedings of IEEE International Conference on Fuzzy Systems, pp. 2291–2298 (2008)

[39] Lander, S.E.: Issues in multiagent design systems. IEEE Expert 12(2), 18–26 (1997)

[40] Flores-Mendez, R.A.: Towards a standardization of multi-agent system framework. Crossroads 5(4), 18–24 (1999)

[41] Cengeloglu, Y.: A Framework for dynamic knowledge exchange among intelligent agents. In: AAAI Symposium, Control of the Physial World by Intelligent Agents (1994)

[42] Genesereth, M., Fikes, R.: Knowledge interchange format, Version 3.0 Reference manual. Technical report, Computer Science Department, Stanford University, USA (1992)

[43] Ginsberg, M.: The Knowledge interchange format: The KIF of death. AAAI Magazine 12(3), 57–63 (Fall 1991)

[44] Finin, T., Labrou, Y., Mayfield, J.: KQML as an agent communication language. In: Software Agents, pp. 291–316. AAAI Press, Menlo Park (1997)

[45] Gibbons, R.: Game theory for Applied Economists. Princeton University Press, Princeton

[46] Greenwald, A.: The search for equilibrium in markov games. Synthesis Lectures on Artificial Intelligence and Machine Learning (2007)

[47] Nash, J.F.: Equilibrium points in n-person games. Proceedings of the National Academy of Sciences 36, 48–49 (1950)

[48] Gibbons, R.: An introduction to applicable game theory. The Journal of Economic Perspectives 11(1), 127–149 (1997)

[49] Nwana, H.S., Lee, L., Jennings, N.R.: Co-ordination in multi-agent systems. In: Nwana, H.S., Azarmi, N. (eds.) Software Agents and Soft Computing: Towards Enhancing Machine Intelligence. LNCS, vol. 1198, pp. 42–58. Springer, Heidelberg (1997)

[50] Chavez, A., Maes, P.K.: An agent marketplace for buying and selling goods. In: Proceedings of the First International Conference on the Practical Application of Intelligent Agents and Multi-Agent Technology, PAAM 1996 (1996)

[51] Tsvetovatyy, M., Gini, M., Mobhasher, B., Wieckowski, Z.: MAGMA: An agent-based virtual marketplace for electronic commerce. Applied Artificial Intelligence 11(6), 501–542 (1997)

[52] Huhns, M., Singh, M.P.: CKBS-94 Tutorial: Distributed artificial intelligence for information systems, DAKE Center, University of Keele (1994)

[53] Smith, R.G.: The contract net protocol: High-level communication and control in a distributed problem solver. IEEE Transactions on Computers 29(12), 1104–1113 (1980)

[54] O'Brien, P.D., Nicol, R.C.: FIPA-Towards a standard for software agents. BT Technology Journal 16(3), 51–59 (1998)

[55] Guestrin, C., Koller, D., Parr, R.: Multiagent planning with factored MDPs. In: Advances in Neural Information Processing Systems, vol. 14. MIT Press, Cambridge (2002)

[56] Kuywe, L., Whiteson, S., Bakker, B., Vlassis, N.: Multiagent reinforcement learning for urban traffic control usin coordination graphs. In: Daelemans, W., Goethals, B., Morik, K. (eds.) ECML PKDD 2008, Part I. LNCS (LNAI), vol. 5211, pp. 656–671. Springer, Heidelberg (2008)

[57] Srinivasan, D., Choy, M.C.: Neural networks for real-time traffic signal control. IEEE Trans. Intelligent Transportation Systems 7(3), 261–272 (2006)

[58] Lhotska, L.: Learning in multi-agent systems: Theoretical issues. In: Moreno-Díaz, R., Pichler, F. (eds.) EUROCAST 1997. LNCS, vol. 1333, pp. 394–405. Springer, Heidelberg (1997)

[59] Gomez, F., Schmidhuber, J., Miikkulainen, R.: Efficient non-linear control through neuro evolution. In: Fürnkranz, J., Scheffer, T., Spiliopoulou, M. (eds.) ECML 2006. LNCS (LNAI), vol. 4212, pp. 654–662. Springer, Heidelberg (2006)

[60] Jiu, J.: Autonomous Agents and Multi-agent Systems. World Scientific Publication, Singapore

[61] Vassiliades, V., Cleanthous, A., Christodoulou, C.: Multiagent reinforcement learning with spiking and non-spiking agents in the iterated prisoner's dilemma. In: Alippi, C., Polycarpou, M., Panayiotou, C., Ellinas, G. (eds.) ICANN 2009. LNCS, vol. 5768, pp. 737–746. Springer, Heidelberg (2009)

[62] Gabel, T., Riedmiller, M.: On a successful application of multi-agent reinforcement learning to operations research benchmarks. In: IEEE International Approximate Dynamic Programming and Reinforcement learning, pp. 68–75 (2007)

[63] Sutton, R.S., Barto, A.G.: Reinforcement Learning: An Introduction. MIT Press, Cambridge

[64] Schneider, J., Wong, W.K., Moored, A., Riedmiller, M.: Distributed Value functions. In: The Proceedings of the sixteenth International Conference on Machine Learning, pp. 371–378 (1999)

[65] Busoniu, L., Babuska, R., De Schutter, B.: A comprehensive survey of multiagent reinforcement learning. IEEE Trans. On Systems, Man and Cybernetics Part C: Applications and Reviews 38(2), 156–212 (2008)

# Chapter 2

# Hybrid Multi-Agent Systems

D. Srinivasan and M.C. Choy

Department of Electrical & Computer Engineering,
National University of Singapore,
10 Kent Ridge Crescent,
Singapore 119260

**Abstract.** Hybrid systems have grown tremendously in the past few years due to their abilities to offset the demerits of one technique by the merits of another. This chapter presents a number of computational intelligence techniques which are useful in the implementation of hybrid multi-agent systems. A brief review of the applications of the hybrid multi-agent systems is presented.

## 1 Overview

This chapter provides a detailed review of existing hybrid multi-agent systems. It also includes the relevant computational intelligent techniques [1-5] and other algorithms that have been used to implement them. The first section of this chapter introduces some well known computational intelligent techniques. It also explains how they have been integrated and hybridized by researchers [6-7]. The second section includes a description of how the hybrid computational intelligent techniques and other relevant algorithms such as Stochastic Approximation Algorithms can be applied. They can be used to develop major functional aspects of multi-agent systems [8-11]. The functional aspects include Decision-Making [12-16], Policy Formulation and Knowledge Acquisition and representation. Online Learning [17] and Cooperation are discussed. Some following sections will then deal with how the techniques and algorithms have been applied in previous research works. The design of the functional aspects of multi-agent systems are then dealt with in a hybrid manner. A summary of this review is presented at the end of the chapter.

## 2 Hybrid Computational Intelligent Techniques

Many researchers have recognized and defined four main components of computational intelligence. Each of these dominates an area of AI. They are (1) Fuzzy Logic [18-12], (2) Neural networks [23-26], (3) Evolutionary Algorithms (these include genetic algorithms and genetic programming), and (4) Machine learning and data mining.

Fuzzy logic [18] is a language, which uses syntax and local semantics in order to provide qualitative knowledge about the problem. This can provide a strong scheme

D. Srinivasan & L.C. Jain (Eds.): Innovations in MASs and Applications – 1, SCI 310, pp. 29–42.
springerlink.com                                      © Springer-Verlag Berlin Heidelberg 2010

which is useful for knowledge representation, reasoning and inference. Artificial Neural Networks (ANN) were introduced in [2] and [3]. Essentially they are computational structures which can be trained to learn from examples using Supervised Learning algorithms. The term Evolutionary Algorithm is a general term used to describe Computer Based Problem Solving Systems. These use computational models of Evolutionary Processes as key elements in their design and implementation. Two major algorithms in this area are Genetic Algorithms and Genetic Programming [4], [5]. Machine Learning [6], [7] was developed four decades ago and used for the Development of Computational Methods that could implement various forms of learning. This applies in particular to mechanisms capable of inducing knowledge from examples or data [8].

In recent times, AI oriented computer science has moved towards finding combinations of intelligent tools and techniques that are sound both on a theoretical and a practical basis.

At this time, some of the most effective and popular hybrid intelligent schemes are probably the Neuro-Fuzzy Systems. The main reason for this is that they are both fast and efficient as a methodology and can be easily understood. They may be designed and implemented in automated computing environments (for example, they may be developed within the MATLAB environment). Neuro-fuzzy systems are usually superior to simple neural networks due to the fact that a Neural Network may be affected by noisy data, whereas the Neuro-Fuzzy System has the ability to "account" for the noise through the use of embedded membership functions. There are many such systems now that use, apply and modify Neuro-Fuzzy algorithms. Reference [9] describes how neural networks can be used for fuzzy decision making problems. Reference [10] uses a neuro-fuzzy algorithm for coordinated traffic responsive ramp metering. Reference [11] adopts a neuro-fuzzy approach for material processing. Reference [12] uses neuro-fuzzy systems to control unknown industrial plants. This shows that neuro-fuzzy systems are well-suited to solve ill-defined problems. They can also be used for decision-making and policy formulation.

Besides neuro-fuzzy systems, there are also fuzzy-genetic systems and neuro-evolutionary systems. The fuzzy-genetic systems are essentially a synergistic combination of fuzzy logic and fuzzy sets using various versions of evolutionary algorithms. Fuzzy-genetic systems are often preferable to simple fuzzy systems. This is due to the fact that generally fuzzy-genetic approaches do not need to define the rule base for the control system. Fuzzy-genetic systems have been used collaboratively for control engineering applications, knowledge acquisition, knowledge representation in addition to many other complex optimization problems [13]. The fuzzy logic driven evolutionary approaches [14] primarily deal with the use of fuzzy logic. This is either for tuning the evolutionary parameters, or for fuzzy encoding of the chromosomes. Evolutionary driven fuzzy logic systems based on such schemes [15] usually consist of fuzzy rule-based systems which use an evolutionary approach for the determination and updating of the rule base.

Neuro-evolutionary systems basically apply evolutionary algorithms for determination of the various neural network parameters. These include the Neural Weights or the Neural Network Architecture or both of these methods. In the first case, the neural network is tuned by the evolutionary algorithm. An example is given in [16] where Genetic Algorithm is integrated with an adaptive conjugate gradient neural network learning algorithm for training the multilayer feed forward neural networks. In the second case, the neural network is often generated by use of the evolutionary algorithm. In [17], different coevolving sub-networks combined to form whole networks. There are also researches that develop hybrid models involving neuro-fuzzy as well as evolutionary algorithms.

The final part of this section will cover briefly hybrid systems that use machine learning and a combination of some of the above mentioned techniques. Machine learning and fuzzy logic can be integrated. Here machine learning is used to assist in the formation of fuzzy membership functions. This done by defining the fuzzy boundaries between the neighbouring linguistic areas. An example can be found in [18], where a hybrid intelligent system is used for diagnosing Coronary Stenosis. The authors combine fuzzy generalized operators with decision rules generated by machine learning algorithms. Machine learning can also be integrated with evolutionary algorithms. This is often done  using machine learning for feature selection or extraction from large collections of data. This is done prior to the application of evolutionary methods. In this way, machine learning works as a mechanism which reduces the complexity of the task. This can be extremely important when using time consuming methods such as evolutionary computation. Machine learning can also be combined with neural networks for tuning  the neural network parameters.

Essentially, the list of viable hybrid intelligent models is extensive. There are many more hybrid models that use techniques other than the four areas mentioned earlier in this section. Successful implementations of hybrid intelligent models in recent years show that they have potential for further explorations and applications in this area.

The following section describes how some of the hybrid methods can be applied to the design of various major functional aspects of multi-agent systems.

## 3   Hybrid Multi-Agent Systems

A Multi-agent system is a form of Distributed Artificial Intelligence (DAI). It generally refers to a group of intelligent agents that collaborate to solve common tasks within a dynamic environment. The major functional aspects of an agent in a multi-agent system can be grouped into four categories in the following way.

### 1)  Knowledge Representation and Acquisition

Assuming a correspondence epistemology [19]. Here it is a mapping of the external world. In order for an agent to behave "intelligently", it is essential for the agent to relate to the assigned task in a similar way to how a human being sees it. Knowledge Representation and Acquisition is a crucial functional aspect of an intelligent agent.

### 2)  Decision-Making and Policy Formulation

Based on its knowledge base, an intelligent agent must make decisions and formulate its policies and goals in an autonomous manner. In order to obtain this functional aspect, an intelligent agent must possess a rule base.

### 3) Online Learning

The "intelligent" behaviour of an agent may be limited given the extreme difficulty in developing knowledge structures and rule bases which completely describe the task of the agent if the task by nature is complex. This problem can be partially overcome by causing the agent to learn on its own during this task. This learning is often known as online learning.

### 4) Cooperation

Cooperative behaviours and collaborative actions are also ways in which an agent can be made to be more intelligent. This is done by gaining and exchanging perspectives and information with other agents in the system.

Given the abstract and complex nature of these functional tasks, they often cannot be implemented effectively by the use of a single technique. In this case it is necessary to us several techniques . These are connected together to realize the required functions.

Hybrid multi-agent systems are multi-agent systems where some or all major functional aspects are designed and developed by integrating various computational intelligent techniques. Other relevant algorithms are used in a hybrid manner so as to increase their synergistic relationships. Consequent of their hybridization, hybrid multi-agent systems are capable of displaying highly advanced behavior. They are able to show cooperative online learning and dynamic collaborative decision-making for example. The price to achieve this performance is that the architectures of such systems may become very complex.

In the next sections of this chapter it is described in detail how different techniques have been applied in a hybrid manner as reported in various by previous research works to implement these four major functional aspects of such hybrid multi-agent systems.

## 4   Knowledge Representation and Acquisition in Hybrid Multi-Agent Systems

There are many different ways in which knowledge can be represented in hybrid multi-agent systems. Some well known methods include Grid Model, Vector Model and Graph Model. Examples are visibility graphs or Voronoi diagrams [20], [21]. In addition to these methods, various other computational intelligent techniques can also be used for knowledge representation.

In reference [21], a multi-agent system is used for image processing for aircraft detection. The multi-agent system works asynchronously with a separate set of multi-layer perceptrons (MLPs). It can detect and identify aircraft based on available images. Reference [22] uses a distributed fragmented approach for knowledge representation. The knowledge required for aircraft detection is shared using a multi-agent system and MLPs. Each of the agents is coded using the knowledge of outline detection. The MLPs are trained to identify the aircraft based on the outlines produced using the agents. The advantage of this method is that knowledge representation in individual entities of this system has been made simpler. This is based on the knowledge that it is distributed and fragmented. Given that each entity only needs to know a section of the problem and does not require knowledge of the complex structure. A knowledge representation scheme that is too fragmented may incur a high communication overhead. This is because each entity does not have sufficient knowledge to support its actions in an autonomous manner.

Given fuzzy logic's strong scheme for knowledge representation, it is not surprising that fuzzy logic and fuzzy sets have been used in numerous research works for many years for the representation of knowledge in hybrid multi-agent systems.

In [23], the multi-agent system is in the form of multiple robots and the problem of motion planning is considered. The vector model is used to generate a static map of the known environment for each robot. A polygonal obstacle is represented as a set of vertices. The unknown environment contains the positions of other robots in the existing known environment. Each robot acquires a knowledge of the unknown environment iteratively. This knowledge is expressed using fuzzy logic.

Fuzzy logic uses a set of fuzzy IF-THEN rules to represent knowledge. Various parameters in fuzzy rules such as membership functions can be tuned to modify the way in which knowledge is expressed for the agent. In [24], agents are implemented for the game of soccer. Knowledge bases for problems, such as that of ball interception, are implemented using sets of fuzzy IF-THEN rules. This is similar to [23]. The knowledge bases are adjusted iteratively when new knowledge is acquired and existing knowledge bases are updated.

A good knowledge model, using an appropriate level of abstraction, is crucial for effective learning [25]. For example, in [25], a multi-agent system consisting of 22 agents is used to simulate the game of soccer. The information about the players directly influences the size of the search space of each agent. If all the information for all of the players is to be considered, the search space must have a dimension of approximately $3.74e10^{50}$. The knowledge model may be simplified significantly by considering only the environment that may interact directly with the agent. The search space required may be reduced by several orders of magnitude to approximately $43e10^{6}$. Details how the model can be simplified are given in [25].

Knowledge representation in hybrid multi-agent systems can also be accomplished by using genetic programming. Genetic Programming is an automatic programming method that finds the best computer programs using ideas from natural selection and genetics [26]. Computer programs or the individuals in genetic programming are usually represented as trees or LISP S-expressions, symbolic knowledge may be readily represented. In [27], a robot control program is expressed in the form of a genetic programming tree. In [28], genetic programming is used to represent knowledge by software agents that are used for e-commerce. Given the ease by which knowledge can be represented using genetic programming, advanced knowledge exchange features can be realized using genetic programming operators.

Summing up this section, the knowledge model or the way in which knowledge is represented in hybrid multi-agent system is a crucial factor in determining how effective the system is in problem solving. This is given that the knowledge model has direct effect on the agent's decision-making and learning process [25]. The next section describes in detail the various techniques used to design decision-making and policy formulation aspects of hybrid multi-agent systems.

## 5  Decision-Making and Policy Formulation in Hybrid Multi-Agent Systems

In order for an agent in a multi-agent system to make autonomous effective decisions and formulate their own policies. Comprehensive decision-making models need to be provided within the structure of the agent. Usually, decision-making and policy formulation models are in the form of a set of rules or a rule base. Using their rule bases, the agents generate their responses based on the inputs provided. The process of designing a rule base is not easy given that a good rule base should be comprehensive enough to enable the agent to effectively deal with different exceptions in the problem. A good rule base should not contain more detail than is necessary. It must not hamper the responsiveness of the agent.

A Fuzzy Rule Base has been generally used in recent years in order to facilitate the process of decision-making and policy formulation in multi-agent systems. In [29], fuzzy rule bases are integrated with neural networks to produce a hybridized fuzzy neural network rule structure. It is noted in [29] that a major disadvantage of fuzzy methods of implementing rule bases is that there are no clear guidelines for fine-tuning the fuzzy membership functions. The fuzzy rule base needs to be updated whenever it is deployed in a new environment which affects the existing rules. A neuro-fuzzy technique is used so that the fuzzy rules can be fine tuned using the neural weight update process. The structure of most neuro-fuzzy rule bases is in the form of a five or six layered MLP. This is dependent on how the particular researcher defines the layers.

Inputs from the environment/problem are received in first layer where they are fuzzified and the degree of membership in the respective fuzzy sets determined. The outputs obtained from these layers are forwarded through the rest of the Connectionist Rule System. In the last layer of the network, the appropriate decision/policy is determined using the process of defuzzification. Such rule bases can be designed to use complex rules used for solving complicated problems.

Artificial neural networks (ANN) have also been used by some researchers to implement the rule bases for the agents. In [30], ANN is used as the decision-making systems of agents for iterative game playing. The environment chosen is that for a non-cooperative repeated market selection game. It has been shown that the agents perform reasonably well using ANN. In [31], a multi-agent system is designed for controlling an industrial hard-coal combustion process in a power plant. In this multi-agent system, ANN is used to approximate a function to find the appropriate control function.

In [21], ANN is used to facilitate decision-making in a multi-agent system intended to control mobile robots. For this research work, the concept of ANN is combined with Cellular Automata (CA). The structure of a neuron is consequently modified. Each neuron is also considered as a cell and within each neuron there is an additional set of expert rules. These rules have been formulated and based on existing databases. These contain records of previous paths taken by the robots. They can then adjust the weights of the links between the neurons for possible conflict resolutions.

Evolutionary computation presents another approach for implementing decision-making functions in hybrid multi-agent systems. In [32], a multi-agent system is implemented for the control of traffic signals at nine traffic intersections. The constraints of the traffic signal control problem are encoded in the genetic algorithm. Each chromosome represents a possible solution or decision of the agent. In [33], a multi-agent approach is used to train Beta Basis Function Neural Networks. Similar to that in [32], the constraints of the problem are encoded in the genetic algorithm of each of the agents and the possible decisions of an agent take the form of chromosomes in the genetic algorithm.

Given that most hybrid multi-agent systems are used to solve complex and possibly dynamic problems, a comprehensive set of rules is often not available to assist the agents in decision-making and policy formulation. In order to overcome this limitation, advanced functions such as Online Learning and Cooperation have to be implemented in the multi-agent systems. The following section describes in detail the concept of online learning and how this function has been implemented in hybrid multi-agent systems of earlier research works.

# 6  Online Learning in Hybrid Multi-Agent Systems

Online learning is essential for hybrid multi-agent systems. This is necessary in order that their perceptions and knowledge structures can be updated when dealing with dynamic problems. Online learning essentially concerns the application of machine learning algorithms to enable learning to be done in an unsupervised manner when the agents are doing their pre-assigned tasks. In order to determine the suitability of a machine learning algorithm for the purposes of online learning, it is necessary to know the characteristics of the algorithm. The following section gives a brief review of some well known machine learning algorithms and describes the relationship between some algorithms.

## 6.1  Machine Learning Algorithms for Online Learning

Machine learning algorithms can be organized into a taxonomy, based on the desired outcome of the algorithms. The most common types of machine learning algorithms are as follows:

1) **Supervised learning:** when the algorithm generates a function that maps a given input to a desired output. Such an algorithm requires the existence of a set of desired outputs from labeled examples. Connectionist learning structures such as neural networks use supervised learning to train and update the neural parameters.

2) **Unsupervised learning:** where the algorithm models a set of inputs. The learning takes place without any labeled examples.

3) **Reinforcement learning:** where the algorithm learns the correct policy based on its observation of its world. Each action chosen by the algorithm has an impact on the environment and the environment provides feedbacks. These are known as reinforcements in this context and they guide the learning algorithm.

For complex application domains, labeled examples are often not readily available. Most complex application domains are dynamic in nature and supervised learning will often yield outdated solutions. Unsupervised learning algorithms may not be responsive enough for dynamic problems since they do not take in feedback from the environment. The outputs are varied only by a re representation of the inputs in a more efficient manner. Reinforcement learning is a cross between Supervised and Unsupervised learning. In doing so, reinforcement learning algorithms avoid the inflexibility of supervised learning algorithms. Simultaneously, they can be used to design effective online learning mechanisms for use in solving dynamic problems.

Over the years, a number of reinforcement learning algorithms have been developed by machine learning researchers. For some research works, the environment is being modeled as a finite-state Markov decision process (MDP) or semi-Markov decision process (SMDP) [34], [35]. Reinforcement learning algorithms in this context are closely related to dynamic programming techniques. Others are based on model-free reinforcement learning such as temporal difference (TD) [36], and the Actor-critic methods [37].

One of the most important breakthroughs in reinforcement learning algorithms to date is Q-learning, which was developed by [38]. Many recent reinforcement learning techniques, such as those reviewed in [18], have been developed based on Q-learning.

Certain Dynamic Programming Based Learning Algorithms such as Q-learning and temporal difference method are related to Stochastic Approximation Theorems. In particular, some of the convergence properties of these algorithms can be proven using selected forms of Stochastic Approximation Theorems. In [39], the authors present a rigorous proof of the convergence for the Temporal Difference method and the Q-learning Algorithm. The authors claim that these dynamic programming based algorithms can be motivated heuristically as approximations to dynamic programming. As such, the authors have shown in this paper the relationship between these dynamic programming based algorithms and the powerful techniques of stochastic approximation theorem via a new convergence theorem. Given the relationship of stochastic approximation theorem and some of these dynamic programming based reinforcement learning methods, it can be shown that the stochastic approximation theorem can be modified and applied directly to facilitate online learning processes.

The following section describes how some of these algorithms have been applied to facilitate online learning in hybrid multi-agent systems.

## 6.2 Applications of Machine Learning Algorithms in Hybrid Multi-Agent Systems

Reinforcement learning has been frequently used in recent years for learning online. It has been shown in many research works that reinforcement learning can be used to improve the performance of the decision-making function of an agent and to update its rule structure.

In [40], the authors develop a method known as Evolutionary Learning of Fuzzy rules (ELF) to enable fuzzy rules to learn. They have successfully applied ELF to develop autonomous agents. The set of fuzzy rules in an agent evolves by means of reinforcement learning. The learning algorithm of ELF cycles through episodes and at each episode, ELF estimates the current state of the problem. Based on this information, ELF distributes rewards to the rules. The rewards are proportional to their contribution to the results obtained. The condition is indicated by the condition of the current state. For each episode, the rules compete with each other on the basis of the rewards received, to be selected by ELF. Their recommended actions are then implemented by the agent.

Reference [41] describes a hybrid learning approach for a multi-agent system by combining the reinforcement learning and the genetic algorithm. Reinforcement learning is in use for online local optimization and the genetic algorithm is used for asynchronous global combinatorial optimization. In each agent, the genetic algorithm search modifies the reinforcement learning's search direction so as to develop a coordinated plan which maximizes the global objective function. This concept has been successfully tested in the multi-agent system in [32]. There the combination of reinforcement learning and the genetic algorithm is used to obtain the objectives of distributed learning and coordination among agents.

In [42], a multi-agent system is developed to solve the problem of distributed dynamic load balancing. This is specifically in the form of dynamic web caching in the Internet. Each agent represents a mirrored piece of content that tries to move closer to the areas of the network having a high demand for this item. A fuzzy rule-base is implemented within each agent to enable the agent to choose the optimal direction of motion. In order to improve each agent's choice of action when the dynamics of the problem change, reinforcement learning is used to update the parameters of the fuzzy rule base. A combination of the Actor-critic method as well as Q-learning is used for reinforcement learning. Reference [24] presents a similar approach and it uses reinforcement learning to update the fuzzy rule bases of soccer agents. In [31], reinforcement learning is used to update the neural function approximator in each agent. It is clear that reinforcement learning can be used for updating the parameters of various rule structures.

Besides machine learning algorithms, genetic algorithms can also be used to update rule structures and to facilitate the online learning process. In [43], a multi-agent system is constructed to simulate the hunter-prey scenario, which is analogous to a pursuit game in a real-world scenario. For this research work, the authors assume that comprehensive knowledge of the problem is not available. They therefore intend to let the controllers evolve while they are engaged in handling the problem. A genetic algorithm is used to generate and update a fuzzy logic controller autonomously. In [27], the robot control programs are represented as trees using genetic programming methods. These control programs evolve by using genetic programming operators. Consequently, the robots are updated with the fittest control program.

Often, an agent may not be able to evolve quickly enough to adapt to the changing problem if it can only perform learning on its own. Hence, it is essential to examine ways to enhance the online learning function of multi-agent systems. One way to do this is by cooperative learning. The next section examines the various concepts of cooperation and describes in details how cooperation has been included as an integral function in some advanced hybrid multi-agent systems.

## 7   Cooperation in Hybrid Multi-Agent Systems

A variety of definitions have been offered as a means of providing cooperation in previous published works [44] and from [45]. The definition is as follows: "Cooperation occurs when two or more agents work together in a common environment in order to more effectively reach the maximal union of their goals". The Cooperative Distributed Problem Solving approach (CDPS) arises when the concepts of cooperation are integrated with multi-agent system. This is a form of DAI.

CDPS essentially concerns the development of knowledge and reasoning techniques so that a network of problem-solving nodes or agents can use to cooperate effectively in solving a distributed, complex problem. In many cases, each node in the network does not have sufficient expertise, resources and information to solve a particular problem. However, different nodes might have the necessary expertise to solve parts of the problem. In addition, different nodes might have different information or viewpoints on the problem due to the nature of the particular network architecture. A good understanding of the various cooperation mechanisms is necessary before the concept of cooperation can be applied to multi-agent systems. The following section gives a taxonomy of cooperation mechanisms and includes a brief discussion of each of them.

### 7.1   Cooperation Mechanisms

Cooperation mechanisms may be classified into three main categories: **System Inherent/Structural-based Mechanism (SSCM), Learning and Inference-based Mechanism (LICM)** and **Communication-based Mechanism (CCM)**.

**LICM** concerns cooperative inference and learning. Using LICM, the Learning and Inference process of an agent can be affected by the learning and inference processes of other agents in the multi-agent systems. An example to this would be a group of agents, each learning a different part of the problem, and sharing their knowledge with one another during the learning processes.

**CCM** involves the setting up of Communication Channels between agents. Such channels can take the form of Horizontal Communication (or communication among agents that belong to the same group) or Vertical Communication between agents (or communication among agents that belong to different groups). CCM can be used for agent negotiation, conflict resolution, etc. It can be seen that CCM needs to be present in order for LICM to work.

**SSCM** refers to the Structures of the environment or in-built features in a multi-agent architecture/model that affords cooperative behavior among the agents. Modifications to the environment or an existing Multi-agent architecture//model can be made to implement such cooperation mechanisms. An example to this would be a Multi-agent system where individual agents are programmed to consider the current states of the problems that assigned to other agents in addition of their own problems. This can be viewed as a form of implicit cooperation where the agents do not perform group learning and do not explicitly communicate with other agents.

The choice of cooperation mechanisms may vary depending on the complexity of the problem which the multi-agent system is handling. At times, it is not essential for all three types of cooperation mechanisms to be present in a multi-agent system. A brief review of recent research works that use cooperation in hybrid multi-agent systems is given in the following section.

## 7.2 Applications of Cooperation in Hybrid Multi-Agent Systems

In [46], Q-learning based agents are used to navigate through a deterministic maze. These agents are heterogeneous in nature. That is, one agent is trained to be an expert for the problem while the other learns from the expert. At the same time, both agents are given the same goal of navigating through the maze and they are designed to do cooperative learning together to achieve that goal. The agent with the least obtains the Q-table. This is a table which contains the Q values [38] of the expert agent. It then uses it to assist in its inference process. Such a feature is a form of LICM. It has been shown in [46] that cooperation during learning, even in its simplest form, can have crucial beneficial effect on the learning process of the team of heterogeneous agents. In [46], it has been empirically verified that the speed of convergence for the learning process of the agents improved with such cooperation.

In [47], a coordination-based cooperation protocol was proposed for an Object-Sorting Task in multi-agent robotic systems. The protocol developed coordinates the agents to move objects to their destination efficiently and effectively. Every agent autonomously makes a subjective optimal decision.Subsequently, the coordination algorithm resolves their conflicts by a consideration of the global results. Each agent sends its decision to a centralized coordinator and does not broadcast. The coordinator runs the algorithm and sends the results to the others so they can update their parameters. The CCM is a form of vertical communication. Every agent runs the same algorithm in order to obtain common results without further communication with other agents. The cooperation mechanism can be classified as a SSCM.

The research work described in [48] illustrates the advantage of cooperation when it is applied in multi-agent systems. The agents in [48] are designed using Q-learning and fuzzy logic. They are applied to solve a variation of the problem known as Tile-World [49]. This was designed to evaluate agent architectures. Reference [49] provides detailed information on the Tile-world problem. Similar to [46], a form of cooperative Q-learning is used by the agents. From the results of these experiments, it has been found that $K$ cooperative agents, each learning $N$ time steps outperform $K$ independent agents, where each is learning in a separate world for $K*N$ time steps.

The advantage of cooperation is also illustrated in [23]. The total cost of the genetic algorithm based robots when used for Motion Planning is reduced after each robot has considered the interactions with other robots during its own planning process.

## 8  Summary

This chapter provides a comprehensive review of hybrid multi-agent systems. It also considers the application of different computational intelligent techniques when applied to previous research works. The design of four major functional aspects: knowledge representation, acquisition, decision-making/policy formulation, online learning and cooperation of such multi-agent systems are explained. Based on the review, it can be seen that for most research works, two or more techniques are often necessary in order to fully implement the features of these major functional aspects.

# References

[1] Zadeh, L.A.: Fuzzy Sets, Information Control, vol. 8, pp. 338–353 (1965)
[2] Rosenblatt, F.: Two theorems of statistical separability in the perceptron, Mechanization of Thought Processes, London HM Stat.Office, pp. 421–456
[3] Widrow, B., Hoff, M.E.: Adaptive switching circuits, IRE WesternElectric Show and Convention Record - Part 4, pp. 96–104 (1960)
[4] Holland, J.H.: Adaptation in Natural and Artificial Systems. MIT Press, Cambridge, MA (1975)
[5] Koza, J.R.: Genetic Programming – On the Programming of Computers by Means of Natural Selection. MIT Press, Cambridge (1992)
[6] Michalski, R.S., Carbonell, J.G., Mitchell, T.M.: Machine Learning: An Artificial Intelligence Approach. Morgan Kaufmann, San Francisco (1983)
[7] Mitchell, T.M.: Machine Learning. McGraw-Hill, New York (1997)
[8] Kubat, M., Bratco, I., Michalski, R.S.: A Review of Machine Learning Methods. In: Michalski, R.S., Bratco, I., Kubat, M. (eds.) Machine Learning And Data Mining–Methods and Applications, pp. 3–69. Wiley, Chichester (1997)
[9] Francelin, R.A., Gomide, F.A.C.: A neural network for fuzzy decision making problems. In: Proceedings of Second IEEE International Conference on Fuzzy Systems – FUZZ-IEEE 1993, San Francisco, USA, vol. 1, pp. 655–660 (1993)
[10] Bogenberger, K., Keller, H., Vukanovic, S.: A Neuro-Fuzzy Algorithm for Coordinated Traffic Responsive Ramp Metering. In: Proceedings of IEEE Intelligent Transportation Systems, pp. 94–99 (1994)
[11] Arafeh, L., Singh, H., Putatunda, S.K.: A neuro fuzzy logic approach to material processing. Proceedings of IEEE Transactions on Systems, Man and Cybernetics, Part C 29(3), 362–370 (1999)
[12] Li, C.S., Lee, C.: Self-organizing neuro-fuzzy system for control of unknown plants. Proceedings of IEEE Transactions on Fuzzy Systems 11(1), 135–150 (2003)
[13] Mishra, S., Dash, P.K., Hota, P.K., Tripathy, M.: Genetically optimized neuro-fuzzy IPFC damping modal oscillations of power system. Proceedings of IEEE Transactions on Power Systems 17(4), 1140–1147 (2002)
[14] Chan, K.C.C., Lee, V., Leung, H.: Generating fuzzy rules for tracking using a steady-state genetic algorithm. Proceedings of IEEE Transactions on Evolutionary Computation 1(3), 189–200 (1997)
[15] Perneel, C., Thernlin, J.M., Renders, J.M., Archeroy, M.: Optimization of fuzzy expert systems using genetic algorithms and neural networks. Proceedings of IEEE Transactions on Fuzzy Systems 3(3), 300–312 (1995)
[16] Hung, S.L., Adeli, H.: A parallel genetic/neural network learning algorithm for MIMD shared memory machines. Proceedings of IEEE Transactions on Neural Networks 5(6), 900–909 (1994)
[17] Garcia-Pedrajas, N., Hervas-Martinez, C., Munoz-Perez, J.: COVNET: a cooperative coevolutionary model for evolving artificial neural networks. Proceedings of IEEE Transactions on Neural Networks 14(3), 575–596 (2003)
[18] Cios, K.J., Goodenday, L.S., Sztandera, L.M.: Hybrid intelligence system for diagnosing coronary stenosis. Combining fuzzy generalized operators with decision rules generated by machine learning algorithms. Proceedings of IEEE Engin. In Medicine and Biology Mag. 13(5), 723–729 (1994)
[19] Russell, S., Norvig, P.: Artificial Intelligence A Modern Approach, Section 26.2, pp. 821–822. Prentice Hall, Upper Saddle River, New Jersey

[20] Timofeev, A.V., Kolushev, F.A., Bogdanov, A.A.: Hybrid algorithms of multi-agent control of mobile robots. In: The Proceedings of the International Joint Conference on Neural Networks, IJCNN 1999, July 10-16, 1999, vol. 6, pp. 4115–4118 (1999)

[21] Yabja, A., Stem, A., Singh, S., Brumitt, B.L.: Framed-Quadtree Path Planning for Mobile Robots Operating in Sparse Environments. In: The Proceedings IEEE Conference on Robotics and Automation, May 1998, vol. 1, pp. 650–655 (1998)

[22] Cozien, R., Querrec, R., Douchet, F., Lopin, F.: Neural networks to depict the knowledge of multi-agents systems: application to image processing. In: The Proceedings of the International Joint Conference on Neural Networks, IJCNN 1999, July 10-16, 1999, vol. 4, pp. 2729–2734 (1999)

[23] Shibata, T., Fukuda, T.: Coordinative behavior by genetic algorithm and fuzzy in evolutionary multi-agent system. In: The Proceedings of IEEE International Conference on Robotics and Automation, May 2-6, 1993, vol. 1, pp. 760–765 (1993)

[24] Nakashima, T., Udo, M., Ishibuchi, H.: Knowledge acquisition for a soccer agent by fuzzy reinforcement learning. In: The Proceedings of IEEE International Conference on Systems, Man and Cybernetics, 2003, October 5-8, 2003, vol. 5, pp. 4256–4261 (2003)

[25] Bonarini, A.: Evolutionary learning, reinforcement learning, and fuzzy rules for knowledge acquisition in agent-based systems. Proceedings of the IEEE 89(9), 1334–1346 (2001)

[26] Koza, J.R.: Genetic Programming: On the Programming of Computers by Means of Natural Selection. The MIT Press, Cambridge (1992)

[27] Lee, K., Zhang, B.: Learning robot behaviors by evolving genetic programs. In: The Proceedings of 26th Annual Conference of the IEEE Industrial Electronics Society, IECON 2000, October 22-28, 2000, vol. 4, pp. 2867–2872 (2000)

[28] Zhu, F.M., Guan, S.: Evolving software agents in e-commerce with GP operators and knowledge exchange. In: The Proceedings of IEEE International Conference on Systems, Man, and Cybernetics, October 7-10, 2001, vol. 5, pp. 3297–3302 (2001)

[29] Pham, D.T., Awadalla, M.H.: Neuro-fuzzy based adaptive co-operative mobile robots. In: The Proceedings of IECON 2002, IEEE 2002 28th Annual Conference of the Industrial Electronics Society, November 5-8, 2002, vol. 4, pp. 2962–2967 (2002)

[30] Ishibuchi, H., Seguchi, T.: Successive adaptation of neural networks in a multi-agent model. In: The Proceedings of the 2002 International Joint Conference on Neural Networks, IJCNN 2002, May 12-17, 2002, vol. 3, pp. 2454–2459 (2002)

[31] Stephan, V., Debes, K., Gross, H.-M., Wintrich, F., Wintrich, H.: A reinforcement learning based neural multiagent system for control of a combustion process. In: The Proceedings of the IEEE-INNS-ENNS International Joint Conference on Neural Networks, 2000. IJCNN 2000, July 24-27, 2000, vol. 6, pp. 217–222 (2000)

[32] Mikami, S., Kakazu, Y.: Genetic reinforcement learning for cooperative traffic signal control. In: The Proceedings of the First IEEE Conference on Evolutionary Computation, IEEE World Congress on Computational Intelligence, June 27-29, 1994, vol. 1, pp. 223–228 (1994)

[33] Kallel, I., Jmaiel, M., Alimi, A.M.: A multi-agent approach for genetic algorithm implementation. In: The Proceedings of 2002 IEEE International Conference on Systems, Man and Cybernetics, October 6-9, 2002, vol. 7, p. 6 (2002)

[34] Barto, A.G., Mahadevan, S.: Recent Advances in Hierarchical Reinforcement Learning. Special Issue on Reinforcement Learning, Discrete Event Systems Journal 13, 41–77 (2003)

[35] Abe, N., Pednault, E., Wang, H., Zadrozny, B., Fan, W., Apte, C.: Empirical comparison of various reinforcement learning strategies for sequential targeted marketing. In: Proceedings of IEEE International Conference on Data Mining, pp. 3–10 (2002)

[36] Sutton, R.S.: Learning to Predict by the Method of Temporal Difference. Machine Learning 3(1), 9–44 (1988)

[37] Mizutani, E., Dreyfus, S.E.: Totally Model-Free Reinforcement Learning by Actor-Critic Elman Networks in Non-Markovian Domains. In: Proceedings of the IEEE World Congress on Computational Intelligence (WCCI 1993), pp. 2016–2021 (1998)

[38] Watkins, C.J.C.H., Dayan, P.: Q-Learning. Machine Learning 8, 279–292 (1992)

[39] Jaakkola, T., Jordan, M.I., Singh, S.P.: On the convergence of stochastic iterative dynamic programming algorithms. Neural Computation 6(6), 1185–1201 (1994)

[40] Bonarini, A.: Evolutionary learning of general fuzzy rules with biased evaluation functions: competition and cooperation. In: The Proceedings of the First IEEE Conference on Evolutionary Computation, IEEE World Congress on Computational Intelligence, June 27-29, 1994, vol. 1, pp. 51–56 (1994)

[41] Mikami, S., Wada, M., Kakazu, Y.: Combining reinforcement learning with GA to find co-ordinated control rules for multi-agent system. In: The Proceedings of IEEE International Conference on Evolutionary Computation, May 20-22, 1996, pp. 356–361 (1996)

[42] Vengerov, D., Berenji, H.R., Vengerov, A.: Adaptive coordination among fuzzy reinforcement learning agents performing distributed dynamic load balancing. In: The Proceedings of the 2002 IEEE International Conference on Fuzzy Systems, FUZZ-IEEE 2002, May 12-17, 2002, vol. 1, pp. 179–184 (2002)

[43] Jeong, I., Lee, J.: Evolving fuzzy logic controllers for multiple mobile robots solving a continuous pursuit problem. In: The Proceedings of IEEE International Fuzzy Systems Conference, FUZZ-IEEE 1999, August 22-25, 1999, vol. 2, pp. 685–690 (1999)

[44] Li, C.H., Li, M.Q., Kou, J.S.: Cooperation structure of multi-agent and algorithms. In: The Proceedings of IEEE International Conference on Artificial Intelligence Systems, ICAIS 2002, September 5-10, 2002, pp. 303–307 (2002)

[45] Jone, P.M., Jacobs, J.L.: Cooperative Problem solving in Human-Machine Systems: Theory, Models, and Intelligent Associate Systems. IEEE Transactions on System, Man and Cybernetics, Part C: Applications and Reviews 30(4) (November 2000)

[46] Reza MirFattah, S.M., Ahmadabadi, M.N.: Cooperative Q-learning with heterogeneity in actions. In: The Proceedings of IEEE International Conference on Systems, Man and Cybernetics, 2002, October 6-9, 2002, vol. 4, p. 5 (2002)

[47] Lin, F.-C., Hsu, Y.-J., Jane: Coordination-based cooperation protocol in multi-agent robotic systems. In: The Proceedings of IEEE International Conference on Robotics and Automation, April 22-28, 1996, vol. 2, pp. 1632–1637.

[48] Berenji, H.R., Vengerov, D.: Advantages of cooperation between reinforcement learning agents in difficult stochastic problems. In: The Proceedings of The Ninth IEEE International Conference on Fuzzy Systems, FUZZ IEEE 2000, May 7-10, 2000, vol. 2, pp. 871–876 (2000)

[49] Pollack, M.E., Ringuette, M.: Introducing the Tile-world: Experimentally Evaluating Agent Architecture. In: The Proceedings of the 8th National Conference on Artificial Intelligence (AAAI 1990), pp. 183–189 (1990)

# Chapter 3

# A Framework for Coordinated Control of Multi-Agent Systems

Howard Li[1,*], Fakhri Karray[2], and Otman Basir[2]

[1] Department of Electrical and Computer Engineering,
University of New Brunswick, Fredericton, New Brunswick, Canada
howard@unb.ca
[2] Department of Electrical and Computer Engineering, University of Waterloo,
Waterloo, Ontario, Canada

**Abstract.** In this chapter, the Coordinated Hybrid Agent (CHA) framework is introduced for the distributed control and coordination of multi-agent systems. In this framework, the control of multi-agent systems is regarded as achieving decentralized control and coordination of agents. Each agent is modeled as a CHA which is composed of an intelligent coordination layer and a hybrid control layer. The intelligent coordination layer takes the coordination input, plant input and workspace input. The intelligent coordination layer deals with the planning, coordination, decision-making and computation of the agent. The hybrid control layer of the framework takes the output of the intelligent coordination layer and generates discrete and continuous control signals to control the overall process. In order to verify the feasibility of the framework, experiments for multi-agent systems are implemented. The framework is applied to a multi-agent system consisting of an overhead crane, a mobile robot and a robot manipulator. The agents are able to cooperate and coordinate to achieve the global goal. In addition, the stability of systems modeled using the framework is also analyzed.

## 1 Introduction

In this section, we give a brief introduction of the proposed research in coordinated control of multi-agent systems.

### 1.1 Agents

Agents are one of the most effective software design paradigms. An agent is anything that can be viewed as perceiving its environment through sensors and acting upon that environment through actuators [1]. Human agents have eyes, ears, and other organs as sensors, and body parts as actuators. An agent's choice

---

* Corresponding author.

D. Srinivasan & L.C. Jain (Eds.): Innovations in MASs and Applications – 1, SCI 310, pp. 43–67.
springerlink.com                                          © Springer-Verlag Berlin Heidelberg 2010

of actions at any moment depends on the perception of the agent's environment. An agent's function maps the perception of an agent to an action. Agents' functions can be implemented as programs. A performance measure can be chosen to evaluate the behaviors of an agent. A rational agent will do the right thing to achieve good performance. It is pointed out in [1] that the rationality of an agent depends on:

- The performance measure;
- Agents' previous knowledge of the environment;
- Agents' perception of the environment;
- Agents' actions.

In order to make agents more flexible, it is desirable to have autonomous agents that do not rely on the designer's prior knowledge. Agents should be able to make decisions based on their own perception of the environment. An agent should learn as much as possible from their perception. Learning can be based on simple events in the environment or past experience of certain behaviors.

There are different ways of classifying agents [2]. It can be based on agents' functional features, tasks they carry out, and so on. The Nwana classification method defines four classes of agents according to their ability to cooperate, learn, and act autonomously. The four classes are collaborative learning agents, interface agents, smart agents, and collaborative agents. The Davis classification method describes three types of agents' behaviors, reflective, reactive, and mediative.

There are a few key features that an agent should possess. Communication is an essential ability of agents. An agent should be able to communicate with other agents. Coordination is an important ability of agents. Coordination enables the combination of multiple entities in a synchronous and mutual beneficial manner. Inter-operability enables agents to operate device and components on other agents. An agent system should also be fault tolerant. In case of the failure of an agent, other agents should be able to replace the agent and the whole system is still functional. Although replacing an agent with other agents in case of failure is possible, the functionality of the system will degrade. When we design multiple agents, graceful degradation of the system in case of failure should be considered.

## 1.2 Multi-Agent Systems

Multi-Agent Systems (MASs) represent a group of agents that cooperate to solve common tasks in a dynamic environment. Multi-agent control systems have been widely studied in the past few years [3] [4] [5] [6] [7] [8] [9] [10]. The control of MASs relates to synthesizing control schemes for systems which are inherently distributed and composed of multiple interacting entities. Because of the wide applications of multi-agent theories in large and complex control systems, it is necessary to develop a framework to simplify the process of developing control schemes for MASs.

## 1.3   Literature Review of Related Work

In [11], an architecture is introduced for an agent. There are a few major subsystems in the architecture: perception, mission planning, behavioral executive, and motion planning. The perception subsystem processes sensor data from the vehicle and provides a collection of semantically rich data products. The mission planning system computes the fastest route to reach the next way point. The behavioral executive system combines the value function provided by mission planning system with local information to generate motion goals. The motion planning system is responsible for the safe and timely execution of the motion goals. In [12], a framework is proposed for the design and analysis of control algorithms for autonomous ground vehicles on outdoor terrains. The proposed framework is developed using MIRO and Matlab. In the proposed framework, the simulated robot navigates in the simulation environment which is a mirror of the real world. Sensors in the real world can also send signals through MIRO to the sensed environment. The robot in the sensed environment can "think" about the simulated environment and make control decision. In [13], a collection of local feedback control policies offer continuous guarantees. They are combined with discrete automata. The approach automatically creates a hybrid feedback control policy that satisfies a given high-level specification without planning a specific configuration space path. Most of the above architecture focuses on specific applications of MASs. It is necessary to develop a generic framework to model agents and MASs. In [14], a generic framework for integrated modeling, control and coordination of multiple multi-mode dynamical systems is developed. This framework of distributed control of multi-agent systems is called Hybrid Intelligent Control Agent (HICA). In this framework, a certain form of knowledge-based deliberative planning is integrated with a set of verified hybrid control primitives and coordination logic to provide coordinated control of systems of agents. This work gives the basis for analyzing MASs as hybrid control systems. Although in this framework, coordination factors have been defined as input coordination factors and output coordination factors, there is no generic coordination mechanism defined for the HICA agents. Only some Pseudo codes are given for the coordination problem. Defining an abstract and domain independent coordination mechanism is necessary for this framework. Furthermore, because this framework was based on the multiple unmanned ground vehicle/unmanned air vehicle pursuit-evasion problem, not all essential primitives are defined.

## 1.4   Discussion of the Proposed Work

In [15], a framework is proposed for the distributed control and coordination of MASs. In this chapter we will introduce this framework for MASs. Each agent in the framework is responsible for controlling subsystems of the whole system. Each subsystem is viewed as a hybrid system that includes time-driven and event-driven dynamics. In this framework, the control of MASs is considered as a decentralized control scheme involving coordination of agents. Each agent is modeled as a Coordinated Hybrid Agent (CHA) which is composed of an intelligent coordination control layer and a hybrid control layer. The intelligent coordination control layer takes the coordination, plant and workspace inputs. After processing the coordination primitives, the intelligent coordination control layer outputs the desired action to the hybrid layer. The intelligent coordination control layer deals with the planning, coordination, decision-making and computation of the agent. The hybrid control layer of the CHA framework then takes the output of the intelligent coordination layer and generates discrete control signals and continuous control signals to control the plant. It is demonstrated that the CHA framework is not only capable of modeling homogeneous MASs [15], but also heterogeneous MASs [16], for example, a multi-agent system consisting of an overhead crane, a mobile robot, and a robot manipulator. The objectives of this research are to develop a generic framework for the control of a system consisting a collection of agents. More specifically, the objectives are:

1. Proposing a generic framework for the control of a multi-agent system where agents cooperate, coordinate and interact with each other.
2. Guaranteeing the stability of the control scheme for the multi-agent system.
3. Applying the framework to homogeneous and heterogeneous systems. The feasibility of the proposed generic framework for the control of multi-agent systems is demonstrated by experiments and/or numerical simulations.
4. Optimizing a multi-agent system modeled using the CHA framework. In order to optimize the performance and time for the MAS, both the event-driven dynamics and time-driven dynamics are formulated.

The rest of this chapter is organized as follows. Section 2 gives background knowledge on this research work. Section 3 describes the CHA framework for the control of multi-agent systems. In addition, the optimization problem of MASs modeled by the CHA framework is formulated. The direct identification algorithm for the optimization of CHA MASs is introduced. Section 4 gives experimental results to illustrate the feasibility of the CHA framework. By using the CHA framework, the control schemes are developed for multi-agent systems. It is demonstrated that the CHA framework is generic and could be applied to homogeneous and heterogeneous multi-agent systems. In Section 5, we conclude by providing insight of what has been introduced.

## 2   Background

In this section, we introduce the background knowledge of various areas.

### 2.1   Multi-Agent Systems

The two most important fields of multi-agent systems are Distributed Artificial Intelligence (DAI) and Artificial Life (AL) [17]. The purpose of DAI is to create systems that are capable of solving problems by reasoning based on dealing with symbols. The purpose of AL is to build systems that are capable of surviving and adapting to the environments. The research into agents was originated in 1977 [18] by Hewitt. He proposed the actor model of computation to organize programs in which the intelligence is modeled using a society of communicating knowledge-based problem-solving experts. Since then, the research in agents has continued and developed. The research of sharing data among agents dates back to 1980 [19]. In this work, the model of the blackboard system was developed. Objects in the working area were put down, modified and withdrawn in a common area called the blackboard.

An agent within a multi-agent system can be thought of as a system that tries to fulfill a set of goals within the complex, dynamic environment. Agents have only a partial representation of the environment. In recent years, there has been a growing interest in control systems that are composed of several interacting autonomous agents instead of a single agent. In order to deal with highly complex control systems, it is important to have systems that operate in an autonomous decentralized manner. One way to do this is to distribute the control/decision making to the local controllers which makes the control of the multi-agent system simpler.

## 2.2   Centralized Control and Decentralized Control

The centralized control paradigm is characterized by a complex central processing unit that is designed to solve the whole problem. The central unit must gather data from the whole system. The solution algorithms are therefore complex and problem specific. The processing unit is able to check whether or not a solution is the globally optimal solution, which is not easily achieved in a decentralized control paradigm. However, utilizing complex algorithms and analyzing all information in a centralized controller always cause slower responses than a decentralized control system.

Decentralized control paradigms are based on distributed control in which individual components react to local conditions simultaneously. These individual components interact with neighboring components to exhibit desired adaptive behaviors. The complex behaviors are a resultant property of the system of connections. The decentralized nature of information in many large-scale systems, requires the control systems to be decentralized. Decentralized control of discrete-event systems, in the absence of communication, has been well studied [16].

## 2.3   Continuous Systems and Discrete Event Systems

Discrete Event Systems (DES) are dynamical systems which evolve in time by occurrence of events at possibly irregular time intervals. Some examples include flexible manufacturing systems, computer networks, logic circuits, and traffic systems [20]. In [21], Passino introduces a logical DES model and defines stability in the sense of Lyapunov and asymptotic stability for logical DES. He shows that the metric space formulation can be used for the analysis of stability for logical DES by employing appropriate Lyapunov functions.

Modern systems involve both discrete and continuous states. Systems of interest in this study are typically governed by continuous dynamic equations at particular discrete states. Systems like these are considered as hybrid systems. In order to study the multi-agent systems consisting of hybrid systems, we need to include the hybrid system concept to model the controlled processes that have both discrete and continuous variables. Hence it is necessary to develop a framework that deals with both the discrete and continuous states.

### 2.4   Hybrid Intelligent Control Agent

Many control problems involve processes that are inherently distributed, complex or that operate in multiple modes. Agent-based control is an emerging paradigm within the sub-discipline of distributed intelligent control. In [14], Fregene proposes the HICA architecture as a conceptual basis for the synthesis of intelligent controllers in problem domains which are inherently distributed, multi-mode and that may require real-time actions.

The key idea of HICA is to combine concepts from hybrid control and multi-agent systems to build agents which are especially suitable for multi-mode control purposes. HICA conceptually wraps an intelligent agent around a core that is itself a hybrid control system. Fregene [14] illustrates how HICA might be used as a skeletal control agent to synthesize agent-based controllers for inherently distributed multi-mode problems.

## 3   The Coordinated Hybrid Agent

Agent-based control is an emerging paradigm within the sub-discipline of distributed intelligent control. In this section, the CHA framework is introduced for the distributed control and coordination of multi-agent systems. In the CHA framework, the control of multi-agent systems focuses on decentralized control and coordination of agents. Each agent is modeled as a CHA which is composed of an intelligent coordination layer and a hybrid control layer as shown in Figure 1. The core of the CHA framework is on developing coordinated agents for the control of hybrid multi-agent systems. A robust and generic control architecture is developed to control a homogeneous multi-agent system or a heterogeneous multi-agent system. The framework is able to model the cooperation, coordination and communication among the members of the multi-agent system. The control scheme is able to control a multi-agent system where agents cooperate, coordinate and interact with each other. The stability of the control scheme for the multi-agent system modeled by the framework is also proved.

### 3.1   The Agent Workspace

Agents can either work within the same workspace or have their own workspace. In order to execute a common task, two or more agents might need to cooperate and coordinate within the same workspace. For other tasks, agents may need to work in their own workspace and communicate with each other to achieve a global goal.

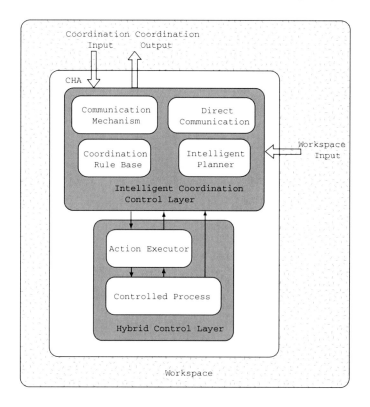

**Fig. 1.** The Internal Structure of a CHA Agent

## 3.2 The Hybrid Control Layer

In this sub-section, we introduce the trajectories of the system, the controlled process, the action executor and the execution of hybrid actions.

**Trajectories of the System:** Let $T$ denote the time axis. Since a hybrid system evolves in continuous time, we assume an interval $V$ of $T \subseteq \mathcal{R}$ to be $V = [t_i, t_f] = \{t \in T | t_i \leq t \leq t_f\}$. The variables of the system evolve either continuously or in instantaneous jumps. The addition of $T$ is also allowed. For an interval $V$ and $t_0 \in T$, we have $t_0 + V = \{t_0 + t' | t' \in V\}$. Using the concepts from [22], we have the following definitions.

If we denote the discrete evolution space of a hybrid system as $Q$ and the continuous evolution space of a hybrid system as $X$, a *trajectory* of a hybrid system can be defined as a mapping $V \rightarrow Q \times X$.

The evolution of the continuous state in each sub-interval of $V$ is described as $f : Q \times X \times U \rightarrow Q \times TX$, where $U$ represents the continuous control signal space, $TX$ represents the tangent space of space $X$. Thus for every sub-interval of $V$, we have $\dot{x}(t) = f(q(t), x(t), u(t))$, in which $f$ is the vector field. We assume the existence and uniqueness of solutions to the ordinary differential equation on $f$.

The application of the continuous control signal $u \in U$ and the discrete control signal $m \in M$ is defined as a *hybrid action* which is denoted by $a \in A$. In each sub-interval, $q(t)$ is a constant.

We define $\diamond$ as the *restriction* of trajectory $E$ to a subset of its domain $d(E)$ in which no discrete state transition occurs rather than at the starting and the ending point. There is no discrete transition at the starting point or the ending point if the interval is left-open or right-open, respectively. $E \diamond [t_1, t_2]$ means the subset of trajectory $E$ over $t_1 \leq t \leq t_2$. It can also be denoted as $E \diamond V$, which means the subset of trajectory $E$ over $[t_i, t_f]$.

If $E_1$ is a trajectory with a right-closed domain $V_1 = [t_i, t_j]$, $E_2$ is a trajectory with domain $V_2 = [t_j, t_f]$, we define the *trajectory link* of $E_1$ and $E_2$ to be the trajectory over $[t_i, t_f]$ as

$$E_1 \propto E_2(t) = \begin{cases} E_1(t) & \text{if } t \in V_1; \\ E_2(t) & \text{otherwise.} \end{cases}$$

For a countable sequence of trajectories, if $E_i$ is a trajectory with domain $V_i$, while all $V_i$ are right-closed, if $1 \leq i \leq \infty$ and $i \in \mathcal{Z}$, the infinite trajectory link can be written as $E_1 \propto E_2 \propto E_3 \ldots$ over $V_1 \cup V_2 \cup V_3 \ldots$.

**The Controlled Process in the Framework:** The controlled process for each agent is essentially a hybrid system whose dynamics are controlled by the coordinated hybrid agent. The evolution of the controlled process is given by

$$I_p \subset Q_p \times X_p \tag{1}$$

$$Y_p \subset Q_a \times X_a \tag{2}$$

$$E_p = E_{p_1} \propto E_{p_2} \propto \ldots \propto E_{p_k} \tag{3}$$

$$\eta_p : Q_p \times X_p \times M \to \mathcal{P}(Q_p \times X_p) \tag{4}$$

$$\gamma_p : Q_p \times X_p \to \mathcal{P}(Q_p \times X_p) \tag{5}$$

$$f_p : Q_p \times X_p \times U \to TX_p \tag{6}$$

$$h_p : Q_p \times X_p \times M \times U \to Y_p \tag{7}$$

- $I_p$ is the initial state of the controlled process that gives both the initial discrete state $Q_p$ and the initial continuous state $X_p$.
- $Y_p$ is the output space of the controlled process which is a subset of the space $Q_a \times X_a$, where $Q_a$ is the discrete state of the hybrid system read by the sensors, $X_a$ is the continuous state of the hybrid system read by the sensors.
- $E_p = E_{p_1} \propto E_{p_2} \propto \ldots \propto E_{p_k}$ is the trajectory of the controlled process. $E_p$ is determined by the discrete state evolution and the continuous state evolution of the controlled process.
- $\eta_p$ is a function that governs the controlled discrete transition of the controlled process. $\mathcal{P}(.)$ represents the power set. $\forall V = [t_i, t_f]$, the controlled discrete jumps of the controlled process is given by

$$q_p(t') = \eta_p(q_p(t), x_p(t), m) \qquad (8)$$

where $q_p \in Q_p$, $x_p \in X_p$ and $m \in M$ represents the discrete control signal.

- $\gamma_p$ is the function that governs the autonomous discrete transition of the process. $\forall V = [t_i, t_f]$, the autonomous discrete jumps of the controlled process is given by

$$q_p(t') = \gamma_p(q_p(t), x_p(t)) \qquad (9)$$

where $q_p \in Q_p$ and $x_p \in X_p$.

- $f_p$ is the vector field determined by the evolution of the continuous state $(x_p \in X_p)$ of the controlled process at a certain discrete state $(q_p \in Q_p)$ of the controlled process (i.e., within the sub-interval of $V$ while the discrete state $q_p(t)$ is a constant or a set of constants). The evolution of the continuous state is given by

$$\dot{x} = f_p(q_p(t_i), x_p(t_i), u(t_i)) \qquad (10)$$

where $q_p \in Q_p \subset \mathcal{Z}^m$, $x_p \in X_p \subset \mathcal{R}^n$ and $u \in U$ represents the continuous control signal.

- The output $y_p(t) \in Y_p \subset \mathcal{Z}^m \times \mathcal{R}^n$ is the feedback of the controlled process. The output is read by the sensors and is given by

$$y_p(t) = h_p(q_p(t), x_p(t), m(t), u(t)) \qquad (11)$$

**The Action Executor:** For each single agent, the evolution of the discrete and continuous state of the system is regarded as the execution of a hybrid action. The action executor has two functions $f_e$ and $\eta_e$:

$$f_e : A \times Y_p \times X_r \to U \qquad (12)$$
$$\eta_e : A \to M \qquad (13)$$

- $f_e$ is the continuous action execution function that takes the desired hybrid action $a \in A$, the output $y_p \in Y_p$ of the process, and the reference value $x_r \in X_r$ as input, then generates the continuous control signal $u \in U$ for the process.
- $\eta_e$ is the discrete action execution function that takes the desired hybrid action $a \in A$ as input, then generates the discrete control output to the process.

The selection of appropriate actions and sequence of the actions are handled by the intelligent coordination control layer which will be introduced later. Because the action executor deals with all the local control problems, in the view of the intelligent coordination control layer, the controlled process can be considered as a discrete event system.

An *execution sequence* is defined as $\beta = E_{p_1} a_1 E_{p_2} a_2 E_{p_3} a_3 \ldots$, where $E_{p_i}$ is the restriction $E_p \diamond V_i$ and $a_i$ is the hybrid action that occurs between $E_{p_i}$ and $E_{p_{i+1}}$.

The execution sequence $\beta$ of the hybrid actions determines the trajectory $E_p$. $E_p$ represents the evolution of the discrete states of the hybrid system, and the evolution of the continuous states in between the discrete transitions.

### 3.3   The Intelligent Coordination Control Layer

In the CHA framework, local hybrid dynamics are considered as hybrid actions. The intelligent coordination control layer has full control of the agent in an abstract way. The intelligent coordination layer plans the sequence of control primitives and selects appropriate hybrid actions without violating the coordination rules. The intelligent coordination control layer is built upon the action executor.

At the intelligent coordination control layer, we define the states of the agent in an abstract way, which we call *coordination states* of the CHA. We denote the set of coordination states as $R$.

Agents repeatedly and simultaneously take actions, which lead them from their previous states to new states. The intelligent coordination control layer interacts with other agents through the *communication mechanism*. In addition, the intelligent coordination control layer takes $Q_a$ and $X_a$ as feedback from the controlled plant, then it outputs the desired action $a \in A$ and reference value $x_r \in X_r$ to the action executor.

In order to coordinate the agents while planning, we introduce the concept of coordination rule base which is inspired by social laws defined in [23]. The coordination rules can be considered as optimal choices and constraints for the actions of agents. The constraints specify which of the actions are in fact not allowed in a given state. The optimal choices in general are optimal actions that are available for a given state.

Given a set of coordination states $R$, a set of rules $L$, and a set of actions $A$, an *optimal choice* is a pair $(a, l_o)$ where $a \in A$ and $l_o \in L$ is a rule that defines an optimal action that results in a transition with the maximum distance along the path of $R$ in the metric space at the given coordination state $r \in R$.

Given a set of coordination states $R$, a set of rules $L$, and a set of actions $A$, a *constraint* is a pair $(a, l_c)$ where $a \in A$ and $l_c \in L$ is a rule that defines a constraint at the given coordination state $r \in R$.

A *coordination rule set* is a set of optimal choices $(a, l_{oi})$ and constraints $(a_i, l_{ci})$. We denote the coordination rule set as $C$. The coordination rule set defines which action should be taken at a given coordination state $r \in R$.

A set of rules $L$ is used to describe what is true and false in different coordination states of the agent. Given a coordination state $r \in R$ and a rule $l \in L$, $r$ might satisfy or not satisfy $l$. We denote the fact that $r$ satisfies $l$ by $r \models l$. The meaning of $(a_i, l_i)$ will be that $l_i$ is the most general condition about coordination states which optimally chooses or prohibits the action $a_i$.

A *coordination rule base* for the intelligent coordination control layer of a CHA is a tuple $(R, L, A, C, T)$ in which $C$ is a coordination rule set, and $T$ is the transition function $T : R \times A \times L \rightarrow \mathcal{P}(R)$ such that: For every $r \in R$, $a \in A$, $c \in C$, if $r \models l_c$ holds and $(a, l_c) \in C$, then $T(r, a, l) = \emptyset$, the empty set, which means the desired transition is prohibited; For every $r \in R$, $a \in A$, $c \in C$, if $r \models l_o$ holds and $(a, l_o) \in C$, then $T(r, a, l) = \check{r}$, where $\check{r}$ is the coordination state after the optimal action is taken. The coordination rule base is used to define agents' behaviors so that agents do not violate any constraints in the workspace.

The coordination rule base provides a skeleton for the agents to coordinate with others. Agents in a multi-agent system with coordination rule base share the set of abstract states, the rule for describing states, the set of potential actions and the transition functions.

Without violating the coordination rule base, the intelligent coordination control layer can have built-in intelligent planners to generate actions as the input to the action executor. Following the next coordination state $r$, the selected action is determined by $T : R \times A \times L \to \mathcal{P}(R)$. The AI approaches for planning tasks such as potential field methods, artificial neural networks, fuzzy logic and knowledge based planning schemes can be implemented as possible intelligent planners.

For a given present state in $R$, denoted by $r_p$, the next state $r_n$ is obtained by

$$r_n \Leftarrow \left( x_{r_n} = \max\{x_i, i = 1, 2, \dots, k\} \right), \tag{14}$$

where $x$ is the degree of fitness of the coordination state, $i$ is the number of the neighboring coordination states including itself (i.e. all the possible next states).

Common Object Request Broker Architecture (CORBA) provides all the abstractions and relevant services for developing a distributed application on heterogeneous platforms. It is illustrated that the CORBA architecture can seamlessly integrate distributed systems. In this study, this architecture is used for the communication among the agents.

In addition to the communication mechanism, agents have access to direct communication to coordinate their behaviors. This is necessary for applications in which agents need to react very fast. Instead of using a network-based communication mechanism, agents interact with each other through sensors and actuators in order to cooperate and coordinate. Implicit communication through actions could be implemented so that agent can share information in the common workspace.

## 3.4   Lyapunov Stability

In order to analyze the stability of the agent in our CHA framework, we apply the stability analysis method proposed by Passino in [21].

According to the model of the intelligent coordination control layer introduced above, the stability properties of the CHA systems can be accurately modeled with $G = (R, A, f_c, g, \mathcal{E}_v)$ where $R$ is the coordination states, $A$ is the set of hybrid actions, $f_c : R \to R$ for $a \in A$ is the transition function. $g : R \to \mathcal{P}(A) - \{\emptyset\}$ is the enable function and $\mathcal{E}_v$ is the set of valid event trajectories for the coordination states $R$. Note that the events we are discussing here are the hybrid actions that the agent will take. It is also possible that, at some states, no actions should be taken; this is represented by a null action. In this way, the system that terminates can also be studied in the Lyapunov stability theoretic framework.

Let $r_k \in R$ represent the $k$th coordination state of the CHA and $a_k \in A$ represent an enabled action for $r_k$ (i.e. $a_k \in g(r_k)$). As described above, at state

$r_k \in R$, action $a_k \in A$ is taken, the next coordination state $r_{k+1}$ is given by the transition function $f_c$. Thus, $r_{k+1} = f_c(r_k)$. Each valid event trajectory $\mathcal{E}_v$ represents a physically possible event trajectory. If $r_k \in R$ and $r_k \in g(r_k)$, $a_k$ can be taken if it lies on a valid event trajectory that leads the state to $r_{k+1} = f_c(r_k)$.

In the CHA framework, we model the intelligent coordination control layer as $G$. First, we model the system via $R$, $A$, $f_c$ and $g$. Then, the possible trajectories $\mathcal{E}_v$ are given. The allowed event trajectories are denoted as $\mathcal{E}_a \subset \mathcal{E}_v$. In the CHA framework, $\mathcal{E}_a$ is governed by the coordination rule base. The allowed event trajectories that begin at state $r_0 \in R$ is denoted by $\mathcal{E}_a(r_0)$. If we use $\mathcal{E}_k = a_0 a_1 a_2 \ldots a_{k-1}$ to denote an event sequence of $k$ events, the value of function $\mathcal{R}(r_0, \mathcal{E}_k, k)$ to denote the coordination state reached at time $k$ from $r_0 \in R$ by the application of event sequence $\mathcal{E}_k$, then $\mathcal{R}(\mathcal{R}(r_0, \mathcal{E}_k, k), \mathcal{E}_{k'}, k') = \mathcal{E}(r_0, \mathcal{E}_k \mathcal{E}_{k'}, k + k')$. In order to guarantee the stability of the CHA system, we need to define the coordination rule base properly to generate desired action sequences that will make the system stable.

A closed invariant set $R_m \subset R$ is called *stable in the sense of Lyapunov* w.r.t. $\mathcal{E}_a$ if for any $\epsilon > 0$, it is possible to find a quantity $\delta > 0$ such that when the metric $\rho(r_0, R_m) < \delta$ we have $\rho(\mathcal{R}(r_0, \mathcal{E}_k, k), R_m) < \epsilon$ for all $\mathcal{E}_k$ such that $\mathcal{E}_k \mathcal{E} \in \mathcal{E}_a(r_0)$ and $k \in \mathcal{Z}^+$ where $\mathcal{E}$ is an infinite event sequence, and $\mathcal{Z}^+$ is the set of positive integers.

Given a coordination state $r \in R$ and a rule $l \in L$, $r$ might satisfy or not satisfy $l$. Recall that we denote the fact that $r$ satisfies $l$ by $r \models l$. Based on the definitions and the CHA model we described, we give the definition of the stability of a CHA system.

A CHA multi-agent system is called *stable* if

1. The action executor can accomplish the hybrid actions so that the coordination states can transition according to $f_c$.

2. All the actions taken are on the allowed event trajectories $\mathcal{E}_a$ that lead the system to the goal set, and for $r \in R$, $a \in A$, $c \in C$, we have $r \models l_o$ holds and $(a, l_o) \in C$, $r \models l_c$ holds and $(a, l_c) \in C$ respectively. $l_o \in L$ defines an optimal action and $l_c \in L$ defines a constraint respectively. $L$ is a set of coordination rules.

3. The invariant set $R_m \subset R$ is stable in the sense of Lyapunov w.r.t. $\mathcal{E}_a$.

In order to satisfy the stability requirements of a CHA system, we need to give the necessary and sufficient condition for closed invariant set $R_m$ to be stable. In [21], the necessary and sufficient condition for a closed invariant set to be stable is given as:

For a closed invariant set $R_m \subset R$ to be stable in the sense of Lyapunov w.r.t. $\mathcal{E}_a$, it is necessary and sufficient that in a sufficiently small r-neighborhood of the set $R_m$ there exists a specified functional $V$ with the following properties:

1. For all sufficiently small $c_1 > 0$, it is possible to find a $c_2 > 0$ such that $V(r) > c_2$ for $r \in$ r-neighborhood of $R_m$ and $\rho(r, R_m) > c_1$.

2. For any $c_4 > 0$ as small as desired, it is possible to find a $c_3 > 0$ so small that when $\rho(r, R_m) < c_3$ for $r \in$ r-neighborhood of $R_m$, we have $V(r) \leq c_4$.

3. $V(\mathcal{R}(r_0, \mathcal{E}_k, k))$ is a non increasing function for $k \in \mathcal{Z}^+$, as long as $\mathcal{R}(r_0, \mathcal{E}_k, k) \in$ r-neighborhood for all $\mathcal{E}_k$ such that $\mathcal{E}_k \mathcal{E} \in \mathcal{E}_a(r_0)$.

*Proof:*
The necessity can be proved by letting the closed invariant set $R_m \subset R$ be stable in the sense of Lyapunov w.r.t. $\mathcal{E}_a$ for some r-neighborhood of $R_m$.

The sufficiency can be proved by letting there exist a specified functional $V$ with the three properties in a r-neighborhood of $R_m$. We can show that the closed invariant set $R_m \subset R$ is stable in the sense of Lyapunov w.r.t. $\mathcal{E}_a$.

The details of the proof can be found in [21].

## 3.5   Optimization

In this sub-section, the optimization of MASs modeled by the CHA framework is discussed. In the CHA framework, each agent is modeled as a hybrid control layer and an intelligent coordination control layer. For a single agent, the controlled plant is at some initial physical state $x_{r_0}(t_0)$ at time $t_0$ and subsequently evolves according to the time-driven dynamics

$$\dot{x}_{r_0} = f_{p_{r_0}}(x_{r_0}, u_{r_0}, t), \tag{15}$$

where $f_p(\cdot)$ represents a continuous function, the subscript $r_0$ represents the initial abstract state. $x$ is the continuous state, $u$ is the continuous control signal, and $t$ represents time.

At time $t_{r_0}$, an event takes place. The abstract state becomes $r_1$ and the physical state becomes $x_{r_1}(t_{r_0})$. There might be a jump of the physical state at $t_{r_0}$. Therefore it is possible that $x_{r_1}(t_{r_0}) \neq x_{r_0}(t_{r_0})$. Then the physical state subsequently evolves according to new time-driven dynamics with this initial condition. The time $t_{r_0}$ at which this event happens, is called the temporal state of the agent. It depends on the event-driven dynamics of the form

$$t_{r_0} = w_{r_0}(t_0, x_{r_0}, u_{r_0}). \tag{16}$$

Let $r_k \in R$ represent the $k$th coordination state of a single agent. In general, after the abstract state switches from $r_{k-1}$ to $r_k$ at time $t_{r_{k-1}}$, the time-driven dynamics are given by

$$\dot{x}_{r_k} = f_{p_{r_k}}(x_{r_k}, u_{r_k}, t), \tag{17}$$

where the initial condition for $x_{r_k}$ is $x_{r_k}(t_{r_{k-1}})$. The event-driven dynamics are given by

$$t_{r_k} = w_{r_k}(t_{r_{k-1}}, x_{r_k}, u_{r_k}). \tag{18}$$

Both the physical state $x_{r_k}$ and the next temporal state $t_{r_k}$ are affected by the choice of the control schemes at the abstract state $r_k$. Note that in order to solve the optimization problem, $t_{r_0}, t_{r_1}, t_{r_2}, \ldots, t_{r_k}$ are considered as temporal states intricately connected to the control of the system.

In a CHA MAS, events corresponding to the actions of one agent can be indexed as $k = 0, 1, \ldots, N_i - 1$, where subscript $i$ represents the $i$th agent in the system. Each agent can be considered as a multi-stage process modeled as a single-server queuing system. The objective for the $i$th agent is to finish $N_i$ actions. In the CHA framework, once the agent takes an action, it cannot be interrupted, and continues its task until it finishes it. Let $a_k \in A$ represent an enabled action for $r_k$. As an agent takes an action $a_k$, the physical state, denoted by $x_{r_k}$, evolves according to the time-driven dynamics of the form

$$\dot{x}_{r_k} = f_{p_{r_k}}(x_{r_k}, u_{r_k}, t), \qquad (19)$$

where the initial condition for $x_{r_k}$ is $x_{r_k}(t_{r_{k-1}})$. The continuous control variable $u_{r_k}$ is used to attain a desired physical state.

If the time required to finish the $k$th action is $s_{r_k}$ and $\Gamma_{r_k}(u_{r_k}) \subset \mathcal{R}^n$ is a given set that defines $x_{r_k}$ satisfying the desired physical state, then the control signal $u_{r_k}$ can be chosen to satisfy the criteria

$$s_{r_k}(u_{r_k}) = \min \Big( t \geq 0 : \big( x_{r_k}(t_{r_{k-1}} + t) $$
$$= x_{r_k}(t_{r_{k-1}}) + \int_{t_{r_{k-1}}}^{t_{r_{k-1}}+t} f_{p_{r_k}}(x_{r_k}, u_{r_k}, \tau) d\tau \Big) $$
$$\in \Gamma_{r_k}(u_{r_k}) \Big) \qquad (20)$$

where we can assume that under the best circumstance (i.e. without any disturbance), $u_{r_k}$ is a fixed constant value at the abstract state $r_k$. The temporal state $t_{r_k}$ of the $k$th action represents the time when the action finishes.

In an MAS, we have two or more agents interacting with each other in order to achieve a global goal. Therefore, when the $i$th agent finishes its $k$th action $a_{k_i}$, it might have to wait for the $j$th agent finishes its $l$th task $a_{l_j}$ before the $i$th agent can start its $(k+1)$th task $a_{(k+1)_i}$. Assume that agent $i$'s tasks will depend only on agent $j$'s tasks. Let $t_{a_{(k+1)_i}}$ represent the starting time of the $(k+1)$th action for the $i$ agent. In this case $t_{a_{(k+1)_i}} \neq t_{r_{k_i}}$. Instead, $t_{a_{(k+1)_i}} = t_{r_{l_j}}$ where the temporal state $t_{r_{l_j}}$ represents the time when the $l$th action of the $j$th agent finishes. Therefore, the event-driven dynamics of the temporal state $t_{r_{k_i}}$ of the $i$th agent can be acquired by

$$t_{r_{k_i}} = \max(t_{r_{(k-1)_i}}, t_{r_{l_j}}) + s_{r_{k_i}}(u_{r_{k_i}}) \qquad (21)$$

where $l = 0, 1, 2, \ldots, N_j - 1$, $j$ is the index of the agent and $j \neq i$. Thus if action $a_{k_i}$ does not depend on the completion of any actions of any other agents, (21) can be simplified as

$$t_{r_{k_i}} = t_{r_{(k-1)_i}} + s_{r_{k_i}}(u_{r_{k_i}}). \qquad (22)$$

One may notice that we need to set $t_{0_i} = 0$ to make sure that $t_{r_{0_i}} = t_{r_{l_j}} + s_{r_{0_i}}(u_{r_{0_i}})$ and $t_{r_{0_i}} = s_{r_{0_i}}(u_{r_{0_i}})$ in case action $a_{0_i}$ does not depend on the completion of other actions.

In order to simplify the optimization problem for the $i$th agent, we assume that the temporal states $t_{r_{l_j}}$ of the $l$th action of the $j$th agent are known. Then we can see that when $t_{r_{l_j}} > t_{r_{k_i}}$, there is an idle period in the interval $[t_{r_{k_i}}, t_{r_{l_j}}]$ during which the physical state of the $i$th agent does not change.

Therefore, the optimization problem for the $i$th agent of the CHA framework becomes the optimization problem of the hybrid control layer (i.e., the optimization of the hybrid system that combines the time-driven dynamics in (18) and the event-driven dynamics in (21)).

The optimization problem to be solved for the $i$th agent has the general form

$$\min_{u_{r_0},\dots,u_{r_{N_i-1}}} \left( \sum_{k=0}^{N_i-1} L_{r_k}(t_{r_{k_i}}, u_{r_{k_i}}) \right) \tag{23}$$

where $L_{r_k}(t_{r_{k_i}}, u_{r_{k_i}})$ is the cost function defined for the $k$th action of the $i$th agent in the system. The cost function has been defined without including $x_{r_{k_i}}$ because $x_{r_{k_i}}$ is supposed to reach the desired value as defined in (20) which gives the $s_{r_k}(u_{r_k})$ for the $i$th agent here.

Notice that for the optimization problem defined in (23), the index $k = 0, 1, 2, \dots, N_i - 1$ does not count time steps, but rather asynchronous actions. Rewrite (23), we can represent the optimization problem using the following form

$$\min_{u_{r_0},\dots,u_{r_{N_i-1}}} \left( \sum_{k=0}^{N_i-1} \left( \phi(t_{r_{k_i}}) + \theta(u_{r_{k_i}}) \right) \right). \tag{24}$$

We propose the direct identification algorithm for the optimization of a CHA MAS as listed in Table 1.

Table 1. The Direct Identification Algorithm

| |
|---|
| for $i = 0$ to $N - 1$ |
| Step 0 initialize $k = 0$, $n = 0$; |
| while $n \leq N_i - 1$ do |
| Step 1 solve the sub-optimization problem $Q(k,n)$; |
| Step 2 identify busy periods: |
| if $t_{r_n} < t_{r_{l_j}^n}$ then |
| $k \leftarrow n + 1$ |
| end if |
| Step 3 increment index n |
| $n \leftarrow n + 1$ |
| end while |
| end for |

In the algorithm, $N$ represents the number of agents in the MAS. $Q(k,n)$ represents the sub-optimization problem of a sequence of actions that starts from state $r_k$ and ends at state $n$. In the algorithm, we assume that agent 0 depends on some external events, and agent $i$ depends on actions of previous agents, $i = 1, \dots, N - 1$. By applying the direct identification algorithm for the optimization of the CHA framework, we are able to optimize the performance and time of agents in a CHA MAS based on the given cost function.

# 4    Implementations

This section gives some experimental and simulation results for systems modeled using the CHA framework. The goal is to implement the tools we have proposed to develop the control algorithm for multi-agent systems. It is demonstrated that the framework is generic and can be applied to the control of both homogeneous and heterogeneous multi-agent systems.

The CHA framework is applied to control a heterogeneous multi-agent system. The control systems involved in this system are:

1. A mobile robot, iRobot ATRV-mini, which is a flexible, robust platform for either indoor or outdoor experiments and applications.
2. An overhead crane.
3. A robot manipulator, CRS F3, which can provide six degrees of freedom.

The goal for this control system is to develop cooperative tasks among the overhead crane, the mobile robot and the robot manipulator. As shown in Figure 2, the mobile robot picks up an object in the overhead crane's workspace (zone 1) and carries it to the manipulator's workspace (zone 2). The robot manipulator is mounted on a track which is an extra axis of control for the robot manipulator. The robot manipulator picks up the object from the mobile robot (zone 2) and delivers it to the other end of the track (zone 3). There are four landmarks set in the workspace to guide the mobile robot to move along the desired trajectory.

Since the analysis of this multi-agent system can be found in [16], we only briefly give the results of this multi-agent system.

## 4.1    The Mobile Robot

In this multi-agent system, the mobile robot is a nonholonomic mobile robot with kinematic constraints in the two dimensional workspace.

### The Action Executor
In order to pick up an object from the crane and deliver it to the robot manipulator, the mobile robot needs to execute the following desired actions:

(*search*) - turn the servo motor of the CCD camera to scan the environment in order to find the landmark;

(*align*) - align the robot body to the target;

(*vision navigation*) - move toward the landmark using the output of the fuzzy controller;

(*turn left*) - turn left $90^o$;

(*turn right*) - turn right $90^o$;

(*turn back*) - turn $180^o$;

(*move back*) - move backward into the loading area of the manipulator.

These actions are guaranteed to be executed by the hybrid action executor.

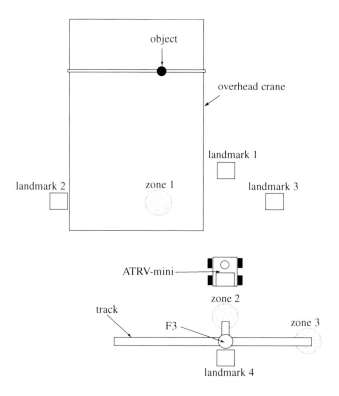

**Fig. 2.** The Setup of the Multi-Agent System

## The Coordination States

For the mobile robot, we have the following coordination states:

$1_m$ idle;
$2_m$ first landmark located;
$3_m$ aligned;
$4_m$ first landmark reached;
$5_m$ second landmark located;
$6_m$ second landmark reached;
$7_m$ loaded;
$8_m$ third landmark located;
$9_m$ third landmark reached;
$10_m$ fourth landmark located;
$11_m$ fourth landmark reached;
$12_m$ ready to be unloaded.

### 4.2 The Robot Manipulator

The robot manipulator is made by CRS Robotics. The robot is connected to the robot server that processes all the requests from the clients that command the robot manipulator. The robot server gets the control of the robot manipulator through the manipulator's controller. Then the robot server controls the joints through the functions described in the table.

**The Action Executor**
In order to pick up an object from the mobile robot and deliver it to the other side of the track, the robot server needs to send out the following desired actions to the robot manipulator:

(*approach*) - the tip of the manipulator approaches the object;
(*close gripper*) - the manipulator grabs the object;
(*move up*) - the tip moves up in order to pick up the object;
(*move left*) - the manipulator moves to the left end of the track;
(*turn left*) - the manipulator turns left $90^o$;
(*drop*) - the gripper opens in order to drop the object;
(*turn right*) - the manipulator turns right $90^o$;
(*move right*) - the manipulator moves right and goes back to the initial position.

These actions are guaranteed to be executed by the controller.

**The Coordination States**
For the robot manipulator, we have the following coordination states:

$1_r$ ready to pick up;
$2_r$ picked up.

## 4.3    Modeling the System Using the Proposed Framework

In this subsection, the CHA framework is applied to model the control of the multi-agent system with the mobile robot, the robot manipulator and the overhead crane.

**The Coordination Rule Base**
Based on the nature of this multi-agent system, the following coordination rule base is defined.
The optimal choices for the crane:

$1_c$ (*pick up*) $2_c$
$2_c$ (*move to*) $3_c$
$3_c$ (*put down*) $4_c$
$4_c$ (*pick up*) $5_c$
$5_c$ (*move to*) $6_c$
$6_c$ (*put down*) $1_c$

The constraints are:
If mobile robot is not at state $6_m$, (*put down*) is not allowed for $3_c$
The optimal choices for the mobile robot:

$1_m$ (*search*) $2_m$

$2_m$ (*align*) $3_m$
$3_m$ (*vision navigation*) $4_m$
$4_m$ (*turn left*) $5_m$
$5_m$ (*vision navigation*) $6_m$
$6_m$ (*null*) $7_m$
$7_m$ (*turn back*) $8_m$
$8_m$ (*vision navigation*) $9_m$
$9_m$ (*turn right*) $10_m$
$10_m$ (*vision navigation*) $11_m$
$11_m$ (*turn back*)(*move back*) $12_m$
$12_m$ (*null*) $1_m$

The constraints are:

If crane is not at state $5_c$, only (*null*) is allowed for $6_m$, and the transition is prohibited

If manipulator is not at state $2_r$, only (*null*) is allowed for $12_m$, and the transition is prohibited

The optimal choice for the robot manipulator:

$2_r$ (*move left*)(*turn left*)(*drop*)(*turn right*)(*move right*) $1_r$ $1_r$ (*approach*)(*close gripper*)(*move up*) $2_r$

The constraint for the manipulator is:

If the mobile robot is not at state $12_m$, (*approach*)(*close gripper*)(*move up*) is not allowed for $1_r$

Note that since there is no states defined between a series of actions for the manipulator, several actions can be executed in consequences. This series of actions can be thought as one single action.

The coordination rule base defines the optimal choices and the constraints for the agents to cooperate and coordinate with each other.

## 4.4    Simulation and Experimental Results

Before the agents are developed, simulation is implemented to verify the feasibility of the framework. In Figure 3, the simulation results for the cooperation and coordination between the mobile robot and the robot manipulator are given. The simulation is implemented using Matlab. The dimensions of the overhead crane, the mobile robot and the robot manipulator are measured to program the multi-agent system. In the figure, the round object represents the load of the overhead crane, while the square object represents the mobile robot. The trajectories of both the overhead crane and the mobile robot are given. From the figure, we can see that the overhead crane starts from the initial position and delivers the object to the loading area to wait for the mobile robot to pick up the object. The mobile robot follows the landmarks into the loading area and picks up the object. Then, the mobile robot turns around. For clarity, the path of the robot returning to the robot manipulator's track, which is shown as a long solid bar in the figure, is omitted. Note that even with a push, the robot can still follow the landmark and finish the desired task.

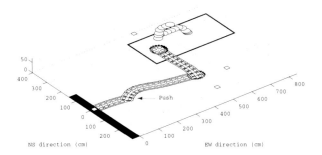

**Fig. 3.** Simulation Results for the Heterogeneous Multi-Agent System

The simulation result shows that the CHA framework can model the control of this multi-agent system. In addition, experiments also verify that the multi-agent system can achieve the desired goal successfully, for example, the overhead crane delivers an object in its workspace to the designated area, then with the vision navigation control, the mobile robot picks up the object from the crane's workspace and delivers it to the robot manipulator. The robot manipulator then picks up the object and transports it to its own workspace. The whole process involves cooperation, coordination and communication among multiple agents. By applying the CHA framework to the control of this multi-agent system, we are able to achieve coordinated control of the heterogeneous multi-agent system. The agents cooperatively work together to achieve the desired global goal.

### 4.5   The Stability of the Multi-Agent System

Based on the definition of the stability of CHA multi-agent systems, the multi-agent system with the mobile robot, the robot manipulator and the overhead crane is stable.

Proof:

1. Because of the design of the hardware and the software of the overhead crane, the mobile robot and the robot manipulator, the action executors can accomplish the hybrid actions. After the actions have been executed, the coordination state transitions to the next coordination state according to the coordination rule base.

2. As described above, all the actions taken by the overhead crane, the mobile robot and the robot manipulator are on the allowed event trajectories $\mathcal{E}_a$ which is governed by the coordination rule base. For $r \in R$, $a \in A$, $c \in C$, we have $r \models l_o$ holds and $(a, l_o) \in C$, $r \models l_c$ holds and $(a, l_c) \in C$ respectively. Recall that for the overhead crane, coordination states $1_c$ and $7_c$ represent state 'idle' and state 'put down without load' respectively. For the mobile robot, coordination states $1_m$ and $12_m$ represent 'idle' and 'ready to be unloaded' respectively. For the robot manipulator, coordination state $1_r$ represents 'ready to pick up'. The goal set is

the region around state $(7_c, 12_m, 1_r)$ for (crane, mobile robot, robot manipulator) and the origin set corresponds to the coordination state $(1_c, 1_m, 1_r)$. $\mathcal{E}_a$ leads the system to the goal set.

3. We wish to show that for this multi-agent system, the invariant set $R_m \subset R$ is stable in the sense of Lyapunov w.r.t. $\mathcal{E}_a$.

We use the metric defined by the Euclidean distance between each agent and the goal region along the allowed event trajectories $\mathcal{E}_a$, which is

$$\rho(R, R_m) = \Sigma_{i=1}^3 \{|x_i - \bar{x}_i| + |y_i - \bar{y}_i| + |z_i - \bar{z}_i|\} \tag{25}$$

in which the goal region is defined as $R_m = \{(7_c, 12_m, 1_r)\}$ which corresponds to $\{(\bar{x}_1, \bar{y}_1, \bar{z}_1), (\bar{x}_2, \bar{y}_2, \bar{z}_2), (\bar{x}_3, \bar{y}_3, \bar{z}_3)\}$. Subscript 1 is used to represent the overhead crane, 2 represents the mobile robot, and 3 represents the robot manipulator. Note that for the mobile robot $z_2 = \bar{z}_2$. We choose

$$V(R) = \rho(R, R_m), \tag{26}$$

then we need to show that in a sufficiently small r-neighborhood of the set $R_m$ the Lyapunov function $V$ has the required properties.

(1) If we choose $c_2 = c_1$, it is obvious that for all sufficiently small $c_1 > 0$, when $V(r) > c_2$ for $r \in$ r-neighborhood of $R_m$, $\rho(r, R_m) > c_1$.

(2) Same as above, if we choose $c_3 = c_4 > 0$ as small as desired, when $\rho(r, R_m) < c_3$ for $r \in$ r-neighborhood of $R_m$, we have $V(r) \leq c_4$.

(3) By design, all the agents only move toward the next goal along the allowed event trajectories $\mathcal{E}_a$, they don't go backward. So we have $V(\mathcal{R}(r_0, \mathcal{E}_k, k))$ a non-increasing function for $k \in \mathcal{Z}^+$, as long as $\mathcal{R}(r_0, \mathcal{E}_k, k) \in$ r-neighborhood for all $\mathcal{E}_k$ such that $\mathcal{E}_k \mathcal{E} \in \mathcal{E}_a(r_0)$.

## 4.6   Optimization

The mobile robot needs to finish various tasks in order to coordinate and cooperate with the overhead crane to achieve the final goal of the multi-agent system. The "quality" of the work of the mobile robot has to be maintained otherwise the cooperation would not be possible. For example, if the mobile robot turns too early or too late, the mobile robot would not stay on track when it gets into the overhead crane's workspace and it will fail the task. As a result, the whole system would not complete the mission. In particular, the mobile robot's actions involve searching the landmark, aligning its body to the target, landmark following, turning left, turning right, and so on. The goal of the whole system is that the mobile robot needs to move into the overhead crane's workspace and wait there, until the overhead crane finishes its dropping action. Then, the mobile robot takes the object that the overhead crane has dropped and transports the object out of the overhead crane's workspace. In [24], jobs are done for different products by the single-stage process. However, in our framework, each agent has to take various actions on a single object in order to accomplish the overall task.

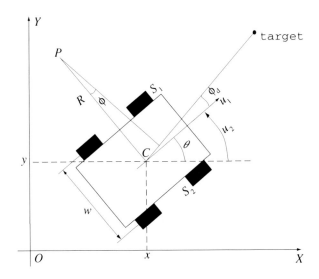

**Fig. 4.** The model of the 4-wheeled mobile robot

Of course, we can also apply the model to systems that require the same task to be done on multiple products.

The mobile robot needs a certain amount of time to finish the task. The position of the mobile robot is critical for the cooperation. The distances between the mobile robot and the landmarks are used to determine the quality level of the actions. In a sufficiently large empty space, a mobile robot can be driven to any position with any orientation, hence the robot's configuration space has three dimensions, two for translation and one for rotation. The physical state of the $k$th action of the mobile robot is denoted by $(x_{r_k}, y_{r_k}, \theta_{r_k})$ and represents the translational and rotational position of the mobile robot. Thus the mobile robot can be illustrated as the model shown in Figure 4, the wheels are aligned with the vehicle. The kinematic model of the mobile robot can be represented as

$$
\begin{aligned}
\dot{x}_{r_k} &= u_{1_{r_k}} \cos \theta_{r_k}, \\
\dot{y}_{r_k} &= u_{1_{r_k}} \sin \theta_{r_k}, \\
\dot{\theta}_{r_k} &= u_{2_{r_k}},
\end{aligned}
\tag{27}
$$

where $u_{1_{r_k}}$ corresponds to the forward velocity of the vehicle and the angle of the vehicle body with respect to the horizontal line is $\theta_{r_k}$, the angular velocity of the vehicle body is $u_{2_{r_k}}$, $(x_{r_k}, y_{r_k})$ is the location of the center point of the robot. The forward velocity $u_{1_{r_k}}$ and the angular velocity $u_{2_{r_k}}$ are used to control the motion of the mobile robot.

The path of the mobile robot can be obtained by integrating (27). Since the mobile robot needs to take a series of actions to achieve the global goal for the multi-agent system, we use the subscript $r_k$ to represent the abstract state of the mobile robot which indicates which action the robot is taking.

Next, the temporal state of the $k$th action of the mobile robot represents the time when the mobile robot starts the next action. Let $t_{r_{l_j}}$ be the ending time of the $l$th action of the $j$th agent that the mobile robot depends on in order to finish its own action, the event-driven dynamics describing the evolution of the temporal states of the mobile robot are given by

$$t_{r_k} = \max(t_{r_{(k-1)}}, t_{r_{l_j}}) + s_{r_k}(u_{r_k}), \tag{28}$$

where $s_{r_k}(u_{r_k})$ is the time for the mobile robot to finish the $k$th action. Notice that we have omitted subscript $i$ for simplicity. In this system, we consider two control objectives: 1) Increasing the performance of the mobile robot, and 2) Reducing the time for the mobile robot to finish all the tasks. Thus, the optimal control problem of interest can be expressed as:

$$\min_{u_{r_0},\ldots,u_{r_{N_i-1}}} \left( \sum_{k=0}^{N_i-1} \big(\phi(t_{r_k}) + \theta(u_{r_k})\big) \right). \tag{29}$$

The function $\phi(t_{r_k})$ above is the cost related to the time an action is finished and the time its depended task is finished. Generally, if the robot moves slower, its performance is better. The function $\theta(u_{r_k})$ is the cost function to penalize lower speed since we want the robot finishes its tasks faster. The cost function should be chosen by the designer based on the problem to be solved. As an example, we can choose $\phi(t_{r_k}) = |t_{r_k} - t_{r_{l_j'}}|$ and $\theta(u_{r_k}) = \frac{1}{u_{r_k}}$.

## 5   Conclusion and Future Work

In this chapter, the CHA framework is introduced for the control of multi-agent systems. In this framework, the control of multi-agent systems is considered as decentralized control and coordination of agents. In this framework, we include the concept of coordination states, a coordination rule base, an intelligent planner and a direct communication module in the intelligent coordination layer. The hybrid control layer of the framework takes the output of the intelligent coordination layer and generates discrete control signals and continuous control signals to control the overall process. The stability of the framework is also analyzed. We also include a direct identification algorithm in the framework for the optimization of multi-agent systems.

Although we have analyzed the optimization problem of the framework, it is unclear how to optimize MASs modeled using the CHA framework where actions are not sequential and agents do not wait for other agents to complete a specific task. A more generic optimization approach should be investigated and proposed. In addition, for the hybrid control layer, we need to define observers properly so that the intelligent coordination control layer will be able to generate actions accordingly. We need to solve problems related to partial observation or limited views of the workspace.

# References

[1] Russell, S., Norvig, P.: Artificial Intelligence - A Modern Approach. Prentice-Hall, Englewood Cliffs (2003)

[2] Genco, A.: Mobile Agents - Principles of Operation and Applications. WIT Press, Southampton, UK (2008)

[3] Brennan, W., Fletcher, M., Norrie, D.H.: An agent-based approach to reconfiguration of real-time distributed control systems. IEEE Transactions on Robotics and Automation 18(4), 444–449 (2002)

[4] Baeza, J., Gabriel, D., Bejar, J., Lafuente, J.: A distributed control system based on agent architecture for wastewater treatment. Computer-Aided Civil and Infrastructure Engineering 17(2), 93–103 (2002)

[5] Earl, M.G., D'Andrea, R.: Modeling and control of a multi-agent system using mixed integer linear programming. In: Proceedings of 41st IEEE Conference on Decision and Control, vol. 1, pp. 107–111 (2002)

[6] Crespi, V., Cybenko, G., Rus, D., Santini, M.: Decentralized control for coordinated flow of multi-agent systems. In: Proceedings of the 2002 International Joint Conference on Neural Networks (IJCNN), pp. 2604–2609 (2002)

[7] Garza, L.E., Cantu, F.J., Acevedo, S.: Faults diagnosis in industrial processes with a hybrid diagnostic system. In: Coello Coello, C.A., de Albornoz, Á., Sucar, L.E., Battistutti, O.C. (eds.) MICAI 2002. LNCS (LNAI), vol. 2313, pp. 536–545. Springer, Heidelberg (2002)

[8] Indrayadi, Y., Valckenaers, H.P., Van Brussel, H.: Dynamic multi-agent dispatching control for flexible manufacturing systems. In: Proceedings 13th International Workshop on Database and Expert Systems Applications, pp. 489–493 (2002)

[9] Pereira, G.A.S., Pimentel, B.S., Chaimowicz, L., Campos, M.F.M.: Coordination of multiple mobile robots in an object carrying task using implicit communication. In: Proceedings of the 2002 IEEE International Conference on Robotics & Automation, Washington DC, USA, pp. 281–286 (2002)

[10] Dasgupta, P.: A multiagent swarming system for distributed automatic target recognition using unmanned aerial vehicles. IEEE Transactions on Systems, Man, and Cybernetics - Part A: Systems and Humans 38(3), 549–563 (2008)

[11] Baker, C.R., Dolan, J.M.: Street Smarts for Boss. IEEE Robotics & Automation Magazine, 78–87 (March 2009)

[12] Broten, G.S., Mackay, D., Monckton, S.P., Collier, J.: The Robotics Experience - Beyond Components, and Middleware. IEEE Robotics & Automation Magazine, 46–54 (March 2009)

[13] Kress-Gazit, H., Conner, D.C., Choset, H., Rizzi, A.A., Pappas, G.J.: Courteous Cars - Decentralized Multiagent Traffic Coordination. IEEE Robotics & Automation Magazin, 30–38 (March 2008)

[14] Fregene, K., Kennedy, D., Wang, D.: HICA: A Framework for Distributed Multiagent Control. In: Proceedings of IASTED International Conference on Intelligent Systems and Control, pp. 187–192 (2001)

[15] Li, H., Karray, F., Basir, O., Song, I.: A framework for coordinated control of multi-agent systems and its applications. IEEE Transactions on Systems, Man, and Cybernetics - Part A: Systems and Humans 38(3) (2008)

[16] Li, H., Karray, F., Basir, O., Song, I.: A coordinated hybrid agent framework as applied to heterogeneous multi-agent Control. In: Proceedings of the 2nd International Conference on Humanoid, Nanotechnology, Information Technology, Communication and Control, Environment, and Management, Philippines (2005)

[17] Ferber, J.: Multi-Agent Systems - An Introduction to Distributed Artificial Intelligence. Addison-Wesley, Reading (1999)

[18] Hewitt, C.: Viewing Control Structures as Patterns of Passing Messages. Artificial Intelligence 8(3), 323–364 (1977)

[19] Erman, L., Hayes-Roth, F., Lesser, V., Reddy: The hearsay-ii speech understanding system: Integrating knowledge resolve uncertainty. ACM Computing Surveys 12 (1980)

[20] Passino, K.M., Burgess, K.L.: Stability Analysis of Discrete Event Systems. John Wiley & Sons, Inc., Chichester (1998)

[21] Passino, K.M., Michel, A.N., Antsaklis, P.J.: Lyapunov Stability of a Class of Discrete Event Systems. IEEE Transactions on Automatic Control 39(2), 269–279 (1994)

[22] Lygeros, J.: Hierarchical, Hybrid Control of Large Systems. PhD Thesis, University of California, Berkeley, CA (1996)

[23] Shoham, Y., Tennenholtz, M.: On social laws for artificial agent societies: off-line design. Artificial Intelligence 73(1–2), 231–252 (1995)

[24] Cho, Y.C., Cassandras, C.G., Pepyne, D.L.: Forward decomposition algorithms for optimal control of a class of hybrid systems. International Journal of Robust and Nonlinear Control 11 (2001)

# Chapter 4

# A Use of Multi-Agent Intelligent Simulator to Measure the Dynamics of US Wholesale Power Trade: A Case Study of the California Electricity Crisis

Toshiyuki Sueyoshi[1,2]

[1] Department of Management, New Mexico Institute of Mining and Technology, Socorro, NM 87801 USA
Tel.: 575-835-6452; Fax: 575-835-5498
toshi@nmt.edu
[2] National Cheng Kung University, College of Business, Department of Industrial and Information Management, Tainan, Taiwan

**Abstract.** During the summer (2000), wholesale electricity prices in California were approximately 500% higher than those during the same months in 1998-1999. This study proposes a practical use of a reengineered Multi-Agent Intelligent Simulator (MAIS) to numerically examine several reasons on why the crisis has occurred during May 2000-Janurary 2001. The proposed MAIS generates artificially numerous trading agents equipped with different learning capabilities and duplicates their bidding strategies in the California electricity markets during the crisis period. In this study, we confirm the methodological validity of MAIS by comparing the estimation accuracy of MAIS with those of the three well-known computer science techniques (Support Vector Machines, Neural Networks and Genetic Algorithms). This study also investigates the dynamic change on agent composition in a time horizon. This investigation finds that all agents gradually shift to multiple learning capabilities so as to adjust themselves to the price fluctuation of electricity. Finally, we apply the sensitivity analysis of MAIS to identify economic rationales concerning the crisis. The sensitivity analysis results in the estimation accuracy (91.15%) during the crisis period. This study finds that 40.46% of the price increase during the crisis period was due to an increase in marginal production cost, 17.85% to traders' greediness, 5.27% to a real demand change and 3.56% to market power. The remaining 32.86% came from other unknown market fundamentals and an estimation error. This numerical result indicates that the price hike has occurred due to an increase in fuel prices and real demand. The change of the two market fundamentals explained 45.73% (= 40.46% + 5.27%) of the price increase and fluctuation during the crisis. The responsibility of energy utility firms was 21.41% (= 17.85% + 3.56%).

**Keywords:** Agent-based Approach, Partial Reinforcement Learning, Electricity, Auction.

D. Srinivasan & L.C. Jain (Eds.): Innovations in MASs and Applications – 1, SCI 310, pp. 69–111.
springerlink.com © Springer-Verlag Berlin Heidelberg 2010

# 1  Introduction

An agent-based approach has been applied to investigate various complex systems (Tesfatsion, 2001). The applicability of the agent-based approach can be found in not only computer science but also social science. For example, Samuelson (2005) discussed the applications of agent-based approach to various social science systems from a perspective of optimization. Similarly, Makowski *et al.* (2005) assembled seventeen articles, all of which discussed various linkages between the agent-based approach and complex systems from the perspective of optimization. In a same vein, Jiang and Leyton-Brown (2007) discussed automated bidding agents within machine learning and Vorobeychik *et al.* (2007) explored the learning of agents in game theory.

An important feature of the agent-based approach is its role in modeling and simulation. The structure of a complex system is modeled with many different types of agents equipped with various learning capabilities. Their learning processes are usually characterized by adaptive learning through which agents determine their decisions by interacting with an environment. The proposed model is numerically expressed and examined by a simulation-based investigation. The incorporation of a problem structure, along with a simulation study, provides us with a numerical capability to handle a large complex system where many components have interactions among them. Consequently, the agent-based approach is gradually recognized as a new promising approach among researchers in natural and social sciences.

The agent-based approach has been recently applied to investigate a dynamic change of wholesale power trading. For example, Jacobs (1997), Bagnall (2000), Bunn and Oliveira (2001), Morikiyo and Goto (2004) developed multi-agent adaptive systems that incorporated a dynamic bidding process for power trading. See Axelrod (1997) that discussed a general view on complexity of agent's cooperation. Unfortunately, the previous studies described only the development of agent-based modeling and simulations. Almost no research discussed the development of an agent-based approach from the perspectives of machine learning.

To deal with such an issue, Sueyoshi (2010) as well as Sueyoshi and Tadiparthi (2005, 2007, 2008a,b,c) have recently explored the agent-based approach applied to investigate a dynamic change of wholesale power trading. Their first study (2005) discussed how to incorporate machine learning techniques into the agent-based approach. The second research (2007) extended their study (2005) by considering two groups of adaptive agents. One of the two groups incorporated multiple learning capabilities. The other group incorporated limited learning capability. Their research (2007) confirmed that agents with multiple learning capabilities did not have an advantage over agents with limited capability in predicting the market price of electricity. However, a theoretical extension of multiple learning capabilities had a potential for developing the agent-based approach for power trading, because experience in power trading needed other types of learning capabilities in addition to their prediction capability. The authors also developed an agent-based simulator, referred to "Multi Agent Intelligent Simulator (MAIS)," based upon the adaptive behaviors and algorithms discussed in the two studies (2005 and 2007). To extend the applicability of MAIS further, Sueyoshi and Tadiparthi (2008a) considered various influences of a transmission capacity limit on the wholesale electricity price. As a result of incorporating such a transmission issue in power trading markets, the study (2008a) could enhance the practicality and applicability of the proposed MAIS. The computer software and its operation on a computer monitor were documented in the fourth study (2008b). They applied the proposed MAIS to investigate an occurrence of the California electricity crisis (2008c) as a primary study. Sueyoshi (2010) extended the study (2008c) by incorporating additional market fundamentals and a dynamic process of power trading.

To extend the previous studies, this research needs to restructure the use of MAIS from the perspective of "partial reinforcement learning" in economics and machine learning in computer science (not conventional "reinforcement learning" as discussed by well-known researchers such as Sutton and Barto, 1999). After the completion of the MAIS reengineering, this study applies the software to investigate why the California electricity crisis has occurred in 2000–2001. In this study, the performance of MAIS is compared with the three well-known approaches in computer science (i.e., Support Vector Machines, Neural Networks and Genetic Algorithms), using a real data set on power trading related to the California electricity crisis (2000–2001). The methodological comparison examines whether the proposed MAIS performs as well as the other well-known approaches in predicting a dynamic price fluctuation of wholesale electricity during the crisis. After confirming the methodological validity of MAIS, this study investigates the dynamic change on agent composition in a time horizon. This investigation finds that all agents gradually shift to multiple learning capabilities so as to adjust themselves to the dynamic price fluctuation of electricity during the crisis. Finally, we apply the sensitivity analysis of MAIS to identify rationales regarding why the crisis has occurred in California. We compare economic implications obtained from the MAIS's sensitivity analysis with those of the very well known economic studies (i.e., Joskow and Kahn, 2002, Borenstein et al., 2002 and Wolak, 2003) regarding the California electricity crisis. The application of MAIS provides new evidences and policy implications regarding the crisis, all of which cannot be found in the previous economic studies. That is the research task of this study.

It is important to note that the previous studies (Sueyoshi, 2010; Sueyoshi and Tadiparthi, 2008c) investigated the California electricity crisis, using the agent-based approach. Unfortunately, their studies could not document the whole computation process and results because of a page limit in journal publication. This study reorganizes the whole research process and adds new computational results (i.e, a learning rate), not published in the previous studies. Thus, this research is the final version of a series of studies on the agent-based approach applied to the California electricity crisis.

The remaining structure of this article is organized as follows: The next section briefly reviews underlying economic concepts used for the proposed MAIS reengineering. Section 3 describes the market clearing scheme in the US wholesale power markets. A numerical model for wholesale power trading is incorporated in the proposed MAIS. Section 4 discusses adaptive behaviors and algorithms incorporated in the MAIS. Section 5 applies the MAIS to a data set related to California electricity markets in the crisis period. The performance of the MAIS is compared with those of the other well known approaches in computer science that estimate a dynamic fluctuation of wholesale electricity price. The MAIS-based sensitivity analysis investigates rationales regarding why the electricity crisis has occurred in California. Section 6 summarizes this research along with future research agendas.

## 2   Economic Concepts for Agent-Based Learning Systems

In computer science, reinforcement learning is a subarea of machine learning that is concerned with how an agent takes actions in an environment to maximize some notion of a reward. Reinforcement learning algorithms attempt to find a policy that maps states of the world to the actions that the agent should to take in those states. An application example of reinforcement learning in power trading can be found in Nanduri and Das (2007) and the other studies described in Section 1. See Sutton and Barto (1999) for a general framework of reinforcement learning and applications. When applying the reinforcement learning to investigate the dynamic fluctuation of electricity price, this study needs to mention the following four concerns:

*Multiple Learning Rate*: Conventionally, reinforcement learning incorporated a single "learning rate" that made a linkage from a reward at step t to the one at step t+1. The previous research paid attention to the convergence of the learning rate. The use of a single learning rate was due to computational convenience. The single learning rate is not sufficient for representing a complex problem such as US wholesale power markets. The US wholesale power markets (like PJM: a large regional transmission controller covering the region of Pennsylvania-New Jersey-Maryland) consist of Day Ahead (DA), Hour Ahead (HA) and Real Time (RT) markets. Moreover, in a spot market of electricity, each software agent (being as an artificial trader) must consider his bidding price and quantity. Thus, it is easily imagined that we need to consider the multiple learning rates of agents in investigating the dynamics of power trading. The proposed MAIS has such a numerical capability to incorporate multiple learning rates as specified by a user. Hence, the proposed MAIS can provide more reliable information on the dynamics of power trading than the previous studies (e.g., Nanduri and Das, 2007), which are based upon the conventional use of reinforcement learning.

*Reinforcement Learning vs. Partial Reinforcement Learning*: In addition to the single learning rate, the conventional use of reinforcement learning in machine learning considers that an agent selects a strategy to maximize some notion of a long-term reward. The reward allocation is usually "deterministic" in the use of reinforcement learning. The concept is inconsistent with the reality of power trading. In a power market, traders make their bidding decisions within a limited time. For example, in the RT market, traders must make their bidding decisions within an interval of 5 minutes. Hence, it is impossible for traders to think about their rewards in a long term horizon. Furthermore, the conventional reinforcement learning assumes that a reward is always given to an agent in each learning process. Such an underlying assumption is questionable because a trader cannot always win in a power market. Sometime, he wins and obtains a reward, but he often loses in the electricity market. Thus, the reward is not always given to each agent and it is allocated by a "stochastic" process. This type of reinforcement learning (a reward in a short time horizon and a limited number of chances to obtain the reward) is referred to as "Partial Reinforcement Leaning". See the reseaerch of Bereby-Meyer and Roth (2006) for a description on the important of partial reinforcement learning from the perspective of game theory. The reengineering of MAIS discussed in this study is based upon the economic concept of partial reinforcement learning.

*Law of Effects and Power Law of Practice*: In addition to the above two concerns, we need to mention that the history of reinforcement (or partial reinforcement) learning originated from animal and human learning psychology, which was first explored in Thorndike (1989). According to the study, "choices that have led to good outcomes in the past are more likely to be repeated in the future" (Erev and Roth, 1998, p.859). The observation was widely known as "Law of Effect" which has been a basic principle for human's adaptive behaviors. Robustness on learning was also found in a learning curve that tended to be steep initially and then be flat. The observation was known as "Power Law of Practice" which dated back at least to Blackburn (1936). See Erev and Roth (1998, p. 859).

In examining the previous research, we find two important lessons for the reengineering of MAIS. First, human intelligence is different from machine intelligence in terms of these learning functions such as a computational capability and a memory space. The concepts of human learning such as "Law of Effect" and "Power Law of Practice" need to be restructured for the development of machine learning. Second, the previous research on machine learning has been focused upon the reinforcement learning of an individual or a single group of people. In contrast, this study is interested in the partial reinforcement learning of multiple adaptive agents who have different preferences on rewards and different bidding strategies in a competitive electricity market. Hence, the partial reinforcement learning discussed in this study needs to be conceptually and computationally reorganized in such a way that it can be fitted for power trading of wholesale electricity.

*Learning from Mistakes*: It is true that we learn many things from our mistakes. In a similar manner, traders adjust their bidding strategies from their mistakes (loses) in power trading. The principle of "Learning from Mistakes" needs to be incorporated into the proposed MAIS. Hence, the partial reinforcement learning is separated into positive and negative experiences in this study. These experiences act as feedbacks for partial reinforcement learning within the proposed MAIS, where the positive experience is originated from a successful bid and the negative experience comes from a failed bid. See, for example, Chialvo and Bak (1999) and Si and Wang (2001) that discussed the principle of "Learning from Mistakes" from the perspectives of active synaptic connections and neural networks. According to their studies, we have a strong desire to avoid making the same mistake within our brains. The aspect on learning from mistakes has been insufficiently considered in the previous studies on agent-based approaches on power trading.

## 3   Market Clearing Scheme

### 3.1   US Wholesale Power Trading and California ISO (Sueyoshi and Tadiparthi, 2008c)

The electric power business is separated into the following four functions: (a) generation, (b) transmission, (c) distribution and (d) retailing. The main parts are generation and transmission, both of which are traded in US wholesale markets of electricity. Generally speaking, two types of transactions can be found in the US wholesale markets. One of the two is a bilateral (usually long-term) exchange contract between a generator(s) and a wholesaler(s). The other is a short-term auction-based transaction. In the short-term transaction, a market operator accepts biddings from both generators and wholesalers. The operator then determines the market price and quantity of electricity. Thus, this type of wholesale market is controlled and coordinated by ISO (Independent System Operator) and RTO (Regional Transmission Organization) such as PJM that operates not only a wholesale market but also a transmission market (Wilson, 2002).

The US wholesale market is functionally separated into (a) a power exchange market and (b) a transmission market. The power exchange market is functionally separated into the four markets: (a1) a Real Time (RT) market, (a2) an Hour-Ahead (HA) market, (a3) a Day-Ahead (DA) market and (a4) a long-term contract market. These markets need to integrate the supply capabilities to satisfy a constantly changing demand on electricity. Coordinated auctions are used for the first three types of exchange market. Bilateral contracts are found in the fourth market (Stoft, 2002).

*California ISO and PX (Power Exchange)*: Before the electricity crisis (2000/2001), utility firms in California use power trading in short-term auction markets: DA market run by PX and HA market run by ISO. Both PX and ISO are market institutions for power trading. The DA market is a "financial and forward" market because all the transactions in the DA market stop one day before real delivery of electricity. The bidding decisions in the DA market are determined by the speculation of traders. Meanwhile, the HA market can be considered as a "physical and spot" market, because the delivery of power in the HA market is not optional like in the DA market. All traders enter the HA market to correspond to actual flows of electricity. Hence, the aspect of financial speculation is very limited in the HA market. Thus, HA is a physical market. In the HA market, traders need to make their decisions within a limited time (one hour). So, it can be considered as a physical spot market in this study.

In this study, a wholesale power exchange market is separated into multiple zones based on the geographical location of nodes and a transmission grid. Each zone consists of several generators and loads. There are two types of transmission connections such as intra-zonal link and inter-zonal link. Intra-zonal links are connections that exist among generators and loads within a zone. Inter-zonal links are connections that exist among the zones. A common MCP (Market Clearing Price) exists if these zones are linked together. However, if those are functionally separated by a capacity limit on an interconnection line, then these zones have different "locational marginal prices" as MCPs.

To explain why we need to consider the influence of capacity limits on transmission on a wholesale power market, we consider the wholesale market operated by California ISO as a real example. The market is divided into three zones for the purpose of pricing; NP-15 is in the north, SP-15 is in the south, and ZP-26 is in the center of the state. The central zone (ZP-26) has only two transmission links, one to northern path (NP-15) and one to southern path (SP-15). The northern path and southern path are not directly connected to each other. If they need excess electricity, they have to obtain it from other states as shown in Figure 1.

**Fig. 1.** Three Zones in California ISO
(Source: http://www.ucei.berkeley.edu/)

## 3.2 Market Clearing Scheme for Multiple Market Zones (Sueyoshi and Tadiparthi, 2008a)

This study considers a wholesale electricity exchange market which consists of Z zones (z = 1, .., Z) and all the zones are connected to each other by transmission links. The subscription "z" indicates the z-th zone. Figure 2 illustrates an algorithm for clearing a wholesale electricity market with multiple zones. In Figure 2, we assume that all links are not limited in the proposed MAIS. Then, the assumption is dropped because we are interested in investigating the influence of a capacity limit on the market clearing process.

The terms used in Figure 2 are summarized as follows: (a) AG (Allocated Generator): a group of generators that are allocated for current generation in all zones. (b) UAG (Unallocated Generators): a group of generators that are not allocated for current generation in all zones. (c) C (Cleared Zones): a set of cleared zones where load requirements are fulfilled. (d) NC (Not Clear Zones): a set of zones that are not cleared. (e) PG (Participating Generators): a group of generators that can participate in the market clearing process of the z-th zone. (f) TCG (Transmission Connected Generators): a group of generators whose transmission lines are connected to the z-th zone. (g) MCPG (Market Clearing Participating Generators): a group of generators who can belong to both PG and TCG.

In addition to these subsets, Figure 2 uses a subscript "t" to indicate the t-th period for real delivery of electricity. Two subscripts "i" and "j" indicate the i-th generator and the j-th wholesaler, respectively. To avoid a descriptive duplication, the study often omits the descriptions on subscripts through this manuscript.

As a preprocessing step in Figure 2, ISO forecasts a total demand ($D_{(z)t}$) for the z-th zone at the t-th period. The total demand is specified as $D_{(z)t} = \sum_j d_{j(z)t}$, where

$d_{j(z)t}$ is the demand forecast for the t-th period by the j-th wholesaler in the z-th zone. The total supply capability ($S_{(z)t}$) for the z-th zone is obtained from all generators and is expressed by $S_{(z)t} = \sum_i s^m_{i(z)t}$, where $s^m_{i(z)t}$ is the maximum generation capacity of the i-th generator in the z-th zone. It is assumed that the total sum of maximum generation capacities ($\sum S_{(z)t}$) is larger than or equal to the total forecasted demand ($\sum D_{(z)t}$) for all periods (t = 1, .., T). Hence, an excess amount of power supply for the zone is specified by $E_{(z)t} = S_{(z)t} - D_{(z)t}$ (for all z and t).

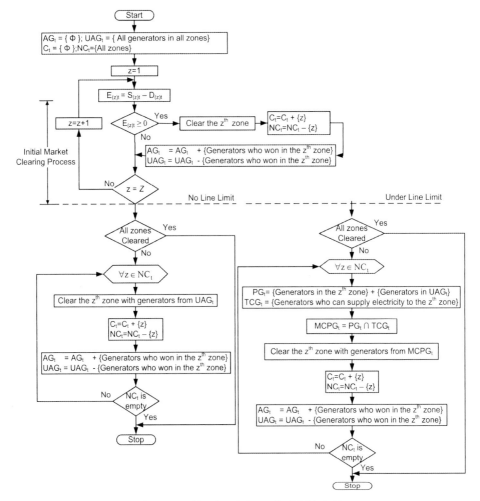

**Fig. 2.** Market Clearing Scheme for Multiple Zones

a) AG (Allocated Generator), UAG (Unallocated Generator), C (Cleared Zone), NC (Not Cleared Zone), PG (Participating Generator), TCG (Transmission Connected Generator) and MCPG (Market-Clear Participating Generator).

In Figure 2, a zone is said to be "cleared" if all the load requirements in a zone are satisfied. That is, if $S_{(z)t} \geq D_{(z)t}$ , then the z-th zone is cleared. Otherwise, ($S_{(z)t} < D_{(z)t}$), the zone is not cleared. In this case, ISO arranges additional electricity by using extra (usually expensive) generators within its own zone and/or obtaining electricity from other linked zone(s). In the former case, ISO needs to reexamine the selection of generators and dispatch scheduling within the zone. In the latter case, ISO needs to examine whether unused generators are available in other zones.

This initial clearing process of Figure 2 continues sequentially for all zones, as depicted in the upper part of the figure. In the initial stage, AG is empty, UAG covers all the generators in a power market (with multiple zones), C is an empty set, and NC consists of all the Z zones. At the end of the initial market clearing process, all the zones are classified into either cleared or not cleared. In Figure 2, the "win" of a generator implies that he bids in the power market and obtains a generation opportunity. So, the generator becomes a member of AG. The market clearing process is repeated Z times because the wholesale market consists of Z zones.

After the initial clearing process is completed, ISO needs to clear all zones where demand is larger than or equal to supply. All these zones belong to NC. Since the market clearing process depends upon whether there is any capacity limit on links, this study first describes an algorithm under no line limit. See the left (no line limit) of Figure 2. To clear the z-th zone in NC, ISO prepares a market for the zone where all unused generators in UAG can participate in the bidding process. Since there is no line limit in transmission, the bidding process of those generators works as a single market entity.

The right hand side of Figure 2 indicates an algorithm within ISO when the z-th zone is not cleared and the links between the not-cleared zone and the other zones have a capacity limit. To clear the zones in NC, ISO identifies not only all generators in NC but also UAG in other zones. Such a group of generators is expressed by PG. In this stage, ISO needs to consider both (a) whether there is a link(s) between the not-cleared z-th zone and the other zones and (b) whether the link has a capacity limit for transmission. The issue regarding whether all generators should be connected to the not-cleared z-th zone through a link is solved by identifying a group of generators whose zones have a link to the not-cleared zone. The group of generators is expressed by TCG. Consequently, a group of generators, which can participate into the market clearing process of the z-th zone, is selected from MCPG that is expressed by an intersection between PG and TCG. The z-th zone is cleared by using all generators in MCPG. Based on the market clearing effort, all the sets (C, NC, AG, UAG, PG, TCG and MCPG) are identified and updated as depicted in Figure 2.

### 3.3  Bidding Strategy of Agents (Sueyoshi and Tadiparthi, 2008a)

Figure 3 depicts the bidding strategy of agents in a market with multiple zones that are connected by links with a capacity limit. Let us select a zone (say, the z-th zone) from NC (a set of zones that are not cleared). Then, n generators for $i(z)t = 1,..,n(z)t$ are selected from MCPG and k wholesalers for $j(z)t = 1,..,k(z)t$ are selected from the z-th zone for the t-th period. In Figure 3, wholesalers are selected from the z-th zone. Meanwhile, generators are selected from not only the z-th zone (as a target zone) but also the z'-th zone where the z'-th zone is a representative of the other zones. So, z' indicates multiple zones in the manner that $z'(\neq z) = 1, 2, ..,Z$. Generators in both the z-th and the other zones as well as wholesalers in the z-th zone can enter a market clearing process of the z-th zone. The zone market consists of DA and HA. The superscripts "1" and "0" indicate DA and HA markets, respectively. We drop the description on the subscripts "z" and "t" to avoid descriptive redundancy. The terms used in Figure 3 are summarized as follows:

$s_{i(z)t}^{m}$      : the maximum amount of power generation capacity of the i-th generator in the z-th zone.

$s_{i(z)t}^{1}$      :  the bidding amount ($s_{i(z)t}^{1} \leq s_{i(z)t}^{m}$) of the i-th generator for DA in the z-th zone.

$\alpha_{i(z)t}$ ($0 \leq \alpha_{i(z)t} \leq 1$): a decision parameter of the i-th generator that expresses the ratio of the bidding amount ($s_{i(z)t}^{1}$) to the maximum generation capacity ($s_{i(z)t}^{1} = \alpha_{i(z)t}s_{i(z)t}^{m}$).

$MC_{i(z)t}^{1}$   : the marginal cost of the i-th generator in DA.

$p_{i(z)t}^{1} [= MC_{i(z)t}^{1}/(1-\beta_{i(z)t})]$ : the bidding price of the i-th generator for DA.

$\beta_{i(z)t}$ ($0 \leq \beta_{i(z)t} < 1$): a mark-up ratio of the i-th generator that indicates how much the bidding price is increased from the marginal cost.

$e_{j(z)t}$      : the demand estimate of the j-th wholesaler at the t-th period of DA.

$w_{j(z)t}^{1}$      : the price estimate the j-th wholesaler from the demand estimate ($e_{j(z)t}$).

$d_{j(z)t}^{1} [= \delta_{j(z)t}e_{j(z)t}]$ : the bidding amount of the j-th wholesaler for DA.

$\delta_{j(z)t}$ ($0 \leq \delta_{j(z)t} \leq 1$): a decision parameter to express the reduction of the bidding amount from the demand estimate ($e_{j(z)t}$).

$p_{j(z)t}^{1} (= \lambda_{j(z)t}w_{j(z)t}^{1})$ : the bidding price of the j-th wholesaler for DA.

$\lambda_{j(z)t}$ ($0 \leq \lambda_{j(z)t} \leq 1$) : a decision parameter to indicate the reduction of the bidding price from the price estimate.

$\hat{p}_{zt}^{1}$      : the market clearing price for the z-th zone in DA.

$\hat{s}_{i(z)t}^{1}$      : the power request to the i-th generator in DA.

$\hat{d}_{j(z)t}^{1}$      : the power allocation to the j-th wholesaler in DA.

$s_{i(z)t}^{0}$      : the bidding amount ($s_{i(z)t}^{0} \leq s_{i(z)t}^{m}$) of the i-th generator for HA.

$\upsilon_{i(z)t}$ ($0 \leq \upsilon_{i(z)t} \leq 1$): a decision parameter of the i-th generator to express

$$s_{i(z)t}^{0} = \upsilon_{i(z)t}s_{i(z)t}^{m}.$$

$p^0_{i(z)t}$ $[= MC^0_{i(z)t}/(1-\eta_{i(z)t})]$: the bidding price of the i-th generator for HA.

$MC^0_{i(z)t}$ : the marginal cost of the i-th generator in HA.

$\eta_{i(z)t}$ : a mark-up ratio ($0 \le \eta_{i(z)t} < 1$) of the i-th generator for HA.

$r_{j(z)t}$ : the real demand to the j-th wholesaler from end-users in the z-th zone at the t-th period.

$d^0_{j(z)t}$ : the bidding amount of the j-th wholesaler for HA.

$\hat{p}^0_{zt}$ : the market clearing price of the z-th zone for HA.

$\hat{s}^0_{i(z)t}$ : the power request to the i-th generator in HA.

$\hat{d}^0_{j(z)t}$ : the power allocation for the j-th wholesaler in HA.

The above description related to the z-th zone can be directly applied to those of the z'-th zone. An exception is found in a line capacity on transmission from the z'-th zone to the z-th zone. Figure 3 includes the following terms related to a line capacity:

$\alpha_{i(z' \to z)t}$ : a decision parameter of the i-th generator in the z'-th that expresses the ratio of the bidding amount ($s^1_{i(z')t}$) in DA to the maximum generation capacity ($s^1_{i(z')t} = \alpha_{i(z' \to z)t} s^m_{i(z')t}$). The generator sends the amount of electricity from the z'-th zone to the z-th zone in DA.

$\upsilon_{i(z' \to z)t}$ : a decision parameter of the i-th generator in the z'-th that expresses the ratio of the bidding amount ($s^0_{i(z')t}$) in HA to the maximum generation capacity ($s^0_{i(z')t} = \upsilon_{i(z' \to z)t} s^m_{i(z')t}$). The generator sends the amount of electricity from the z'-th zone to the z-th zone in HA.

$\ell^m_{(z' \to z)t}$ : the maximum capacity limit on an interlink from the z'-th zone to the z-th zone.

*Supply Side in DA*: All the generators are classified into two groups: (a) generators within the z-th zone and (b) generators in the other zones (say, the z'-th zone in MCPG). In Figure 3, the z-th zone is a specific zone whose market needs to be cleared. Meanwhile, the z'-th zone indicates all the other zones whose transmission lines are connected to the z-th zone ($z' \ne z \in$ MCPG).

The i-th generator selected from the z-th zone bids $s^1_{i(z)t}$ for DA. The bidding amount is expressed by $s^1_{i(z)t} = \alpha_{i(z)t} s^m_{i(z)t}$ where $\alpha_{i(z)t}$ is a decision parameter of the i-th generator to express the ratio of the bidding amount to his maximum generation capacity ($s^m_{i(z)t}$). Meanwhile, the i-th generator selected from the other z'-th zone in MCPG bids $s^1_{i(z')t}$ for DA ($s^1_{i(z')t} \le \min\{\alpha_{i(z' \to z)t} s^m_{i(z')t}, \ell^m_{(z' \to z)t}\}$). In determining his bidding amount, the generator needs to consider both how much he can send electricity to the z-th zone and the maximum transmission limit on an interlink from the z'-th zone to the z-th zone. His bidding strategy is expressed by $\alpha_{i(z' \to z)t} s^m_{i(z')t}$ where $\alpha_{i(z' \to z)t}$ is a decision parameter to express the ratio of the bidding amount to the maximum capacity ( $s^m_{i(z')t}$ ). Furthermore, $\ell^m_{(z' \to z)t}$ stands for the maximum capacity limit on an interlink from the z'-th zone to the z-th zone. Thus, the generator in the z'-th zone must consider not only his generation capacity but also the line capacity for transmission.

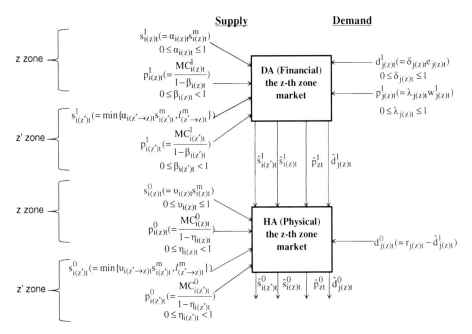

**Fig. 3.** Bidding Strategies for Multiple Zone Markets

Let $MC^1_{i(z)t}$ be the marginal cost of the i-th generator in the z-th zone at the t-th period. The generator determines a bidding price ( $p^1_{i(z)t}$ ) for DA by $p^1_{i(z)t} = MC^1_{i(z)t} / (1 - \beta_{i(z)t})$. Here, $\beta_{i(z)t}$ ($0 \le \beta_{i(z)t} < 1$) is a mark-up ratio of the generator for DA. The mark-up ratio expresses numerically how much the bidding price is inflated from the marginal cost. The mark-up rate reflects the price strategy of the generator toward DA. Considering different magnitudes of the mark-up ratios, the proposed MAIS examines various price strategies for DA. Since a line limit does not influence the pricing strategy, the description on his pricing strategy can be applied to another generator in the z'-th zone.

*Demand Side Strategy in DA*: A wholesaler predicts an expected amount of electricity demanded on a delivery day, using a forecasting method. Let $e_{j(z)t}$ be the demand estimated by the j-th wholesaler in the z-th zone. The wholesaler determines a bidding price ($w^1_{j(z)t}$) by using an inverse function (IF) of demand. That is, $w^1_{j(z)t} = IF(e_{j(z)t})$. See Sueyoshi and Tadiparthi (2005, 2008b) for a visual description on the inverse function. In the proposed MAIS, the wholesaler makes a demand bid ($d^1_{j(z)t}$) whose amount is less than or equal to the demand estimate ($e_{j(z)t}$) so that it can be expressed by $d^1_{j(z)t} = \delta_{j(z)t} e_{j(z)t}$, where $\delta_{j(z)t}$ ($0 \le \delta_{j(z)t} \le 0$) is a parameter to indicate how each bid is strategically reduced from the demand estimate. Similarly, the bidding price is determined by $p^1_{j(z)t} = \lambda_{j(z)t} w^1_{j(z)t}$. Here, $\lambda_{j(z)t}$ ($0 \le \lambda_{j(z)t} \le 1$) is a parameter for price adjustment from the estimated price.

*DA Market*: After ISO obtains their bids from all generators and all wholesalers, the organization allocates $\hat{s}^1_{i(z)t}$ to the i-th generator within the z-th zone and $\hat{s}^1_{i(z')t}$ to the i-th generator in the other z'-th zone. The real allocations are different from their bidding amounts ($s^1_{i(z)t}$ and $s^1_{i(z')t}$). ISO also determines $\hat{d}^1_{j(z)t}$ (a real allocation) to each wholesaler in the z-th zone. The market clearing price ($\hat{p}^1_{zt}$) is equally allocated to both supply and demand sides in the DA market.

*Supply Side in HA*: The i-th generator selected from the z-th zone bids $s^0_{i(z)t}$ for DA where the bidding quantity is expressed by $s^0_{i(z)t} = \upsilon_{i(z)t} s^m_{i(z)t}$ and. $\upsilon_{i(z)t}$ is a decision parameter to express the ratio of the bidding amount for HA to his maximum generation capacity ($s^m_{i(z)t}$). Meanwhile, the i-th generator selected from the other z'-th zone in MCPG bids the amount $s^0_{i(z')t}$ for HA that is expressed by $s^0_{i(z')t} \le \min\{\upsilon_{i(z' \to z)t} s^m_{i(z')t}, \ell^m_{(z' \to z)t}\}$. His bidding amount for HA is determined by comparing his power generation capability supplied to the z-zone with the maximum capacity limit on an interlink from the z'-th zone to the z-th zone. The power generation capability is expressed by $\upsilon_{i(z' \to z)t} s^m_{i(z')t}$ where $\upsilon_{i(z' \to z)t}$ is a decision parameter to express the ratio of the bidding amount to the maximum capacity. Furthermore, $\ell^m_{(z' \to z)t}$ stands for the maximum capacity limit on an interlink from the z'-th zone to the z-th zone. Thus, the generator in the z'-th zone considers both his generation capacity and the line capacity for transmission.

The i-th generator in the z-th zone bids $p^0_{i(z)t} [= MC^0_{i(z)t} / (1 - \eta_{i(z)t})]$, where $\eta_{i(z)t}$ is a mark-up rate ($0 \le \eta_{i(z)t} < 1$). In a similar manner, the generator in the z'-th zone bids $p^0_{i(z')t}$ $[= MC^0_{i(z')t}/(1-\eta_{i(z')t})]$. The mark-up ratio indicates each generator's pricing strategy in HA.

*Demand Side Strategy in HA*: In the HA market, all the wholesalers specify their required quantities on electricity. Since they need to satisfy the actual amount from end users, they must purchase all the necessary electricity from DA and/or HA markets. Let $r_{j(z)t}$ be a real demand to the j-th wholesaler on the delivery day. The wholesaler specifies the purchasing amount, $d^0_{j(z)t}$ $[= r_{j(z)t} - \hat{d}^1_{j(z)t}]$ in the HA market.

*HA Market*: ISO adjusts all the requests from market participants to determine $\hat{s}^0_{i(z)t}$ and $\hat{s}^0_{i(z')t}$ (real requests to all generators), $\hat{d}^0_{j(z)t}$ (a real allocation for the wholesaler) and $\hat{p}^0_{zt}$ (a market clearing price) for the HA market of the z-th zone.

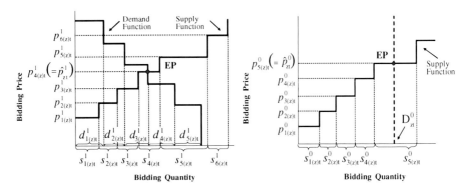

**Fig. 4.** Clearing Scheme for DA          **Fig. 5.** Clearing Scheme for HA

*Auction Process*: Figure 4 visually describes the market clearing mechanism for the DA market. ISO reorders the biddings of generators and wholesalers. In the figure, the supply side combinations ($s^1_{i(z)t}$ and $p^1_{i(z)t}$) for all i = 1, .., n in the z-th zone are reordered according to the ascending order of these bidding prices. The bidding process can be considered as a sealed English auction with acceptance of multiple bids. In contrast, the demand side combinations ($d^1_{j(z)t}$ and $p^1_{j(z)t}$) are reordered according to the descending order of these bidding prices. The bidding process is a sealed Dutch auction with acceptance of multiple bids. In Figure 4, ISO allocates the generation amount ($s^1_{1(z)t}$) of the first generator to satisfy the demand ($d^1_{1(z)t}$) of the first wholesaler. Such a power allocation is continued until an Equilibrium Point (EP) is found in the DA market. The equilibrium point is identified on EP, where the four generators are used to satisfy the demand required by the three wholesalers.

Consequently, $p^1_{4(z)t}$ (the bidding price of the fourth generator) becomes the market price ($\hat{p}^1_{zt}$) for all participating traders in the DA market for the z-th zone.

Figure 5 depicts the market clearing mechanism for the HA market, where generators bid both $s^0_{i(z)t}$ and $p^0_{i(z)t}$. The combination of quantities and bidding prices are reordered according to the ascending order of these bidding prices. Meanwhile, wholesalers submit their demands ($d^0_{j(z)t}$), but not the bidding prices, because the HA market is a physical market where the demand of end users must be always satisfied. In Figure 5, ISO accumulates the generation amounts until the total demand is satisfied. In the figure, $D^0_{zt}$ is such a point, where

$$D^0_{zt} = \Sigma_j d^0_{j(z)t} = \Sigma_j \left( r_{j(z)t} - \hat{d}^1_{j(z)t} \right).$$ In Figure 5, an equilibrium point is identified as EP, where five generators are used to satisfy the total demand required by wholesalers. Consequently, $p^0_{5(z)t}$ (the bidding price of the fifth generator) becomes the market price ($\hat{p}^0_{zt}$) for all participating traders in the HA market of the z-th zone.

*Reward to Agents*: Table 1 summarizes a reward of the i-th generator in the z-th zone at the t-th period. Each cell of Table 1 indicates a winning reward of the generator. For example, if $\hat{p}^1_{zt} < p^1_{i(z)t}$, then he cannot have any chance to generate electricity, so having no reward in DA. Conversely, if $\hat{p}^1_{zt} \geq p^1_{i(z)t}$, then he receives a reward ($\hat{p}^1_{zt} - MC_{i(z)t})\hat{s}^1_{i(z)t}$), as listed in the cell under "DA" and "within the z-zone." In a similar manner, if $\hat{p}^0_{zt} \geq p^0_{i(z)t}$ in HA, then he obtains ($\hat{p}^0_{zt} - MC_{i(z)t})\hat{s}^0_{i(z)t}$.

In addition to the sale, the i-th generator in the z-th zone can sell electricity in the z'-th market. That is, if the generator sells electricity to the z'-th zone, he can obtain a reward from the zone. In this case, the reward becomes ($\hat{p}^1_{z't} - MC_{i(z)t})\hat{s}^1_{i(z \to z')t}$) in DA and ($\hat{p}^0_{z't} - MC_{i(z)t})\hat{s}^0_{i(z \to z')t}$) in HA. Here, $\hat{s}^1_{i(z \to z')t}$ and $\hat{s}^0_{i(z \to z')t}$ are the amount of electricity transmitted in DA and HA, respectively, from the z-th zone to the z'-th zone. The transmission from one zone to another zone is associated with a transmission cost that is listed as $TC_{(z \to z')}$ in Table 1. The total reward ($R_{i(z)}$) for the i-th generator in the z-th zone is determined by subtracting the transmission cost from a sum of these sales. The transmission cost within a same zone is zero. Here, $TC_{(z \to z')}$ stands for a unit transmission cost (\$/MWH), that is associated with physical losses, ancillary services and others related to transmission services) from the z-th zone to z'-th zone. The cost in the table indicates a total transmission cost (\$).

Next, a reward to the j-th wholesaler in the z-th zone can be specified as follows: If $\hat{p}^1_{zt} > p^1_{j(z)t}$, then the wholesaler cannot access electricity through the DA market. So, his reward is zero. In contrast, if $\hat{p}^1_{zt} \leq p^1_{j(z)t}$, then he can obtain electricity from DA and sell the electricity to end users. The reward is listed in the left side of Table 2 where $p^R_{zt}$ is the retail price of the z-th zone at the t-th period. Similarly, if $\hat{d}^0_{j(z)t} \geq 0$, then the wholesaler can access electricity in HA. The reward is listed in the right side of Table 2. An opposite case can be found if $\hat{d}^0_{j(z)t} = 0$. The wholesaler provides end-users with electricity whose retail price is ruled by a regulatory agency.

**Table 1.** Reward for Generators

| Reward = Sale - Cost | | DA | HA |
|---|---|---|---|
| Sale | Within the z − zone | $\left\{\hat{p}^1_{zt} - MC_{i(z)t}\right\}\hat{s}^1_{i(z)t}$ | $\left\{\hat{p}^0_{zt} - MC_{i(z)t}\right\}\hat{s}^0_{i(z)t}$ |
| | Transmission $(z \rightarrow z')$ | $\left\{\hat{p}^1_{z't} - MC_{i(z)t}\right\}\hat{s}^1_{i(z \rightarrow z')t}$ | $\left\{\hat{p}^0_{z't} - MC_{i(z)t}\right\}\hat{s}^0_{i(z \rightarrow z')t}$ |
| Transmission Cost | | $\left\{\hat{s}^1_{i(z \rightarrow z')t} + \hat{s}^0_{i(z \rightarrow z')t}\right\}TC_{(z \rightarrow z')}$ | |

**Table 2.** Reward for Wholesalers

| | DA | HA |
|---|---|---|
| Reward | $\left\{p^R_{zt} - \hat{p}^1_{zt}\right\}\hat{d}^1_{j(z)t}$ | $\left\{p^R_{zt} - \hat{p}^0_{zt}\right\}\hat{d}^0_{j(z)t}$ |

At the end of this section, it is important to describe that the reward allocations summarized in Tables 1 and 2 are based upon the performance in a short term (i.e., DA and HA) horizon. [The speculation capability is incorporated into each agent as his knowledge base in a long term horizon. Thus, there is a difference between the reward allocation (in a short term horizon) and the knowledge base development (in a long term horizon). The former comes from business and the latter comes from computer science.] It is possible to reorganize the rewards into a finite or infinite horizon. In this case, we need to re-compute the rewards in a long term (finite or infinite) horizon. Then, the simulator needs to incorporate such a computation process into the speculation process of each agent so that the agent can compute a discount value from the long term horizon to the short term horizon. In the other aspects are the same in the proposed simulator. An exception is that the proposed transmission cost needs to incorporate a maintenance cost due to the transmission failure in a long term horizon. This study does not consider the transmission failure because the reward allocation belongs to a short term horizon.

## 4 Adaptive Behavior of Agents

### 4.1 Adaptive Behavior

In the proposed MAIS, traders are represented by software agents. Figure 6 illustrates an adaptive learning process and its related knowledge base development, both of which are incorporated into the agents. As depicted in the figure, each agent recognizes that there is an opportunity to obtain a reward by participating into wholesale power markets. He understands that the market participation is always associated with risk and therefore, he tries to obtain risk-hedge ability through trading experience.

*Adaptive Learning of Agents (Partial Reinforcement Learning / Non-Reinforcement Learning):* In the simulator, each market consists of many agents who can accumulate knowledge from their bidding results to adjust their proceeding bidding strategies. The agents operate in two modes: practice and real experience, as depicted in Figure 6. During practice (at the left hand side of Figure 6), the agents use non-reinforcement learning (or self-learning) where a power trading market determines the win or lose of each practice bid.

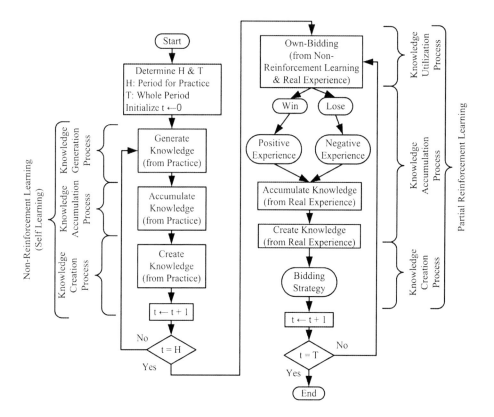

**Fig. 6.** Adaptive Learning and Knowledge Base Development

The non-reinforcement learning is separated into three processes: (a) knowledge generation, (b) knowledge accumulation and (c) knowledge creation. In the knowledge generation, each agent has to generate knowledge by itself. The purpose of "knowledge generation" process is to discover or become familiar with the market as an environment. Thus, the agent bids in the markets (DA & HA) by changing decision parameters and mark-up ratios, using random data generation. After their biddings, the values of decision parameters and their win-lose results are stored in a "knowledge accumulation" process. The accumulated knowledge is further processed by a sigmoid decision rule (for speculation on a winning probability) and an exponential utility function (for risk preference of each agent). This process is referred to as "knowledge creation." This non-reinforcement learning is repeated until the practice period (H) is over. The period can be considered as a training process for each agent. The final knowledge developed at the knowledge creation process becomes a starting basis for the own-bidding process of the partial reinforcement learning (at the right hand side of Figure 6).

After the practice is completed, each agent starts real trading experience. The bidding decisions during the real experience period are based upon the knowledge/information obtained from the previous trading practice period. The real experience period follows "partial reinforcement learning" because the agent reacts according to the feedback obtained from the external environment. The partial reinforcement learning is functionally separated into three sub-processes: (a) "knowledge utilization," (b) "knowledge accumulation" and (c) "knowledge creation." In the knowledge utilization process, each agent fully utilizes both the processed information from the knowledge creation process of both non-reinforcement learning (practice period) and the information from the partial reinforcement learning (real experience period) in order to create a bidding strategy. Based on the positive and negative experience, the agents may change the direction of decision parameters and mark-up ratios according to the partial reinforcement learning algorithm. The positive feedback is generated from successful bidding experience which an agent wins and obtains a reward in the market. In contrast, the negative feedback is generated from a bidding result when he loses and does not obtain any reward in the market. The win-lose results and the corresponding values regarding decision parameters and mark-up ratios are stored in the knowledge accumulation process of each agent. The knowledge accumulation and knowledge creation process is similar to that of the practice period. The partial reinforcement learning is repeated until all iterations (T) are completed.

At the end of this subsection, we need to mention two concerns: One of the two concerns is that the adaptive behavior of agents in the proposed simulator has the non-reinforcement learning (self-learning) as a preparatory process of the partial reinforcement learning. The rationale regarding why the proposed simulator has the non-reinforcement learning is because a knowledge base at the initial stage is often inaccurate in terms of selecting decision parameters. This study considers that all agents in the training period do not have any knowledge base (i.e. experience). They need a practice period to obtain their biding experience. Their practices are not associated with any reward. Thus, the practice period is not formulated by a conventional use of reinforcement learning. The other concern is that both the non-reinforcement learning and the partial reinforcement learning operate under an episodic mode (not a batch mode).

## 4.2 Types of Agents (Type I: Multiple Capabilities and Type II: Limited Capability)

Following the research results of Sueyoshi and Tadiparthi (2007), this study uses two groups of adaptive agents who investigate the dynamics of electric power trading. The first group consists of agents who are equipped with multiple learning capabilities. Their learning capabilities include a risk-averse utility function and a long-term view (using a speculation capability) on how to obtain a reward from electric power trading. This group is referred to as "Type I" in this study. Meanwhile, the other group of agents looks for a short-term gain via limited learning capability. They do not incorporate any utility functions or any speculation capability. Thus, the second group is less informed than Type I. This group of agents is referred to as "Type II" in this research. See Appendix of this article that provides the adaptive behaviors of Types I and II.

In developing the two types of agents, Sueyoshi and Tadiparthi (2007) depended upon the assertions on adaptive behaviors discussed by Roth and Erev (1995) and Erev and Roth (1998). The study (2007) prepared two hypotheses to investigate the adaptive behaviors of agents in power trading. One of the two is that "Type II predicts a price change of electricity more accurately than Type I in a real auction market." The research (2007) concluded that Type II slightly outperformed Type I in terms of estimating electricity price. The other assumption is that "Type I outperforms Type II in a power market where the two groups compete with each other." The research (2007) concluded that Type I outperformed Type II in terms of head-to-head competition because the power trading needed other learning capabilities in addition to the price estimation ability.

The following concerns are important in extending the two assertions into research regarding why the electricity crisis has occurred in California. First, Sueyoshi and Tadiparthi (2008c) assumed that the two groups of agents did not coexist in a power trading market. Furthermore, the type of agents did not depend upon the change of market fundamentals. The two assumptions, not clearly discussed in the study (2008c), do not reflect the reality of power trading. The two groups of agents usually coexist in the auction market for power trading and the agent composition changes constantly along with a dynamic fluctuation of market fundamentals. For example, agents in Type II could survive before the California electricity crisis because the market was stable and they could easily predict a price fluctuation of electricity. However, they needed to shift to Type I during the crisis period because the price fluctuation was drastically changing and consequently they had to speculate carefully the electricity price during the crisis period. Thus, the agent composition between Types I and II depends upon the dynamic change of electricity price and market fundamentals. The previous study (2008c) did not consider the important issue regarding the change of agent composition in a time horizon.

# 5   An Application to the California Electricity Crisis (Sueyoshi, 2010)

## 5.1  California Electricity Market from Data Structure

As depicted in Figure 1, the California market is divided into three zones for the purpose of pricing. A data set on the California electricity market used in this study is available from the University of California Energy Institute website. The data set consists of market information such as time of transaction, date of transaction, price at each zone in DA and HA markets, unconstrained price and quantity of the system, import/export quantities in each zone and prices of various auxiliary services. The DA trading was stopped from the California electricity market after $31^{st}$ January 2001.

**Table 3.** Data Description on California Market Price

| Market | Before crisis | | | | During crisis | | | |
|---|---|---|---|---|---|---|---|---|
| | Mean | Median | Skewness | Kurtosis | Mean | Median | Skewness | Kurtosis |
| SP-15 DA | 26.77 | 26.04 | 3.98 | 29.66 | 136.91 | 92.65 | 5.43 | 71.51 |
| SP-15 HA | 26.84 | 23.99 | 5.53 | 70.96 | 126.01 | 105.03 | 1.86 | 6.92 |
| NP-15 DA | 28.69 | 27.04 | 11.88 | 343.80 | 155.07 | 109.49 | 4.89 | 60.31 |
| NP-15 HA | 29.42 | 25.47 | 4.80 | 35.61 | 155.76 | 150.00 | 1.71 | 6.82 |
| ZP-26 DA | 28.87 | 29.91 | 8.72 | 119.51 | 131.14 | 91.21 | 6.05 | 85.09 |
| ZP-26 HA | 28.38 | 29.31 | 4.80 | 59.14 | 124.35 | 105.13 | 1.79 | 6.96 |
| AVG | 28.16 | 26.96 | 6.62 | 109.78 | 138.21 | 108.92 | 3.62 | 39.60 |

a) Skewness gives an idea about the direction of the tail of a distribution (a degree of asymmetry). A positive value of skewness means that the tail extends more to the right. A negative value of skewness means that the tail extends more to the left. A normal distribution has a skewness of 0.
b) Kurtosis gives an idea about size of the tail of a distribution. A positive value of kurtosis means that the distribution is relatively peaked. A negative value of kurtosis means that the distribution is flat. A normal distribution has a kurtosis of 0.

Table 3 provides a descriptive statistics on the data set regarding the three California zones (DA and HA) over two different periods: before crisis and during crisis. Each sample represents hourly prices representing 24 hours per day. The before crisis period for SP-15 DA and HA and NP-15 DA and HA is from $1^{st}$ April, 1998 to $30^{th}$ April, 2000. The before crisis period for ZP-26 DA and HA is from $1^{st}$ February, 2000 to $30^{th}$ April, 2000. The during crisis period for SP-15 DA and HA, NP-15 DA and HA, and ZP-26 DA and HA is from $1^{st}$ May 2000 to $31^{st}$ January 2001.

For all the DA markets, a maximum price of \$2499.58/MWH was observed at 7 PM on $21^{st}$ January 2001. All the HA markets had a maximum price of \$750/MWH starting from $26^{th}$ June 2000. It was observed that prices started rising steadily from the summer of 2000.

A data set regarding the capacity limits of California transmission links is not available publicly. To estimate a capacity limit on the lines between zones, we calculate the difference between import and export quantities to the wholesale market. A rationale behind choosing such a method is the assumption that a quantity more than the transmission limit has not been transferred on the transmission lines. After observing the data set, the transmission limit was set to a value of 11752 GWH (maximum difference between imports and exports). The limit was applied on both the transmission links: the transmission link between central zone and northern zone and the transmission link between central and southern zones.

*Market Composition*: As mentioned previously, the electricity market consists of generators and wholesalers. Since we do not have any information about an exact composition of the market during the years from 1998 to 2001, we use the information provided by California Energy Commission on their website for year 2005. The website (http://www.energy.ca.gov/maps/electricity_market.html) provides an approximate composition of the generators. Thus, this study considers that 964 generators of which 343 are hydroelectric with 20% market capacity, 44 are geothermal with 3% market capacity, 373 are oil/gas with 58% market capacity, 17 are coal with 6% market capacity, 94 are wind with 4% market capacity, 80 are WTE with 2% market capacity, 2 are nuclear with 7% market capacity, 11 are solar with 1% market capacity. The wholesaler composition is estimated from the website: http://www.energy.ca.gov/ electricity/electricity_consumption_utility.html. There are a total of 48 wholesalers. Pacific Gas and Electric has 30% of the share, San Diego Gas & Electric has 7% of the share, Southern California Edison has 31% of the share, LA Department of Water and Power has 9% of the share. Sacramento Municipal Utility District has 4% of the share, California Dept. of Water Resources has 3% of the share, and other 41 utilities have a 12% share. Self-generating agencies account for 4% of the share.

## 5.2 Alternate Approaches to Predict Market Price

*Evaluation Criterion*: An evaluation criterion is estimation accuracy (%) which is defined as

$$1 - \frac{1}{N} \sum_{t=1}^{N} \left| \frac{\text{Real Market Price}(t) - \text{Estimated Market Price}(t)}{\text{Average Real Market Price}(t)} \right|. \qquad (1)$$

Here, N stands for the number of evaluation periods. This criterion is suggested by Shahidehpour *et al.* (2002).

*Support Vector Machines (SVMs)*: We use the kernlab package of R version 2.5.1 for running our experiments (source: http://www.r-project.org/). We chose an epsilon support vector regression model to predict the market price of electricity. The different parameters of the SVM regression were optimally chosen based on many repetitions of the experiment. The optimal value of C, a parameter to assign penalties to the errors in the support vector regression, was set to 1.5. A Gaussian radial basis kernel function with σ =16 was used as a kernel parameter. A 3-fold cross-validation was applied for training data.

The input vectors for the DA market consisted of day-of-the-week, temperature, and DA demand. The target vector was the DA market price. Similarly, for an HA market, the input vectors consisted of day-of-the-week, temperature, HA demand, and corresponding DA price. The target vector was the HA market price. For SP-15DA, SP-15HA, NP-15DA, and NP-15HA, the first 6216 data points (259 days x 24 hours) are used for training and the next 18672 data points (778 days x 24 hours) are used for predicting the performance of the trained SVM. For ZP-26DA and ZP-26HA, the first 2160 data points (90 days x 24 hours) are used for training and the next 6624 data points (276 days x 24 hours) are used for predicting the performance of the trained SVM.

*Neural Networks (NN)*: The second alternative is NN whose use for price estimation has been recommended by many researchers (e.g., Taylor and Buizza, 2002 and Zhang *et al* 2003). We use a Normalized Gaussian Radial Basis Function Neural Networks to forecast the price. [See, for example, neural v1.4.2 package on R 2.5.1 http://www.r-project.org/).] The network uses a back propagation algorithm for training the network. The input parameters used are as follows: time of day, day of week, power imports, power exports, temperature, system wide load, and previous day's price. These parameters are reported in Gao *et al* (2000). The width of each gauss function is assigned as the Euclidean distance between the two nearest training samples. [Source: Documentation of neural v1.4.2 package]. The learning rate, alpha, is assigned to 0.20. The error condition to stop is specified as 0.001, i.e. the algorithm will stop if the average error between the target vector and the predicted vector is lower than 0.001. As a preprocessing step, the data was normalized to lie in the range between -1 and 1. Each sample has been divided into three sets: training set, validation set and testing set. For SP-15 and NP-15, the first 456 days are used for training, the next 305 days are used for validation, and the next 60 days are used for testing. For ZP-26, the first 216 days are used for training, the next 148 days are used for validation, and the next 31 days are used for testing. To avoid a problem of over fitting, we conducted experiments repeatedly by varying the number of neurons in the hidden layer from 5 to 13. It is found that the number of neurons that gives the least error is 10, 10, and 7 for SP-15, NP-15, and ZP-26, respectively.

*Genetic Algorithm (GA)*: We use "genalg" package (version 0.1.1), an R based Genetic Algorithm to run our experiments. A genetic algorithm essentially consists of three steps: initialize population, evaluate the fitness of a population, and apply genetic operators. We modeled the problem as a parameter estimation problem for a nonlinear regression model. See Pan *et al.* (1995) for more information on the use of a genetic algorithm for nonlinear regression model. The DA and HA prices are modeled as unknown parameters that have to be estimated. The known parameters are day-of-the-week, hour of the day, temperature, and system wide load. Each individual in the population is encoded by two binary strings. The first binary string represents the DA price. The second binary string represents the HA price. We know from the dataset that the maximum value of DA price is 2499.58. Thus, the range of DA price is [0, 2500]. Since we assume a precision of 2 digits after decimal point, the domain of DA price should have 250000 equal divisions. That is, it should be represented by a 18 bit binary string ($131072 = 2^{17} < 250000 < 2^{18} = 262144$). We know that the range of HA price is [0,750]. Assuming a precision of 2 digits after decimal point, the domain of HA price should have 75000 equal divisions. i.e. it should be represented by a 17 bit binary string ($65536=2^{16}<75000<2^{17}=131072$). Thus, an individual in the population is represented by a 35 (18+17) bit binary string.

The fitness value (FV) of an individual in a population in z-th zone is given by the following equation: $FV\,(\hat{p}_{zt}^{1}, \hat{p}_{zt}^{0}) = \sum \left[SWL_{t} - Avg\ Load\ (DW,\ HD,\ Temp,\ \hat{p}_{zt}^{1}, \hat{p}_{zt}^{0})\right]^{2}$ Here, SWL is system-wide load, DW is day of week, HD is hour of day, Temp is a temperature. We use a rank-selection strategy to choose the individuals based on the rank of the fitness function (FV) for the market prices for DA ($\hat{p}_{zt}^{1}$)and HA ($\hat{p}_{zt}^{0}$). The strategy is similar to the one described in Pan *et al.* (1995). The initial population

size is 50. The maximum generation is 25000. The mutation chance is varied from
0.02 to 0.10. Elitism is set to 0.25. The prices for the (t+1)-th period is estimated from
the t-th period.

## 5.3   Estimation Results

Each agent in the MAIS can bid by using either a Type I learning algorithm or a Type
II learning algorithm. Table 4 summarizes the estimation accuracy of MAIS under 11
different combinations of trader composition. There are 6 market zones (2 types of
markets x 3 market zones). Each market zone has 11 agent combinations between
Type I and Type II. For example, the second row indicates that all agents (generators
and wholesalers) belong to Type II (equipped with limited learning capability). There
is no trader in Type I (equipped with multiple learning capabilities). An opposite
agent combination can be found in the last row of Table 4. The row with (60%, 40%)
implies that 60% of traders belong to Type I (multiple learning capabilities) and 40%
of traders belong to Type II (limited learning capability). An agent does not change
his learning strategy during the course of the simulation. The purpose of calculating
the estimation accuracy for different combinations of agents is to find the closest mix
of agents that represent the market composition.

Table 4 indicates that the best agent composition between Type I and Type II is
(60%, 40%) in terms of estimation accuracy. As listed in the last column, the average

**Table 4.** Estimation Accuracy (%) of MAIS with Different Combinations of Traders

| (Type I %, Type II %) | SP-15 DA | SP-15 HA | NP-15 DA | NP-15 HA | ZP-26 DA | ZP-26 HA | Mean |
|---|---|---|---|---|---|---|---|
| (0%, 100%) | 72.68 | 71.24 | 66.28 | 70.34 | 86.19 | 89.12 | 75.98 |
| (10%, 90%) | 72.86 | 71.29 | 67.22 | 70.50 | 85.55 | 89.10 | 76.09 |
| (20%, 80%) | 74.54 | 72.14 | 68.29 | 71.53 | 84.84 | 88.34 | 76.61 |
| (30%, 70%) | 74.32 | 72.68 | 69.47 | 72.91 | 86.33 | 87.19 | 77.15 |
| (40%, 60%) | 76.19 | 75.50 | 69.76 | 72.19 | 84.64 | 88.45 | 77.79 |
| (50%,50%) | 84.98 | 87.15 | 76.91 | 80.33 | 86.32 | 88.67 | 84.06 |
| (60%, 40%) | 86.19 | 88.59 | 79.22 | 82.57 | 88.05 | 90.62 | 85.87 |
| (70%, 30%) | 85.24 | 87.21 | 78.01 | 81.67 | 87.54 | 90.07 | 84.96 |
| (80%, 20%) | 85.93 | 87.48 | 77.65 | 81.20 | 86.90 | 89.07 | 84.71 |
| (90%, 10%) | 86.21 | 87.35 | 78.55 | 79.21 | 88.12 | 89.57 | 84.84 |
| (100%, 0%) | 85.76 | 86.89 | 75.19 | 79.32 | 89.71 | 89.12 | 84.33 |

a)   Estimation accuracy (%) on each zone market is measured by Equation (1). The
     mean, at the bottom of the table, implies the total average (the sum of estimation
     accuracies related to market zones divided by six) of these market zones.

b)   The number of observations (hourly) is 24888 for the two markets in both
     SP-15 and NP-15. The number of observations is 8784 (hourly) for the two
     markets in ZP-26. Hence, the numbers in the table is the averages of these
     hourly observations. The numbers in the last column are the total average of
     each market zone.

estimation accuracy is 85.87(%) under the trader composition. The average estimation accuracy of Table 4 is related to the changes of price fluctuation and market fundamentals before and during the California electricity crisis. The result indicates that agents in Types I and II have coexisted in the power trading markets in the entire observed periods (SP-15 and NP-15: from 1st April 1998 to 31st January 2001 as well as ZP-26: from 1st February 2000 to 31st January 2001). Hence, this study uses the estimation accuracy of the agent combination (60% of traders belong to Type I and 40% of traders belong to Type II) as the computational result of the proposed MAIS.

Table 5 lists the estimation accuracies of MAIS before and during the electricity crisis at the right column. The table compares them with the estimation accuracies of the three other well-known computer science techniques (i.e., SVM: Support Vector Machines, GA: Genetic Algorithms and NN: Neural Networks). Figure 7 visually compares the observed prices of electricity with the estimated ones obtained by MAIS in the SP-15 (DA) market before the electricity crisis. Figure 8 depicts such a comparison during the electricity crisis.

**Table 5.** Estimation Accuracy Comparison (%) among Four Approaches

| Market | SVM | | | GA | | | NN | | | MAIS | | |
|--------|-----|-----|-----|-----|-----|-----|-----|-----|-----|------|------|------|
| SP-15 DA | [82.58] | (69.41) | 79.10 | [68.15] | (49.87) | 63.32 | [87.95] | (47.10) | 77.16 | [87.17] | (83.45) | 86.19 |
| SP-15 HA | [82.73] | (63.47) | 77.64 | [68.37] | (44.53) | 62.10 | [85.37] | (52.46) | 76.67 | [89.67] | (85.59) | 88.59 |
| NP-15 DA | [79.76] | (80.96) | 80.08 | [65.84] | (41.97) | 59.54 | [84.27] | (61.46) | 78.24 | [83.54] | (67.19) | 79.22 |
| NP-15 HA | [79.90] | (80.85) | 80.15 | [65.67] | (42.95) | 59.67 | [83.91] | (61.95) | 78.11 | [85.46] | (74.52) | 82.57 |
| ZP-26 DA | [89.45] | (90.15) | 89.97 | [65.19] | (58.42) | 60.12 | [73.54] | (61.05) | 64.19 | [92.97] | (86.40) | 88.05 |
| ZP-26 HA | [91.53] | (92.29) | 92.10 | [67.53] | (59.79) | 61.74 | [75.46] | (60.09) | 63.95 | [93.96] | (89.50) | 90.62 |
| Mean | [84.33] | (79.52) | 83.17 | [66.79] | (49.59) | 61.08 | [81.75] | (57.35) | 73.05 | [88.80] | (81.11) | 85.87 |

a) [  ] and (  ) indicate the average estimation accuracy before the California electricity crisis and the one during the crisis, respectively. The total average between them is listed at the right hand side of each column.

b) SVM: Support Vector Machines, GA: Genetic Algorithms, NN: Neural Networks and MAIS: Multi-Agent Intelligent Simulator (proposed in this study).

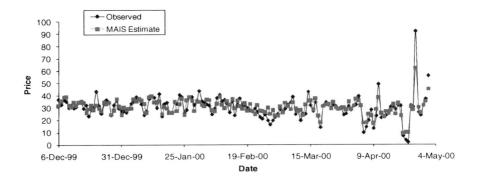

**Fig. 7.** MAIS Price Estimate of SP-15 for DA (the Pre-Crisis)

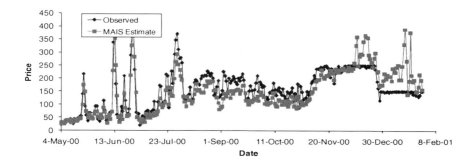

**Fig. 8.** MAIS Price Estimate of SP-15 for DA (During the Crisis)

Table 5 indicates that the proposed MAIS (average estimation accuracy = 85.87%) estimates the dynamic price fluctuation of electricity as well as the other three methods (SVM: 83.17%, GA: 61.08% and NN: 73.05%). The average estimation accuracy of MAIS before the electricity crisis is 88.80%, while its accuracy during the crisis is 81.11%. Furthermore, as depicted in Figure 8, there is a significant difference between observed market prices and MAIS's estimates. Such results indicate that the observed high electricity prices in the crisis period may not be explained as a natural outcome of changes in market fundamentals. Therefore, we need to investigate the issue further in this study. Such a difference (88.80% and 81.11%) can be visually confirmed in Figures 7 and 8, as well. Both figures compare the price fluctuation of observed prices with that of the price estimates in the two periods (before and during the crisis). Note that the price range of Figure 7 (from $0/MWH to $90/MWH) is much smaller than that of Figure 8 (from $25/MWH to $400/MWH).

### 5.4  Learning Speed (Convergence)

Table 6 compares learning speeds (convergences) of decision parameters and mark-up rates before and during the crisis when the market consists of trading agents whose 60% have multiple learning capabilities (Type I) and 40% have limited learning capability (Type II).

The learning speeds (measured by the convergence rates concerning decision parameters and mark-up rates) of trading agents before the crisis are less than those during the crisis. For example, the decision parameter ($\alpha$) needs an average of 41.49 iterations to converge at an average value of 0.42 before the crisis. Meanwhile, the parameter needs 99.40 iterations on average to obtain an average convergence value of 0.46 during the crisis. The observation can be applied to the other decision parameters and mark-up ratios. The result indicates that the learning speeds of agents depend upon the dynamic change of market fundamentals. Before the crisis, agents can adjust themselves to a market change within a shorter time period because the market is stable and predictable. In contrast, they need a long time for their learning during the crisis because a large market fluctuation occurs in the market.

**Table 6.** Learning Speeds of Trading Agents before and during California Electricity Crisis

| | Market zone | Type I | | | | | | Type II | | | | | |
|---|---|---|---|---|---|---|---|---|---|---|---|---|---|
| | | Generator | | | | Wholesaler | | Generator | | | | Wholesaler | |
| | | α | β | v | η | δ | λ | α | β | v | η | δ | λ |
| Before | SP-15 | 35.69 | 54.44 | 35.50 | 62.56 | 30.41 | 81.06 | 81.74 | 94.35 | 81.55 | 88.91 | 55.56 | 114.74 |
| | | (0.45) | (0.68) | (0.55) | (0.79) | (0.53) | (0.81) | (0.43) | (0.67) | (0.57) | (0.79) | (0.52) | (0.77) |
| | NP-15 | 53.52 | 69.79 | 53.50 | 69.01 | 26.42 | 85.94 | 108.36 | 108.08 | 108.54 | 123.12 | 63.14 | 136.61 |
| | | (0.39) | (0.79) | (0.61) | (0.85) | (0.57) | (0.81) | (0.41) | (0.76) | (0.59) | (0.85) | (0.57) | (0.82) |
| | ZP-26 | 35.27 | 55.22 | 35.50 | 57.97 | 20.72 | 72.42 | 97.60 | 96.30 | 98.00 | 115.37 | 48.79 | 123.79 |
| | | (0.43) | (0.84) | (0.57) | (0.84) | (0.61) | (0.74) | (0.45) | (0.84) | (0.55) | (0.84) | (0.61) | (0.72) |
| | AVG | 41.49 | 59.82 | 41.50 | 63.18 | 25.85 | 79.80 | 95.90 | 99.58 | 96.03 | 109.13 | 55.83 | 125.05 |
| | | (0.42) | (0.77) | (0.58) | (0.83) | (0.57) | (0.79) | (0.43) | (0.76) | (0.57) | (0.83) | (0.57) | (0.77) |
| During | SP-15 | 84.94 | 97.25 | 78.94 | 83.96 | 57.44 | 113.49 | 98.52 | 105.20 | 90.22 | 96.23 | 78.60 | 122.81 |
| | | (0.49) | (0.79) | (0.51) | (0.90) | (0.69) | (0.89) | (0.53) | (0.85) | (0.47) | (0.86) | (0.55) | (0.81) |
| | NP-15 | 111.04 | 109.62 | 106.11 | 129.05 | 60.40 | 132.24 | 117.60 | 122.94 | 123.72 | 131.35 | 86.11 | 151.53 |
| | | (0.45) | (0.89) | (0.55) | (0.91) | (0.59) | (0.89) | (0.51) | (0.88) | (0.49) | (0.88) | (0.59) | (0.86) |
| | ZP-26 | 102.21 | 98.00 | 100.77 | 110.89 | 49.69 | 127.40 | 116.37 | 108.71 | 109.06 | 123.65 | 69.07 | 130.72 |
| | | (0.45) | (0.89) | (0.55) | (0.94) | (0.65) | (0.79) | (0.46) | (0.89) | (0.54) | (0.86) | (0.62) | (0.85) |
| | AVG | 99.40 | 101.62 | 95.27 | 107.97 | 55.84 | 124.37 | 110.83 | 112.28 | 107.67 | 117.08 | 77.93 | 135.02 |
| | | (0.46) | (0.86) | (0.54) | (0.92) | (0.64) | (0.86) | (0.50) | (0.87) | (0.50) | (0.87) | (0.59) | (0.84) |

(Type I: 60% Multiple Learning Capabilities; Type II: 40% Limited Learning Capabilities)

a)   The number of iterations, indicating a learning speed, is listed at the top portion of each cell. The convergence value is listed at the bottom portion of each cell and is denoted with ( ).

## 5.5 Dynamic Change in Agent Composition (Type I vs Type II)

In subsection 5.3, we assumed that the composition of the MIAS is initialized as one of 11 different possible combinations between Type I and Type II. To examine the change of agent composition in a time horizon, this study incorporates a time horizon in the adaptive behaviors of agents visually summarized in Figure 6. That is, each agent has an opportunity to change his learning strategy. Figure 9 summarizes a computational flow regarding the adaptive behaviors in a time horizon. The figure describes the flowchart for an adaptive strategy for a generator. A similar algorithm can be extended to a wholesaler. There are n generators (i = 1,.., n) that can participate in a market. The generator is represented by an agent in the proposed MAIS. Each agent has to choose either Type I learning or Type II learning. The premise of the algorithm is that an agent prefers a learning strategy that fetches more reward in the previous learning experience. The agent evaluates its previous learning strategy at the end of every month based on the reward obtained in previous month.

Let x represent the percentage of agents who are of Type I. Then, (100-x) represents the agents who are of Type II. Each agent is assigned two variables: Ownstrategy (the learning strategy of agent) and AltStrategy (the alternate strategy of agent). For example, if Ownstrategy(i) = Type I, then AltStrategy(i) = Type II, and vice versa. As a starting step, the learning strategies of each agent are initialized. Initialize the iteration period t to 1 April 1998. The variable, monthly reward, calculates the total reward obtained by an agent in during each month. The time variable is incremented by one day. During the simulation, the i-th agent calls both Type I and Type II learning. Then, the agent bids based on the value in OwnStrategy(i). If the bidding price is less than the market price, the reward obtained

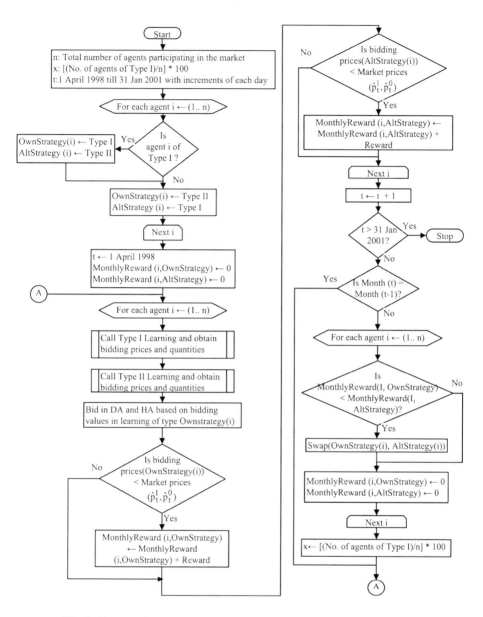

**Fig. 9.** Computational Process to Identify Dynamics of Agent Composition

is added to the monthly reward of OwnStrategy. Similarly, if the bidding price obtained using the alternate strategy is less than the market price, the reward obtained is added to the monthly reward of alternate strategy. See Table 1 and Table 2 for the methodology to calculate rewards for a generator and wholesaler, respectively. Note that the market price is obtained as a result of bids using own strategy, not alternate strategy. The alternate strategy is used to analyze a "what-if" case. Such a simulation is repeated for each month. At the end of each month, the monthly reward obtained by own strategy is compared with the hypothetical monthly reward that could have been obtained if the agent followed an alternate strategy. If the monthly reward of the alternate strategy is greater than the monthly reward of own strategy, then the agent switches strategy for the next month. That is, the agent follows an alternate strategy that could fetch more profit than the own strategy. The value of x is calculated at the end of each month and is plotted in the graph.

Using the algorithm in Figure 9, this study investigates a dynamic change in the agent composition between Type I and Type II in the California market from April 1998 to January 2001. Figure 10 visually summarizes the dynamics. This study considers two initial agent compositions: (a) 50% agents are equipped with Type I adaptive learning and the remaining 50% consists of Type II at April 1998 and (b) all agents are equipped with Type II adaptive learning (so, there is no Type I agent). In Figure 10, the solid line represents a dynamic change in the portion of Type I for the first agent composition and the dotted line represents the dynamic change in the portion of Type I for the second agent composition.

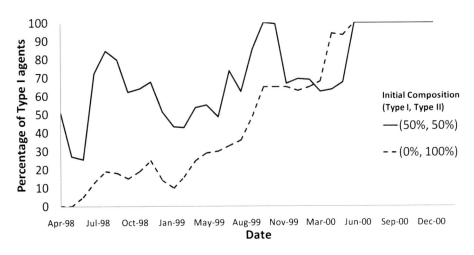

**Fig. 10.** Agent Composition (Type I and Type II) in Time Horizon

Agents in Types I and II coexisted in the California electricity market before the crisis. During the initial period of the simulation and before the crisis period, all agents fluctuated between Type I and Type II. But, there was a gradual increasing trend in the number of agents of Type I. All agents in the electricity market became

Type I during the crisis period (from May 2000 to January 2001). This result was because the price of wholesale electricity increased sharply and fluctuated drastically during the crisis. Agents equipped with Type II adaptive learning needed to adjust themselves to the change in price dynamics during the crisis. Consequently, they became agents of Type I for their survival purposes.

## 5.6  Sensitivity Analysis

As previously mentioned, a significant advantage of MAIS is that we can analyze different market scenarios by sensitivity analysis. To explore a rationale regarding why a price hike has occurred during the California electricity crisis, this study prepares seven economic assertions related to market fundamentals, all of which are examined by the MAIS-based sensitivity analysis.

*Hypothesis 1 (Increase in Marginal Cost)*: An increase in the marginal cost of oil and natural gas has influenced an increase in the wholesale price during the California electricity crisis.

This assertion is due to Joskow and Kahn (2002) and Borenstein *et al* (2002). Borenstein *et al.* (2002) reported that production costs contributed to 21% of the power price increase. A drought in Northwest Pacific region caused a significant shortage of electricity from hydro-electric dams. The price of natural gas started increasing gradually from the beginning of 2000. See Lee (2004) and Joskow and Kahn (2002, pp.5-7 and p.17) that reported a similar description on the fuel increase. The significant increase in the price of natural gas contributed to an increase in the marginal cost of oil/gas power plants. The sensitivity analysis examines the first hypothesis by increasing the marginal cost, $MC_{i(z)}$, of 373 gas-fired generators according to the rates depicted in Figure 1 of Joskow and Kahn (2002, p.7). This study uses marginal costs for hydroelectric, geothermal, oil/gas, coal, wind, and nuclear is $25, $30, $40, $35, $10, and $50, respectively as the initial starting values of the simulation analysis. These marginal costs are all observed before the crisis. The marginal costs, except that of oil/gas, are used to examine the price fluctuation of electricity during the electricity crisis, because a major price change has been not found for those fuels during the observed period.

*Hypothesis 2 (Increase in Real Demand)*: An increase in electricity consumption has influenced an increase in the wholesale price during the California electricity crisis.

The second assertion is due to an observed data set on real demand, $r_{j(z)t}$, in the three zones of the California market. This increase in demand attributes to the temperature and economic growth in California. The summer of 2000 was "unusually hot" (Lee, 2004). While the temperature increase explained the increase in power price for summer, it failed to explain the reasons for the continued increase in power for the rest of the year. The economic growth of the industry due to the opening of many startup companies increased the power demand. The unusually high prices in the summer of 2000 caused financial difficulties for many utilities. See also Joskow and Kahn (2002, p. 28) in which increased demand is considered as one factor of the price hike in the crisis. The data set indicates a load increase during the crisis. Hence, the sensitivity analysis increases the real demand by 20% from the original values. This increase in demand is applied only during the peak load time (from 7:00 to 21:00), because a price hike occurred during the peak load period. Any major price change was not found in the other period. The 20% increase is due to the information that is obtained from the website (www.ucei.berkeley.edu/datamine/uceidata/uceidata.zip). According to the information source, the real demand increased during the period from 4$^{th}$ May 2000 to 31$^{st}$ January 2001. For example, the load was 20,880.40 GWH (at 9:00 on 1$^{st}$ June 1999) and 23,694.10 GWH (at 9:00 on 1$^{st}$ June 2000), respectively. This denotes an increase of 13.48% during that time. An average of all such increase in loads is 19.69% and hence, the proposed sensitivity analysis uses 20% for increase.

*Hypothesis 3 (Greed of Traders)*: Traders have exhibited overwhelming desire for more profit during the California electricity crisis. Generators increased their bidding prices, while wholesalers reduced their bidding prices.

An important numerical capability of the proposed MAIS is that it can examine the level of trader's greed by observing the utility function of each trader. It was observed that the wholesale market price of electricity maintained an increasing trend, but the retail price of electricity did not increase significantly during the crisis. Rather, the retail price almost remained the same under the control of regulatory agencies. The increase in market price implies that the generators obtained a profit and the wholesalers suffered from an economic loss from the power market. Thus, it is assumed that generators were looking for a more risk-taking behavior to more profit by bidding high prices. Meanwhile, the wholesalers were looking for an opposite direction in the manner that they reduced their bidding prices in the power market. In the experiments, the parameter ($\zeta$) represents the level of risk aversion. See Equations from (A-5) to (A-7) in the appendix. For evaluating the economic assertion, the range of $\zeta$ for generators is changed from (0.004, 0.065) to (0.0003, 0.0015). Similarly, the range of $\zeta$ for wholesalers is changed from (0.004, 0.065) to (0.01, 0.07). See Appendix (A) of this article that explains the exponential utility function.

*Hypothesis 4 (Electricity Withholding by Generators)*: Large generators withheld electricity during the California electricity crisis. The exercise of market power attributed to the price hike during the period.

The fourth hypothesis is due to Joskow and Kahn (2002, pp. 19-28) and Borenstein *et al.* (2002). Their empirical analyses suggested that some generation firms deliberately did not make their maximum supply capacities available for the California electricity market. This study reexamines their assertion. The proposed sensitivity analysis uses the mean values of output gap that are summarized in Table 7 of Joskow and Kahn (2002, p. 23). In our experiments, the maximum supply ($s_{i(z)t}^{m}$) was reduced to its adjusted vale (= $s_{i(z)t}^{m}$ - a mean value of output gap).

*Hypothesis 5 (Capacity Limit in Transmission Lines)*: A capacity limit in transmission lines was a source of supplier withholding. The transmission congestion occurred during the California electricity crisis and influenced the price hike of electricity.

Joskow and Kahn (2002, p.8) reported that there was relatively little significant transmission congestion during the crisis period. However, it is important to investigate whether the supplier withholding during the crisis was intentional by generators or was due to an occurrence of congestion in transmission lines (so, the withholding in the supply side was accidental). There is no data available on the transmission line limits in the California transmission grid. We depend upon the information about system-wide imports and exports for both DA and HA market which is available at the source (www.ucei.berkeley.edu/datamine/uceidata/uceidata.zip). To estimate a transmission limit, we calculate the average of the absolute difference between the imports and exports for both DA and HA markets. The average is 11,752 GWH, which is chosen as the line limit for all the transmission lines. For the sensitivity analysis, the line limit is reduced by 25%. Thus, the new line limit is 8,814 GWH.

*Hypothesis 6 (Competitive Rent)*: An increase in competitive rent has influenced an increase in the wholesale price during the California electricity crisis.

The research of Borenstein *et al.* (2002, p.1397) defines competitive rent as the profit that fossil-fuel and reservoir generators make in a competitive market. Since the out-of-state generators cannot participate in the competitive market, those generators supply power at a fixed price. Thus, competitive rents apply only for in-state generators. The in-state generators increase the market price by increasing the competitive rent (profit). In the proposed simulation, this means an increase in bidding price, or an increase in the mark-up ratios, $\beta$ and $\eta$. $\beta = 0$ implies that the competitive rent for DA is 0. Similarly, $\eta = 0$ implies that the competitive rent for HA is 0. Originally, we assume that the parameters ($\beta$ and $\eta$) exist in the range (0, 1). See Borenstein *et al.* (2002, p.1396, Table 3) in which the competitive rent increased 3 times approximately from 708 (in 1999) to 2101 (in 2000). To verify the hypothesis, we change the values of $\beta$ to $\dfrac{3\beta}{(1+2\beta)}$ and $\eta$ to $\dfrac{3\eta}{(1+2\eta)}$.

*Hypothesis 7 (Combination)*: The price hike in the California electricity crisis was not caused by a single source (i.e., a market fundamental that is specified by each hypothesis). Rather, the price hike was caused by some combination of multiple market fundamentals.

**Table 7.** Estimation Accuracy (%) of MAIS under Different Combinations of Market Fundamentals

**Left half**

| No. of MF | MC | RD | G | WC | TL | CR | SP-15 DA | SP-15 HA | NP-15 DA | NP-15 HA | ZP-26 DA | ZP-26 HA | AVG |
|---|---|---|---|---|---|---|---|---|---|---|---|---|---|
| 0 | No Sensitivity | | | | | | 72.45 | 72.93 | 56.72 | 61.91 | 87.66 | 86.67 | 73.06 |
| 1 | ○ | | | | | | 84.19 | 86.27 | 75.18 | 76.82 | 91.24 | 90.07 | 83.96 |
| 1 | | ○ | | | | | 73.28 | 79.64 | 63.15 | 63.48 | 85.19 | 87.67 | 75.40 |
| 1 | | | ○ | | | | 87.56 | 89.27 | 68.27 | 69.48 | 83.19 | 86.17 | 80.66 |
| 1 | | | | ○ | | | 73.11 | 71.66 | 59.37 | 58.12 | 88.10 | 87.27 | 72.94 |
| 1 | | | | | ○ | | 73.19 | 71.04 | 55.17 | 60.14 | 88.10 | 87.94 | 72.60 |
| 1 | | | | | | ○ | 82.67 | 83.54 | 76.82 | 77.02 | 88.32 | 89.14 | 82.92 |
| 2 | ○ | ○ | | | | | 74.35 | 75.62 | 59.59 | 61.16 | 86.32 | 87.85 | 74.15 |
| 2 | ○ | | ○ | | | | 74.62 | 76.09 | 59.63 | 61.77 | 86.90 | 86.27 | 74.21 |
| 2 | ○ | | | | | ○ | 69.07 | 50.27 | 57.18 | 58.37 | 60.24 | 60.86 | 59.33 |
| 2 | | ○ | ○ | | | | 88.67 | 88.93 | 68.77 | 69.97 | 85.56 | 84.84 | 81.12 |
| 2 | | ○ | | ○ | | | 85.77 | 84.91 | 64.07 | 68.34 | 80.17 | 80.74 | 77.33 |
| 2 | | | ○ | ○ | | | 90.17 | 91.71 | 79.21 | 82.04 | 90.77 | 89.69 | 87.27 |
| 2 | | | ○ | | | ○ | 74.33 | 75.49 | 69.79 | 71.22 | 75.68 | 72.93 | 73.24 |
| 2 | | ○ | | | ○ | | 64.08 | 63.27 | 53.01 | 57.14 | 72.08 | 79.65 | 64.87 |
| 2 | | ○ | | | | ○ | 75.18 | 76.27 | 67.17 | 62.08 | 83.90 | 87.19 | 75.30 |
| 2 | | | | ○ | | ○ | 81.24 | 88.88 | 68.12 | 68.90 | 82.90 | 85.19 | 79.21 |
| 3 | ○ | ○ | ○ | | | | 78.53 | 78.54 | 80.14 | 81.74 | 86.33 | 86.68 | 81.99 |
| 3 | ○ | ○ | | ○ | | | 83.19 | 87.27 | 77.18 | 76.98 | 92.01 | 91.13 | 84.63 |
| 3 | ○ | ○ | | | ○ | | 80.12 | 81.32 | 70.18 | 73.19 | 84.29 | 85.57 | 79.11 |
| 3 | ○ | ○ | | | | ○ | 90.08 | 89.93 | 82.06 | 88.97 | 91.34 | 90.21 | 88.77 |
| 3 | ○ | | ○ | ○ | | | 80.17 | 81.64 | 70.24 | 71.69 | 80.37 | 82.67 | 77.80 |
| 3 | ○ | | ○ | | ○ | | 73.96 | 68.61 | 56.32 | 61.71 | 60.82 | 67.92 | 64.89 |
| 3 | ○ | | ○ | | | ○ | 79.59 | 80.87 | 67.52 | 67.43 | 84.57 | 85.31 | 77.55 |
| 3 | ○ | | | ○ | ○ | | 79.92 | 79.64 | 77.48 | 78.40 | 84.84 | 85.98 | 81.04 |
| 3 | ○ | | | ○ | | ○ | 89.21 | 89.38 | 78.34 | 80.21 | 86.63 | 88.51 | 85.38 |
| 3 | | ○ | ○ | ○ | | | 63.95 | 63.73 | 61.87 | 64.40 | 77.49 | 77.19 | 68.11 |
| 3 | | ○ | ○ | | | ○ | 75.95 | 77.33 | 71.43 | 70.88 | 73.90 | 80.15 | 74.94 |
| 3 | | ○ | | ○ | | ○ | 74.28 | 77.17 | 68.27 | 69.31 | 84.27 | 87.19 | 76.75 |
| 3 | | | ○ | ○ | ○ | | 81.46 | 87.73 | 68.25 | 69.13 | 84.87 | 84.35 | 79.30 |
| 3 | | | ○ | ○ | | ○ | 82.07 | 83.09 | 67.18 | 65.28 | 80.27 | 80.29 | 76.36 |
| 3 | | | ○ | | ○ | ○ | 91.27 | 91.66 | 85.09 | 83.27 | 90.10 | 90.12 | 88.59 |

**Right half**

| No. of MF | MC | RD | G | WC | TL | CR | SP-15 DA | SP-15 HA | NP-15 DA | NP-15 HA | ZP-26 DA | ZP-26 HA | AVG |
|---|---|---|---|---|---|---|---|---|---|---|---|---|---|
| 3 | ○ | | | | ○ | ○ | 82.03 | 86.12 | 77.92 | 77.90 | 89.14 | 90.52 | 83.94 |
| 3 | | ○ | ○ | | | ○ | 79.58 | 80.65 | 71.34 | 74.19 | 84.77 | 85.18 | 79.28 |
| 3 | | ○ | | | ○ | ○ | 70.68 | 75.29 | 60.10 | 60.86 | 65.27 | 63.19 | 65.90 |
| 3 | | | ○ | | ○ | ○ | 90.21 | 89.49 | 83.37 | 85.99 | 89.54 | 89.82 | 88.07 |
| 3 | | | | ○ | ○ | ○ | 88.29 | 90.16 | 80.24 | 87.19 | 92.75 | 92.10 | 88.46 |
| 3 | ○ | ○ | ○ | | | ○ | 91.67 | 91.46 | 85.09 | 86.75 | 90.18 | 90.08 | 89.21 |
| 3 | ○ | ○ | | ○ | | ○ | 80.42 | 79.17 | 71.70 | 73.95 | 79.92 | 80.22 | 77.56 |
| 3 | ○ | | ○ | | | ○ | 78.33 | 79.58 | 67.18 | 68.39 | 75.95 | 76.73 | 74.36 |
| 3 | ○ | | | ○ | | ○ | 88.17 | 88.62 | 75.18 | 77.64 | 81.52 | 82.67 | 82.30 |
| 3 | | ○ | ○ | ○ | | ○ | 91.65 | 92.08 | 84.15 | 89.27 | 91.74 | 92.23 | 90.19 |
| 4 | ○ | ○ | ○ | ○ | | | 89.66 | 88.98 | 78.76 | 79.88 | 86.09 | 85.74 | 84.85 |
| 4 | ○ | ○ | ○ | | ○ | | 76.41 | 78.41 | 64.24 | 65.29 | 82.01 | 82.09 | 74.74 |
| 4 | ○ | ○ | | ○ | ○ | | 81.56 | 82.27 | 64.15 | 65.37 | 82.13 | 82.33 | 76.30 |
| 4 | ○ | | ○ | ○ | ○ | | 89.62 | 88.81 | 84.28 | 86.06 | 89.91 | 90.82 | 88.25 |
| 4 | | ○ | ○ | ○ | ○ | | 88.28 | 89.44 | 81.24 | 84.28 | 86.29 | 86.77 | 86.05 |
| 4 | ○ | ○ | ○ | | | ○ | 70.06 | 74.94 | 56.60 | 55.25 | 62.68 | 66.99 | 64.42 |
| 4 | ○ | ○ | | ○ | | ○ | 88.07 | 88.12 | 80.51 | 88.84 | 91.41 | 90.02 | 87.83 |
| 4 | ○ | | ○ | ○ | | ○ | 89.57 | 90.62 | 84.63 | 85.84 | 91.17 | 90.69 | 88.75 |
| 4 | | ○ | ○ | ○ | | ○ | 90.10 | 90.10 | 82.17 | 81.35 | 87.28 | 89.64 | 86.77 |
| 4 | ○ | ○ | | | ○ | ○ | 77.60 | 76.35 | 65.54 | 69.85 | 75.87 | 78.46 | 73.94 |
| 4 | ○ | | ○ | | ○ | ○ | 80.64 | 80.09 | 73.28 | 77.64 | 78.95 | 79.29 | 78.32 |
| 5 | ○ | ○ | ○ | ○ | | ○ | 90.20 | 91.22 | 82.74 | 86.29 | 94.19 | 94.60 | 89.87 |
| 5 | ○ | ○ | ○ | | ○ | ○ | 90.55 | 91.67 | 89.28 | 88.16 | 83.64 | 90.29 | 88.93 |
| 5 | ○ | ○ | | ○ | ○ | ○ | 90.44 | 91.82 | 90.34 | 90.56 | 91.67 | 92.08 | 91.15 |
| 5 | ○ | | ○ | ○ | ○ | ○ | 87.21 | 86.82 | 81.71 | 80.86 | 88.14 | 88.48 | 85.54 |
| 5 | | ○ | ○ | ○ | ○ | ○ | 91.73 | 91.04 | 85.06 | 83.25 | 87.58 | 89.81 | 88.08 |
| 5 | ○ | ○ | ○ | ○ | ○ | | 81.35 | 76.65 | 79.99 | 80.03 | 81.40 | 80.98 | 80.07 |
| 5 | ○ | ○ | ○ | ○ | ○ | ○ | 88.63 | 88.36 | 81.47 | 81.99 | 89.53 | 89.39 | 86.56 |
| 5 | ○ | ○ | ○ | ○ | ○ | ○ | 86.26 | 86.37 | 83.99 | 86.87 | 90.25 | 92.45 | 87.70 |
| 5 | ○ | ○ | ○ | ○ | ○ | ○ | 80.17 | 83.16 | 91.12 | 95.34 | 93.08 | 90.24 | 88.85 |
| 6 | ○ | ○ | ○ | ○ | ○ | ○ | 88.51 | 87.77 | 86.68 | 86.10 | 86.75 | 86.95 | 87.13 |

a) MF: Market fundamentals, MC: Marginal Cost, RD: Real Demand, G: Greed, WC: Withholding Capacity, TL: Transmission Limit, CR: Competitive Rent

Table 7 summarizes the estimation accuracies of MAIS under 64 combinations of different market fundamental changes. The right hand side of the table indicates the average estimation accuracy of MAIS. The simulator starts from the initial values on decision parameters and markup ratios that are obtained from a data set before the crisis and then it computes the estimation accuracy, using a data set during the crisis. The average estimation accuracy of MAIS (at the first row of the left hand side) is 73.06% that serves as a benchmark score for proceeding comparison. The estimation accuracy indicates how much a price fluctuation of electricity during the crisis can be explained by the market fundamentals before the crisis. See the last row of Table 4 that indicates the average estimation accuracy (84.33%) of MAIS when all the agents in the market belong to Type I for the whole simulation period (before and during crisis). The average (84.33%) is computed from 90.35% (the average estimation accuracy before the crisis) and 73.06% (the average estimation accuracy during the crisis).

The second row indicates the result of MAIS-based sensitivity analysis that examines the first hypothesis by increasing the marginal cost of 373 gas-fired generators according to the rate depicted in Joskow and Kahn (2002, p.7). The estimation accuracy increases from 73.06% to 83.96%. This implies that the increase in the marginal cost of oil/natural gas explains the price increase and fluctuation of electricity during the crisis with an estimation accuracy of 83.96%. There is an increase of 10.90% (= 83.96%-73.06%) in its estimation accuracy. Thus, the first hypothesis is confirmed by the MAIS-based sensitivity analysis. A similar result is identified in the second hypothesis (an increase in real demand), the third hypothesis (traders become greedy) and the sixth hypothesis (an increase in competitive rent) because these estimation accuracies (75.40%, 80.66% and 82.92%) are higher than the benchmark estimation accuracy (73.06%), but lower than the highest value (83.96%). Conversely, the fourth hypothesis (withholding by market power in the supply side) and the fifth hypothesis (a capacity limit on transmission lines) cannot be confirmed by the sensitivity analysis because those estimation accuracies (72.94% and 72.60%) are lower than the benchmark score (73.06%). The estimation accuracy of 83.96(%) serves as a benchmark score for the proceeding sensitivity analysis with two market fundamentals.

Under the MAIS-based sensitivity analysis of two market fundamentals, the combination of the first hypothesis (an increase in the marginal cost of oil/natural gas) and the third hypothesis (all traders became greedy for more profit) produces the best estimation accuracy (88.77%). This implies that the changes of the two market fundamentals explain the price increase and fluctuation of electricity during the crisis with the estimation accuracy of 88.77%. There is an increase of 4.81% (= 88.77%-83.96%). The estimation accuracy (88.77%) becomes a benchmark score for the proceeding sensitivity analysis with three market fundamentals.

Under the MAIS-based sensitivity analysis of three market fundamentals, the combination of the previous two hypotheses and the second hypothesis (an increase in real demand) produces the best estimation accuracy (90.19%). This implies that the increase in real demand influences the price increase and fluctuation of electricity during the crisis. There is an increase of 1.42% (= 90.19%-88.77%) in the estimation accuracy. The estimation accuracy (90.19%) becomes a benchmark score for the proceeding sensitivity analysis with four market fundamentals.

Under the MAIS-based sensitivity analysis of four market fundamentals, the combination of the previous three hypotheses and the fourth hypothesis (withholding capacity of generators) produces the best estimation accuracy (91.15%). This implies that the withholding capacity of generators influences the price increase and fluctuation of electricity during the crisis. There is an increase of 0.96% (= 91.15%-90.19%). The estimation accuracy (91.15%) becomes a benchmark score at this stage of sensitivity analysis.

Finally, there is no increase in estimation accuracy under the five and six market fundamentals. Hence, this study considers that the changes in the four market fundamentals are the main sources of the price hike and large fluctuation of electricity during the California electricity crisis.

The California electricity crisis was initiated by an increase in the marginal cost of oil / natural gas. The price increase occurred along with an increase in real demand of electricity. Under such a business circumstance, all traders became greedy for more profit. The three market fundamentals explain the price fluctuation of electricity during the crisis period at the level of 90.19% estimation accuracy. Besides the three market fundamentals, the withholding of large generators had an additional impact (at the level of 1%) to the price hike during the electricity crisis. The estimation accuracy increased to 91.15% under the four combinations. Meanwhile, both a capacity limit on transmission lines (or an occurrence of congestion) and a competitive rent did not have any major influence on the price hike during the crisis. These results indicate that 40.46%[= (83.96%-73.06%) / (100%-73.06%)] of the price increase during the crisis was due to an increase in marginal cost, 17.85% [= (88.77%-83.96%) / (100%-73.06%)] to traders' greediness, 5.27% [= (90.19%-88.77%) / (100%-73.06%)] to a real demand change and 3.56% [= (91.15%-90.19%) / (100%-73.06%)] to market power (withholding electricity). The remaining 32.86% is from other unknown market components and an estimation error.

Consequently, the price hike during the crisis occurred due to an increase in fuel prices and real demand at the level of 45.73% (= 40.46%+5.27%). The responsibility of energy utility firms (their greediness and a use of market power) was 21.41% (= 17.85%+3.56%). The increase of fuel price and real demand was twice more influential than the responsibility of energy utility firms in terms of the price hike and large fluctuation during the crisis.

## 6   Conclusion and Future Extensions

This study proposed a use of MAIS to numerically examine rationales regarding why the crisis occurred during May 2000-Janurary 2001. The proposed MAIS generated numerous trading agents equipped with different learning capabilities and artificially duplicated their bidding strategies in the California electricity markets during the crisis period.

In this study, we confirmed the methodological validity of MAIS by comparing its estimation accuracy with those of the three well-known computer science techniques (Support Vector Machines, Neural Networks and Genetic Algorithms).

The estimation accuracy of MAIS outperformed those of the three techniques. The result indicates that MAIS performs as well as the three other methods in terms of price estimation.

After confirming the methodological validation of MAIS, this study investigated the dynamic change on agent composition in a time horizon. We found that all agents in Types I and II coexisted in the electricity market, but they gradually shift to Type I (equipped with multiple learning capabilities) so as to adjust themselves to the price fluctuation of electricity during the crisis. This study considered two types of market compositions as initial settings: (a) all agents belonged to Type II and (b) 50% of agents were Type I and the remaining 50% were Type II. All agents in the two cases converged to Type I during the crisis period. This result indicated that the market composition depended upon the change of market fundamentals in an electricity trading market. Different types of agents could coexist in a stable market (like the California market before the crisis) where they could easily predict the change of price and market fundamentals. However, all agents needed to speculate more carefully the change of electricity price and market fundamentals in an unstable market (like the California market during the crisis) where they could not predict the dynamics between price and market fundamentals. Consequently, all agents shifted to Type I in such an unstable market.

Finally, the sensitivity analysis of MAIS found that 40.46% of the price increase was due to an increase in marginal production cost, 17.85% to traders' greediness, 5.27% to a real demand change and 3.56% to market power. The remaining 32.86% was attributed to other unknown market fundamentals and an estimation error. This numerical result indicated that the price hike in the crisis occurred due to an increase in fuel price and real demand. The change of two market fundamentals explained 45.73% (= 40.46% + 5.27%) of the price increase during the crisis period. The responsibility of energy utility firms was 21.41% (= 17.85% + 3.56%). Consequently, the price hike during the California electricity crisis was mainly due to the increase in production cost and real demand. This study found the exercise of market power during the crisis, but the influence of the market power was less (half) than that of the increase in fuel price and real demand. The policy implication obtained from this study was different from very well-known economic research (Joskow and Kahn, 2002 and Borenstein *et al.* 2002), which had attributed the crisis to the exercise of market power by large energy utility firms.

As a future research extension, we apply the proposed approach to investigate the fluctuation of price and the market changes of other commodities (e.g., $CO_2$ emission trade) which are traded in international markets. In particular, the $CO_2$ emission trade will be an important application of the proposed MAIS because the $CO_2$ reduction is a major policy issue in the world. The market fundamentals for the $CO_2$ emission trade are considerably different from those of electricity trade explored in this study. Hence, we need to develop a new type of modeling and simulation studies for such an application.

Finally, it is hoped that this study makes a contribution to the agent-based approach applied to power trading. We expect further research extensions as specified in this study.

# References

Axelrod, R.: The Complexity of Cooperation. Princeton University Press, Princeton (1997)

Bagnall, A.J.: A multi-adaptive agent model of generator bidding in the UK market in electricity. In: Proceeding of Genetic and Evolutionary Computation Conference, Las Vegas, Nevada, pp. 605–612 (2000)

Bereby-Meyer, Y., Roth, A.E.: The speed of learning in noisy games: Partial reinforcement and the sustainability of cooperation. American Economic Review 96, 1029–1042 (2006)

Blackburn, J.M.: Acquisition of skill: an analysis of learning curves. IHRB Report. No. 73 (1936)

Borenstein, S., Bushnell, J.B., Wolak, F.A.: Measuring market inefficiencies in California's restructured wholesale electricity market. American Economic Review 92, 1376–1405 (2002)

Bunn, D.W., Oliveira, F.S.: Agent-based simulation: An application to the new electricity trading arrangements of England and Wales. IEEE Transaction on Evolutionary Computation 5, 493–503 (2001)

Chialvo, D.R., Bak, P.: Learning from mistakes. Neuroscience 90, 1137–1148 (1999)

Erev, I., Roth, A.E.: Predicting how people play games: reinforcement learning inexperimental games with unique, mixed strategy equilibria. American Economic Review 88, 848–881 (1998)

Gao, F., Guan, X., Cao, X.R., Papalexpoulos, A.: Forecasting power market clearing price and quantity using a neural network method. In: Proceedings of the IEEE Power Engineering Society Transmission and Distribution Conference, Seattle, WA, pp. 2183–2188 (2000)

Jacobs, J.M.: Artificial power markets and unintended consequences. IEEE Transactions on Power Systems 12, 968–972 (1997)

Jiang, A.X., Leyton-Brown, K.: Bidding agents for online auctions with hidden bids. Machine Learning 67, 117–143 (2007)

Joskow, P.L., Kahn, E.: A quantitative analysis of pricing behavior in California's wholesale electricity market during summer 2000. The Energy Journal 23, 1–35 (2002)

Lee, W.W.: US lessons for energy industry restructuring: Based on natural gas and California electricity incidences. Energy Policy 32, 237–259 (2004)

Makowski, M., Nakamori, Y., Sebastain, H.J.: Advances in complex system modeling. European Journal of Operational Research 166, 593–858 (2005)

Morikiyo, T., Goto, M.: Artificial trading for US wholesale electric power market. Asia Pacific Management Review 9, 751–782 (2004)

Nanduri, V., Das, T.K.: A reinforcement learning model to assess market power under auction-based energy policy. IEEE Transactions on Power Systems 22, 85–95 (2007)

Pan, Z.J., Chen, Y.J., Kang, L.S., Zhang, Y.T.: Parameter estimation by genetic algorithms for nonlinear regression. In: Liu, G.Z. (ed.) Optimization Techniques and Applications, Proceedings of International Conference on Optimization Technique and Applications 1995, vol. 2, pp. 946–953. World Scientific, Singapore (1995)

Roth, A.E., Erev, I.: Learning in extensive-form games: experimental data and simple dynamic models in the intermediate term. Games and Economic Behavior 8, 164–212 (1995)

Samuelson, D.: Agents of change. OR/MS Today 32, 26–31 (2005)

Shahidehpour, M., Yamin, H., Li, Z.: Market Operations in Electric Power System. John Wiley & Sons, Chichester (2002)

Si, J., Wang, Y.-T.: On-line learning control by association and reinforcement. IEEE Transactions on Neural Networks 12, 264–276 (2001)

Stoft, S.: Power System Economics. IEEE Press, Piscataway (2002)

Sueyoshi, T.: Beyond economics for guiding large public policy issues: Lessons from the Bell System divestiture and the California electricity crisis. Decision Support Systems 48, 457–469 (2010)

Sueyoshi, T., Tadiparthi, G.R.: A wholesale power trading simulator with learning capabilities. IEEE Transactions on Power Systems 20, 1330–1340 (2005)

Sueyoshi, T., Tadiparthi, G.R.: An agent-based approach to handle business complexity in US wholesale power trading. IEEE Transactions on Power Systems 22, 532–543 (2007)

Sueyoshi, T., Tadiparthi, G.R.: Wholesale power price dynamics under a capacity limit in transmission. IEEE Transactions on Man, Machine and Cybernetics, Part C 38, 229–241 (2008a)

Sueyoshi, T., Tadiparthi, G.R.: An agent-based decision support system for wholesale electricity market. Decision Support Systems 44, 425–446 (2008b)

Sueyoshi, T., Tadiparthi, G.R.: Why did California electricity crisis occur? A numerical analysis using multi-agent intelligent simulator. IEEE Transactions on Man, Machine and Cybernetics, Part C 38, 779–790 (2008c)

Sutton, R.S., Barto, A.G.: Reinforcement Learning: an Introduction. The MIS Press, Cambridge, Massachusetts (1999)

Taylor, J.W., Buizza, R.: Neural network load forecasting with weather ensemble predictions. IEEE Transactions on Power Systems 17, 626–632 (2002)

Tesfatsion, L.: Agent-based modeling of evolutional economic systems. IEEE Transactions on Evolutionary Computation 5, 437–441 (2001)

Thorndike, E.L.: Animal intelligence: an experimental study of the associative processes in animals. Psychological Monographs 2(8) (1898)

Vorobeychik, Y., Wellman, M.P., Singh, S.: Learning payoff functions in infinite games. Machine Learning 67, 145–168 (2007)

Wilson, R.: Architecture of power markets. Econometrica 70, 1299–1340 (2002)

Wolak, F.A.: Measuring unilateral market power in wholesale electricity markets: The California market, 1998-2000. American Economic Review 93, 425–430 (2003)

Zhang, L., Luh, P.B., Kasiviswanathan, K.: Energy clearing price prediction and confidence interval estimation with cascaded neural networks. IEEE Transactions on Power Systems 18, 99–105 (2003)

# Appendix (Sueyoshi and Tadiparthi, 2007)

## (A) Type I: Adaptive Behavior of Agents Equipped with Multiple Learning Capabilities

In the proposed simulator, each market consists of many artificial traders who can accumulate knowledge from their bidding results in order to adjust their proceeding bidding strategies. As depicted in Figure 6, their adaptive learning process is separated into (a) a non-reinforcement learning (self-learning) process and (b) a partial reinforcement learning process. The former process provides each trading agent with not only a forecasted estimate on wholesale power price and amount, but also a win-loss experience from their biddings. The learning process can be considered as a training process for each trading agent. After the non-reinforcement learning process is completed, each agent starts his bidding decisions based upon previous trading experience. All agents constantly update and accumulate their knowledge (experiences) at each trade. The bidding and learning process is considered as the partial reinforcement learning process. The bidding experience in the learning process is incorporated into his database as updated information.

*Adaptive Sigmoid Decision Rule*: In the adaptive learning process of the proposed MAIS, each agent constantly looks for an increase in an estimated winning probability. In other words, he looks for a combination of unknown decision variables and mark-up rates that can increase a winning probability. The win or lose of each agent is considered as a binary response. To express an occurrence of the binary response, a sigmoid model is widely used to predict a winning probability. Mathematically, the probability cumulative function of the sigmoid model is expressed by $F(\sigma) = \int_{-\infty}^{\sigma} e^{u} / (1 + e^{u})^2 \, du = 1/(1 + e^{-\sigma})$.

The win or loss status of the i-th generator of the z-th zone at the t-th period is predicted by the following linear probability model:

$$R_{i(z)t} = c_0 + c_1 \alpha_{i(z)t} + c_2 \beta_{i(z)t} + c_3 \eta_{i(z)t} + \varepsilon . \qquad (A-1)$$

Here, $R_{i(z)t}$ is a reward obtained by the i-th generator. Parameters to be estimated are denoted by c in (A-1). An observational error is listed as $\varepsilon$. The parameters are unknown. Hence, we need estimate them by OLS (Ordinary Least Squares) regression. The winning probability (Prob) can be specified as follows:

$$\text{Prob(WIN)} = \text{Prob}( R_{i(z)t} \geq 0 ) = \frac{\text{EXP}(\hat{c}_0 + \hat{c}_1 \alpha_{i(z)t} + \hat{c}_2 \beta_{i(z)t} + \hat{c}_3 \eta_{i(z)t})}{1 + \text{EXP}(\hat{c}_0 + \hat{c}_1 \alpha_{i(z)t} + \hat{c}_2 \beta_{i(z)t} + \hat{c}_3 \eta_{i(z)t})} . \qquad (A-2)$$

The symbol ( ^ ) indicates a parameter estimate obtained by ordinary least square method. The above equations suggest that the winning probability can be predicted immediately from the parameter estimates of the sigmoid model. The losing probability is measured by 1- Prob(WIN).

The reward of the j-th wholesaler of the z-th zone at the t-th period can be estimated by the following linear probability model:

$$R_{j(z)t} = c_0 + c_1 \delta_{j(z)t} + c_2 \lambda_{j(z)t} + \varepsilon \, . \tag{A-3}$$

Hence, the winning probability is specified as

$$Prob(\text{ WIN }) = Prob(R_{j(z)t} \geq 0) = \frac{EXP(\hat{c}_0 + \hat{c}_1 \delta_{j(z)t} + \hat{c}_2 \lambda_{j(z)t})}{1 + EXP(\hat{c}_0 + \hat{c}_1 \delta_{j(z)t} + \hat{c}_2 \lambda_{j(z)t})} \, . \tag{A-4}$$

The adaptive learning process of the wholesaler provides three parameter estimates of the sigmoid model. Two ($\hat{c}_1$ and $\hat{c}_2$) of the three parameter estimates are important in determining the bidding strategies of the wholesaler. If the parameter estimate is positive, the wholesaler should increase its corresponding decision variable in order to enhance a winning probability. Conversely, an opposite strategy is necessary if the estimate is negative. Thus, the sign of each parameter estimate provides information regarding which decision variable needs to be increased or decreased. However, the winning probability, obtained from the sigmoid model, does not immediately imply that the agent can always win in a wholesale market with the estimated probability. That is a theoretical guess. The win or lose is determined through the DA and HA market mechanism.

*Exponential Utility function*: It is assumed that all the agents have an exponential utility function. The utility function represents a risk aversion preference. Mathematically, the exponential utility function employed in this study is expressed by $U(R_{j(z)t}) = 1 - EXP(-\zeta R_{j(z)t})$ on $R_{j(z)t} \geq 0$, where $\zeta$ indicates a parameter to express the level of risk aversion. The utility function is a smooth concave function. Different $\zeta$ values represent different risk-hedge behaviors of traders.

Returning to (A-3), the utility value ($\phi_{j(z)t}$) for a reward ($R_{j(z)t}$) of the wholesaler is given by $\phi_{j(z)t} = 1 - EXP(-\zeta R_{j(z)t})$. Hence, given $\phi_{j(z)t}$, the reward is expressed by

$$R_{j(z)t} = -\ln(1 - \phi_{j(z)t})/\zeta = \hat{c}_0 + \hat{c}_1 \delta_{j(z)t} + \hat{c}_2 \lambda_{j(z)t} \, , \tag{A-5}$$

where "ln" stands for a natural logarithm. After obtaining the parameter estimates of the sigmoid model, along with a given utility value or its range; the wholesaler considers a bidding strategy for the next period. In this study, the bidding strategy for the next period $(t+1)$ is specified as follows: $\lambda_{j(z)t+1} \leftarrow \lambda_{j(z)t} + \tau \Delta^\lambda_{j(z)t}$ and $\delta_{j(z)t+1} \leftarrow \delta_{j(z)t} + \tau \Delta^\delta_{j(z)t}$, where $\Delta^\lambda_{j(z)t} = \lambda^U_{j(z)t} - \lambda^L_{j(z)t}$ and $\Delta^\delta_{j(z)t} = \delta^U_{j(z)t} - \delta^L_{j(z)t}$. The prescribed quantities ($\lambda^U_{j(z)t}$ and $\lambda^L_{j(z)t}$) indicate the upper and lower bounds on $\lambda_{j(z)t}$, respectively. The other prescribed quantities ($\delta^U_{j(z)t}$ and $\delta^L_{j(z)t}$) also indicate the upper and lower bounds on $\delta_{j(z)t}$. In this case, we need to identify these quantities from the upper and lower bounds of previous bidding amounts. An unknown parameter ($\tau$) indicates the magnitude of such a bidding change. Along with the changes and given $\phi_{j(z)t+1}$, (A-5) becomes

$$-\ln(1-\phi_{j(z)t+1})/\zeta = \hat{c}_0 + \hat{c}_1(\delta_{j(z)t} + \tau\Delta^{\delta}_{j(z)t}) + \hat{c}_2(\lambda_{j(z)t} + \tau\Delta^{\lambda}_{j(z)t}). \qquad (A\text{-}6)$$

From (A-6), the magnitude variable is determined by

$$\tau = -\left(\ln(1-\phi_{j(z)t+1})\Big/\zeta + \hat{c}_0 + \hat{c}_1\delta_{j(z)t} + \hat{c}_2\lambda_{j(z)t}\right)\Big/\left(\hat{c}_1\Delta^{\delta}_{j(z)t} + \hat{c}_2\Delta^{\lambda}_{j(z)t}\right). \qquad (A\text{-}7)$$

Thus, we can determine the magnitude of a bidding change ($\tau$) along with a previously determined strategic direction. Different utility values produce different magnitudes of $\tau$, consequently generating different bidding prices and amounts for the j-th wholesaler. Note that the description on the utility function of the wholesaler can be extended to the utility function of the i-th generator in a similar manner.

*Algorithm*: Each agent is assumed to have two databases: positive database and negative database. A positive database stores all the decision variables and mark-up ratios of an agent when the bidding result is a win. In the same vein, a negative database stores all the decision variables and mark-up rations of an agent when the bidding result is a loose. Using the partial reinforcement learning depicted in Figure 6 (the right hand side), the j-th wholesaler in the z-th zone has the following bidding strategy (with t = H as the start), where H is the practice period:

Step 1:  Set initial bidding variables from the current knowledge base. A forecasting method (e.g., moving average and exponential smoothing) with different time periods is used to compute the initial bidding variables. Also, set the upper ($\delta^{U}_{j(z)t}$ and $\lambda^{U}_{j(z)t}$) and lower ($\delta^{L}_{j(z)t}$ and $\lambda^{L}_{j(z)t}$) limits from the current knowledge base.

Step 2:  Use OLS to obtain parameter estimates of the sigmoid model from the knowledge accumulation process. Obtain the magnitude of a bidding change ($\tau$) from an exponential utility function.

Step 3:  Based upon the signs of parameter estimates, the decision variables on bidding are changed as follows:

(a) If $\hat{c}_1 > 0$ & $\hat{c}_2 > 0$, then $(\delta_{j(z)t+1}, \lambda_{j(z)t+1}) = \{\delta_{j(z)t} + \tau\Delta^{\delta}_{j(z)t}, \lambda_{j(z)t} + \tau\Delta^{\lambda}_{j(z)t}\}.$

(b) If $\hat{c}_1 > 0$ & $\hat{c}_2 = 0$, then $(\delta_{j(z)t+1}, \lambda_{j(z)t+1}) = \{\delta_{j(z)t} + \tau\Delta^{\delta}_{j(z)t}, \lambda_{j(z)t}\}.$

(c) If $\hat{c}_1 > 0$ & $\hat{c}_2 < 0$, then $(\delta_{j(z)t+1}, \lambda_{j(z)t+1}) = \{\delta_{j(z)t} + \tau\Delta^{\delta}_{j(z)t}, \lambda_{j(z)t} - \tau\Delta^{\lambda}_{j(z)t}\}.$

(d) If $\hat{c}_1 = 0$ & $\hat{c}_2 > 0$, then $(\delta_{j(z)t+1}, \lambda_{j(z)t+1}) = \{\delta_{j(z)t}, \lambda_{j(z)t} + \tau\Delta^{\lambda}_{j(z)t}\}.$

(e) If $\hat{c}_1 = 0$ & $\hat{c}_2 = 0$, then $(\delta_{j(z)t+1}, \lambda_{j(z)t+1}) = \{\delta_{j(z)t}, \lambda_{j(z)t}\}.$

(f) If $\hat{c}_1 = 0$ & $\hat{c}_2 < 0$, then $(\delta_{j(z)t+1}, \lambda_{j(z)t+1}) = \{\delta_{j(z)t}, \lambda_{j(z)t} - \tau\Delta^{\lambda}_{j(z)t}\}.$

(g) If $\hat{c}_1 < 0$ & $\hat{c}_2 > 0$, then $(\delta_{j(z)t+1}, \lambda_{j(z)t+1}) = \{\delta_{j(z)t} - \tau\Delta^{\delta}_{j(z)t}, \lambda_{j(z)t} + \tau\Delta^{\lambda}_{j(z)t}\}.$

(h) If $\hat{c}_1 < 0$ & $\hat{c}_2 = 0$, then $(\delta_{j(z)t+1}, \lambda_{j(z)t+1}) = \{\delta_{j(z)t} - \tau\Delta^{\delta}_{j(z)t}, \lambda_{j(z)t}\}.$

(i) If $\hat{c}_1 < 0$ & $\hat{c}_2 < 0$, then $(\delta_{j(z)t+1}, \lambda_{j(z)t+1}) = \{\delta_{j(z)t} - \tau\Delta^{\delta}_{j(z)t}, \lambda_{j(z)t} - \tau\Delta^{\lambda}_{j(z)t}\}.$

Step 4: Compute $d^1_{j(z)t}$ and $p^1_{j(z)t}$ using ( $\delta_{j(z)t}, \lambda_{j(z)t}$ ), and check whether the bidding variables belong to the negative database (from losing experience). If those belong to the negative database, then go to Step 1. If it does not belong to the negative database, submit the bids to the DA market. If $t = T$, then stop. Otherwise, go to Step 5.

Step 5: If the wholesaler loses, then update the bidding variable in his negative database (form losing experience) to generate a negative feedback. Go to Step 6. If the wholesaler wins, then update the bidding strategy in his positive database (form winning experience) to generate a positive feedback. Go to Step 6.

Step 6: Add information on the current bidding variables into the positive and negative databases and determine the strategy related to Step 1. Go to Step 1.

Note that (a) even if a trading agent keeps the same strategy, his market result may be different from the previous one, because the wholesale markets determines the price and amount of power allocation. (b) In Step 3 for each generator, the generator has 27 (= 3 x 3 x 3) bidding strategies, as structured for the wholesaler. The three parameters need to be considered in the algorithmic steps for the generator. (c) The algorithm proposed for DA can be applied to the bidding price and quantity of a generator for HA in a similar manner.

## (B)  Type II: Adaptive Behavior of Agents Equipped with Limited Learning Capability

In addition to Type I, this study develops agents in Type II who prepare the proceeding bidding strategies only from the current bidding results. So, the bidding strategies are myopic. They are equipped with neither the exponential utility function nor the sigmoid decision rule, both of which are incorporated in Type I.

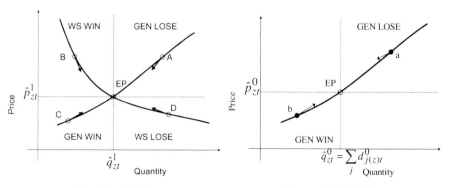

Fig. A-1. DA Market                  Fig. A-2. HA Market

Figures A-1 and A-2 visually describe the bidding strategies of agents in Type II. In the two figures, equilibrium points for DA and HA are depicted on $EP(\hat{q}^1_{zt}, \hat{p}^1_{zt})$

and EP $(\hat{q}_{zt}^0, \hat{p}_{zt}^0)$ respectively, along with supply and demand curves. Centering on each equilibrium point, each figure is separated into four regions. The north-east region indicates that a generator (GEN) loses in the DA market. Let us consider that a current bid for the DA market exists on A along a supply curve in Figure A-1. The generator loses in the DA market because the bidding price is higher than the market price on equilibrium. So, the generator loses an opportunity to supply electricity. The loss generates a "negative feedback" to the agent. As a result of the negative feedback, he reduces both the bidding price and quantity for the proceeding DA market. An opposite strategy is observed on C in the south-west region where the generator wins in the DA market. The north-west region indicates that a wholesaler (WS) on B wins in the current DA market, because the bidding price is higher than the market price on equilibrium. So, the wholesaler obtains electricity through the DA market. The result produces a "positive feedback" to the agent. As a result of the positive feedback, he decreases the bidding price and increases the bidding quantity to obtain more profit in the proceeding DA market. An opposite strategy is found on D in the south-east region where the wholesaler loses in the DA market.

The bidding strategy in the DA market illustrated in Figure A-1 is shifted to the HA market of Figure A-2. As mentioned previously, the HA is a physical and spot market, so a generator's bidding strategy in the HA market is different from that for the DA market. The pricing strategy for the HA market is identified by the two arrows in Figure A-2. A wholesaler enters the HA market with his strategy on quantity. The wholesaler does not have any strategy concerning price, because he must satisfy demand of end-users without any choice about price. As listed at the bottom of Figure A-2, the total amount of demand submitted by all wholesalers is listed as $\sum_j d_{j(z)t}^0$, and it determines the quantity of EP.

The bidding strategy of agents in Type II is separated into the following four strategies for the DA market:

(a)  If the j-th wholesaler wins in the current DA market, then he increases the bidding amount and decreases the bidding price for the proceeding DA market as follows:

$$d_{j(z)t+1}^1 = d_{j(z)t}^1 + \left|q_{j(z)t+1} - d_{j(z)t}^1\right|\delta_{j(z)t+1} \text{ and } p_{j(z)t+1}^1 = p_{j(z)t}^1 - \left|p_{j(z)t}^1 - \tilde{p}_{j(z)t+1}\right|\lambda_{j(z)t+1} \text{ (A-8)}$$

The above strategy indicates that the next bidding amount $\left(d_{j(z)t+1}^1\right)$ is increased from the current bidding amount $\left(d_{j(z)t}^1\right)$ by $\left|q_{j(z)t+1} - d_{j(z)t}^1\right|\delta_{j(z)t+1}$, where $q_{j(z)t+1} - d_{j(z)t}^1$ is a difference between a forecasted amount (for the proceeding DA market) and the current bidding amount. The difference in an absolute value is adjusted by $\delta_{j(z)t+1}$. Similarly, the next bidding price $\left(p_{j(z)t+1}^1\right)$ is reduced from the current bidding price $\left(p_{j(z)t}^1\right)$ by $\left|p_{j(z)t}^1 - \tilde{p}_{j(z)t+1}\right|\lambda_{j(z)t+1}$. Here, the wholesaler compares the current bidding price with the forecasted price estimate

$(\widetilde{p}_{j(z)t+1})$, obtained from the knowledgebase. The difference between them in an absolute value is adjusted by $\lambda_{j(z)t+1}$.

(b) If the j-th wholesaler loses in the current DA market, then he decreases the bidding amount and increases the bidding price for the proceeding DA market as follows:

$$d^1_{j(z)t+1} = d^1_{j(z)t} - \left|q_{j(z)t+1} - d^1_{j(z)t}\right|\delta_{j(z)t+1} \text{ and } p^1_{j(z)t+1} = p^1_{j(z)t} + \left|\widetilde{p}_{j(z)t+1} - p^1_{j(z)t}\right|\lambda_{j(z)t+1} \text{ (A-9)}$$

The difference between (A-8) and (A-9) is identified by the quantity reduction and price increase in (A-9).

(c) If the i-th generator wins in the current DA market, he increases both the bidding amount and price for the proceeding DA market as follows:

$$s^1_{i(z)t+1} = s^1_{i(z)t} + \left|s^m_{i(z)t+1} - s^1_{i(z)t}\right|\alpha_{i(z)t+1} \text{ and } p^1_{i(z)t+1} = p^1_{i(z)t} + \left|\widetilde{p}_{i(z)t+1} - p^1_{i(z)t}\right|\beta_{i(z)t+1} \text{ (A-10)}$$

The above strategy indicates that the next bidding amount $(s^1_{i(z)t+1})$ is increased from the current bidding amount $(s^1_{i(z)t})$ by $\left|s^m_{i(z)t+1} - s^1_{i(z)t}\right|\alpha_{i(z)t+1}$ where $s^m_{i(z)t+1} - s^1_{i(z)t}$ is a difference between a maximum generation capacity (for the proceeding DA market) and the current bidding amount. The difference in an absolute value is adjusted by $\alpha_{i(z)t+1}$. Similarly, the next bidding price $(p^1_{i(z)t+1})$ is increased from the current bidding price $(p^1_{i(z)t})$ by $\left|\widetilde{p}_{i(z)t+1} - p^1_{i(z)t}\right|\beta_{i(z)t+1}$. Here, the generator compares the current bidding price with the price estimate $(\widetilde{p}_{i(z)t+1})$ that is obtained from the current knowledgebase. The difference between them in an absolute value is adjusted by $\beta_{i(z)t+1}$.

(d) If the i-th generator loses in the current DA market, he decreases both the bidding price and amount for the proceeding DA market as follows:

$$s^1_{i(z)t+1} = s^1_{i(z)t} - \left|s^m_{i(z)t+1} - s^1_{i(z)t}\right|\alpha_{i(z)t+1} \text{ and } p^1_{i(z)t+1} = p^1_{i(z)t} - \left|p^1_{i(z)t} - \widetilde{p}_{i(z)t+1}\right|\beta_{i(z)t+1} \text{ (A-11)}$$

The difference between (A-10) and (A-11) can be found in the quantity and price reduction in (A-11).

*Comment*: The HA pricing strategies of a generator is depicted in Figure A-2. The HA strategies correspond to (c) and (d) for the DA market. A wholesaler bids only an amount of electricity in HA.

# Chapter 5

# Argument Mining from RADB and Its Usage in Arguing Agents and Intelligent Tutoring System

Safia Abbas[1] and Hajime Sawamura[2]

[1] Graduate School of Science and Technology, Niigata University
8050, 2-cho, Ikarashi, Niigata, 950-2181, Japan
safia@cs.ie.niigata-u.ac.jp
[2] Institute of Science and Technology, Niigata University
8050, 2-cho, Ikarashi, Niigata, 950-2181, Japan
sawamura@ie.niigata-u.ac.jp

**Abstract.** Argumentation is an interdisciplinary research area that incorporates many fields such as artificial intelligence, multi-agent systems, and collaborative learning. In this chapter, we describe argument mining techniques from a structured argument database "RADB", a sort of relational database we designed specially for organizing argument databases, and their usage in arguing agents and intelligent tutoring systems. The RADB repository depends on the Argumentation Interchange Format Ontology (AIF) using "Walton Theory" for argument analysis. It presents a novel approach that summarizes the argument data set into structured form "RADB" in order to (i) facilitate the data interoperability among various agents/humans/tools, (ii) provide the ability to freely navigate the repository by integrating the data mining techniques gathered in a classifier agent; mine the RADB repository and retrieve the most relevant arguments to the users' queries, (iii) illustrate an agent-based learning environment outline, where the mining classifier agent and the RADB are incorporated together within an intelligent tutoring system (ITS). Such incorporation assists in (i) deepening the understanding of negotiation, decision making, and critical thinking, (ii) guiding the analysis process to refine the user's underlying classification, and improving the analysis and the students' intellectual process.

Later in the chapter, we describe an effective usage of argument mining for arguing agents, which interact with each other in the Internet environment and argues about issues concerned, casting arguments and counter-arguments each other to reach an agreement. We illustrate how argument mining allows to strengthen arguing agent intelligence, resulting in expanding the main concern in formal argumentation frameworks that is to formalize methods in which the final statuses of arguments are to be decided semantically and/or dialectically. In both usages, we yield new forms of argument-based intelligence, which allows establishing one's own argument by comparing diverse views and opinions and uncovering new leads, differently from simple refutation aiming at cutting down other parties.

D. Srinivasan & L.C. Jain (Eds.): Innovations in MASs and Applications – 1, SCI 310, pp. 113–147.
springerlink.com                                    © Springer-Verlag Berlin Heidelberg 2010

# 1   Introduction

Argumentation theory embraces the arts and sciences of civil debate, dialog, conversation, and persuasion. It studies rules of inference, logic, and procedural rules in both artificial and real world settings. Argumentation is concerned primarily with reaching conclusions through logical reasoning, that is, claims based on premises. We argue all the time in our daily life, scientific communities, parliaments, courts, and online discussion boards and so on. That is, we would say "living is arguing in the age of globalization just as living is eating in the days of plenty". Humans' knowledge and wisdom produced there usually have the form of arguments that are built up from more primitive knowledge of the form of facts and rules. Those repositories or treasuries of knowledge are now about to be organized, by different tries, to argument data bases or corpora that can be retrieved, stored, and reused online freely. One of these trials is the argument mapping tools (e.g., Compendium[1], Araucaria[2], Rationale [3], etc.) that aims at improving our ability to articulate, comprehend and communicate reasoning. This can be achieved by supporting to produce diagrams of reasoning and argumentation for especially complex arguments and debates. It is greatly anticipated that it helps end users learning critical thinking methods as well as promoting critical thinking in daily life[20,22]. Although much work has been done in this area, there are still some concerns about formal computational argumentation frameworks; formalizing methods in which the internal statuses of arguments have to be decided semantically and/or dialectically[9]. These concerns can briefly be summarized into the following three general points: (i) both areas for argumentation-conscious works lack means of arguments' search for an argument to be retrieved, classified or summarized. Put it differently, such an idea of argument to be retrieved, summarized or classified in order to gain useful information and get used from a large argument database is missing. (ii) The existing frameworks miss the interoperability among each other, such that instead of producing an appropriate common files type, that enables the user to access the same argument using different tools, special types of files are generated such as AML files that are primarily committed to be used by Araucaria DB only. (iii) The context analysis, presented in the existing frameworks, does not follow empirical regularities, where any one can analyze any document depending on his own thoughts and beliefs and not on a dedicated structure for different argument schemes.

In this chapter, we discuss an effective usage of argument mining for both arguing agents and intelligent tutoring system in order to overcome the mentioned obstacles. We firstly present a novel approach that sustains argument analysis, retrieval, and re-usage from a relational argument database (RADB). The database is considered as a highly structured argument repository managed by a classifier agent. This agent makes use of different mining techniques that

---

[1] http://compendium.open.ac.uk/software.html
[2] http:// araucaria.computing.dundee.ac.uk/
[3] http://rationale.austhink.com/

are gathered and used to:(i) summon and provide a myriad of arguments at the user's fingertips, (ii) reveal more relevant results to users' queries. The chapter mainly concerns with mining arguments through RADB that (i) summarizes the argument dataset, (ii) supports the fast interaction between the different mining techniques and the pre-existing arguments, (iii) and facilitates the inter-operability among various agents/tools/humans. Then, an agent-based learning environment (ALES) outline is illustrated, where the mining classifier agent and the RADB are integrated to an intelligent tutoring system (ITS). Such integration should assist in (i)deepening the understanding of argumentation, decision making and critical thinking, (ii)guiding and tracing the user during the analysis process to refine his/her underlying classification by providing the suitable individualized feedback. Finally, we declare how argument mining and retrieval can strengthen arguing agent intelligence and can enhance computational agents with an ability of argumentation. Where the agents could benefit a huge amount of knowledge for argumentation, and reciprocally argument mapping systems could gain a mechanism for deciding final argumentation statuses with the help of formal argumentation systems.

The chapter is organized as follows. Section 2 provides a background about the AIF ontology[5] and Walton theory of argumentation. Section 3 presents the benefits, design and implementation of the structured/relational argument database "RADB". Section 4 states different mining techniques and strategies in a certain classifier agent, which mines the RADB repository, as a managing tool, to verify some goals. Section 5 outlines the usage of the RADB together with the classifier agent in an agent-based learning environment named "ALES" and provides an illustrative example for student-system interaction. Section 6 describes an effective usage of argument mining for arguing agents that interact with each other. In Section 7, we discuss and compare our work with the most relevant work performed in the same field, particularly from the motivational point of view. Finally, conclusions and future work are presented in Section 8.

## 2    Background: AIF Ontology and Walton Theory of Argumentation

Argumentation theory has provided several sets of forms such as deductive, inductive and presumptive patterns of reasoning [10]. The earliest accounts of argument schemes were advanced and developed in several researches by different authors [10]. Each scheme set unanimously presupposes a particular theory of argument, which in turn, implies a particular perspective regarding the relation between logic and pragmatic aspects of argumentation, and notions of plausibility and defeasibility. However, the most influential theories in argumentation are Toulmin and Walton[21,25,26]. The latter, which is our main interest in this paper, represents a new form of reasoning that depends on fallacious argument, where each argument is considered as a set of premises presented as reasons to accept or reject the conclusion. It could be reformulated based on the AIF ontology[5] and be used thereafter in different context analysis.

AIF is a core ontology, which reveals a way that can be extended to capture a variety of argumentation formalisms and schemes [5][18]. It assumes that every argument can be represented semantically in the form of nodes connected together with directed edges in a directed graph known as argument network [5]. The nodes can be classified as information nodes (I-nodes) that hold pieces of information or data, and scheme nodes (S-nodes) that represent the applications of schemes. In other words, I-nodes contain the content as the declarative aspects of the domain of discourse such as claims, data, evidence, and propositions. On the other hand, S-nodes are applications of schemes that can be considered as patterns of reasoning or deductive patterns associated with argumentative statements. The ontology deals with three different types of scheme nodes namely: rule of inference application nodes (RA-nodes), preference application nodes (PA-nodes) and conflict application nodes (CA-nodes) [5][18].

The Walton theory of argumentation depicts a new aspect in informal logic by using critical questions to evaluate and analyze arguments dialogs. These critical questions help to find the gaps and problems in the argument and evaluate it as weak or strong. This kind of fallacious argumentation reasoning has become the tool of choice in numerous argumentation studies for evaluating arguments [25][26]. Thus, the theory formalized the most common kinds of arguments in structures and taxonomies of schemes, where each scheme type has a name, a conclusion, a set of premises and a set of critical questions bound to it. An example of Walton-style schemes is the 'Argument from Expert Opinion [12] that contains a set of premises, a conclusion and a set of six basic critical questions. Let us see the following example 1.

*Example 1.* The scheme Argument from Expert Opinion has two premises and a conclusion:

- *Premise:*Source E is an expert in the subject domain X containing proposition B.
- *Premise:*E asserts that proposition B in domain X is true.
- *Conclusion:*B may plausibly be taken to be true.

Its basic critical questions consist of:

1. *Expertise Question:* How credible is expert E as an expert source?
2. *Field Question:* Is E an expert in the field that the B is in?
3. *Opinion Question:* Does E's assertions imply B?
4. *Trustworthiness Question:* Is E reliable as source?
5. *Consistency Question:* Does B consistent with the assertions of other experts?
6. *Backup Evidence Question:* Are there any evidences sustain B?

Considering the AIF ontology, if the cyclic problem (the same information node (I-node) refines more than one scheme node (S-node)) is avoided, each scheme structure can be semantically represented in the form of directed tree "argument network". From that, any Walton-style scheme can be represented as a general skeleton as seen in *Fig.*1.

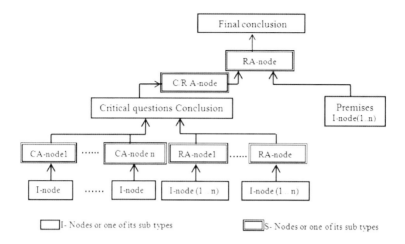

**Fig. 1.** Argument network representation for Walton schemes

- *CA-node:* conflict application nodes.
- *RA-node:* rule of inference application nodes.
- *C/R-node:* either CA-node or RA-node.

# 3   The Relational Argument Database Structure "RADB"

In spite of the different argument repositories, such as Araucaria DB[20,22], that facilitates the data preprocessing and contains different analyzed contexts, the problem of producing appropriate common files types still are left unsolved. Instead, special types of files are generated, such as AML[22] files that are primarily committed to be used by Araucaria only. In addition, the context analysis, as in discourse DB[4], does not follow empirical regularities, and indeed any one can analyze any document depending on his own thoughts and beliefs, not on a dedicated structure for different argument schemes. Furthermore, usual textual data representation is also intractable to be processed, mined or manipulated [8] [13]. As a result, such representations limit the knowledge extraction abilities that could be beneficial and valuable for different end users. So, the relational argument database "RADB" is presented in this paper, as a solution for argument representation, that overcomes the mentioned barriers.

The RADB can be defined as a set of information reformulated and categorized into a set of files (tables) that are general enough to be accessed, gathered (queried), and manipulated in different manners. Therefore, the resulting "clumps" of this organized data is much easier to be understood. The RADB enjoys a number of advantages in comparison with traditional representation. It aims to summarize the high-scale argument dataset, such that all information

---

[4] http://discoursedb.org

required are encoded in an appropriate form in order to (i)support the fast inter-action between the different mining techniques and the pre-existing arguments, (ii)facilitate the interoperability among various argument tools/agents/humans.

In a more abstract view, according to the AIF ontology [5] [18] and Walton theory of argumentation that were discussed in Section 2, any argument can semantically be represented in the form of a directed tree that can be structured as well-established relational argument database "RADB". Not only Walton theory of argumentation is regarded to be an abstract propositional form, but also a pattern instantiated in real dialogs. This property allows the possibility of the relational database conversion. In addition, It provides a space to create a new argumentation scheme, whereas this feature importance will be explained later in Section 3.2.2.

## 3.1   The RADB Main Building Blocks

In the mean time, our repository "RADB" summons a number of arguments that were selected from Araucaria database. However, the RADB design and structure take into account that: (i)Araucaria DB contains general contexts that do not belong to a specific domain, so the RADB structure is general enough to encapsulate multiple domains, (ii)the chosen contexts' analyses do not follow any empirical regularities, so these arguments' context were re-analyzed based on different schemes structures of Walton theory of argumentation [26][25][12], preserving the constraints of the AIF ontology [5] (such that no information node (I-node) refines another I-node), and were visualized using compendium as a mapping tool to reflect the analysis trees.

Accordingly, the current repository can semantically be represented as a forest of a numerous directed trees [7]. Wherein, each directed tree lays out a semantic representation for a specific argument analysis based on specific schemes. This representation is illustrated in the following Example 2 and $Fig2$.

*Example 2.* The following context from Araucaria repository database[27][22][20] was reanalyzed based on the expert opinion scheme [12][25][26], and its directed tree is shown by means of Compendium in $Fig.$ 2.

"*Eight-month-old Kyle Mutch's tragic death was not an accident and he suf-fered injuries consistent with a punch or a kick, a court heard yesterday. The baby, whose stepfather denies murder, was examined by pathologist Dr. James Grieve shortly after his death. Dr. Grieve told the High Court at Forfar the youngest was covered in bruises and had suffered a crushed intestine as well as severe internal bleeding. When asked by Advocate Depute Mark Stewart, prose-cuting, if the bruises could have been caused by an accident, he said "No. Not in a child that is not walking, not toddling and has not been in a motor car." Dr. Grieve said the injuries had happened "pretty quickly" and would be "diffi-cult for an infant to cope with". The lecturer in forensic medicines at Aberdeen University told the jury that the bruises could have been caused by a single blow from a blunt instrument, like a closed hand. Death, not accident, court told*"

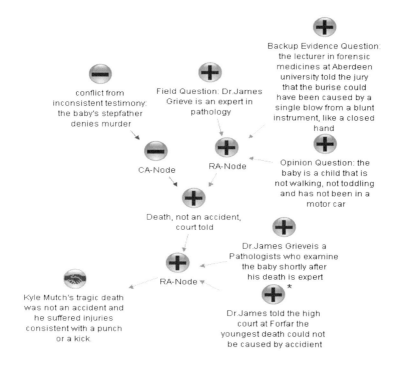

**Fig. 2.** The analysis diagram of the context in Example 2 based on the expert opinion scheme (this is represented by the table form of the RADB in *Fig.* 6)

 Final decision.

Support node approves the final decision.

Conflict node rebuts the final decision.

The main building blocks of the RADB are: (a) the table of schemes "Scheme_TBL", which gathers the names and the indexes for different schemes notations, (b) the schemes structure table "Scheme_Struct_TBL" that assembles the details of each scheme stated in "Scheme_TBL", (c) the transaction table "Data_TBL", which contains the analyses of arguments based on scheme structures and preserves the constraints of the AIF ontology [5]. The relation between those basic tables is shown in *Fig.* 3. In what follow, we describe the various building blocks concerned with "RADB" by the use of the screen shots of our implemented system.

### 3.1.1   The Scheme_TBL Table

This table gathers different scheme kinds, such that schemes are formulated, as seen in *Fig.* 4, in the form of rows and columns. The rows act as records of data for different schemes, and the columns as features (attributes) of records. The Scheme_TBL has two features ($ID$, discrete) and ($SCH\_Name$, text). The

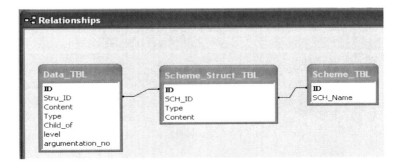

**Fig. 3.** The relation between the basic tables in RADB

$ID$ denotes the identification number proper to $SCH\_Name$ that stands for the different schemes' names. Moreover, this $ID$ plays a role of primary key for this table and foreign key in the others. (Note that any $ID$ attribute/feature mentioned later will stand for the same function)

| | | ID | SCH_Name |
|---|---|---|---|
| | + | 1 | Expert Opinion |
| | + | 2 | Popular Opinion |
| | + | 3 | Verbal Classific |
| | + | 5 | inference |
| | + | 6 | Conflict |
| | + | 7 | Prefrence |
| | + | 8 | Depate Result |
| ▶ | + | 9 | Argument from |

⊞ **Scheme_TBL : Table**

**Fig. 4.** The design and structure of Scheme_TBL table

### 3.1.2    The Scheme_Struct_TBL Table

The details of each scheme stated in the Scheme_TBL is exemplified in this table. Its rows represent records for the different information associated with the different schemes, and columns reveal the features/attributes of these records. The attributes are defined as follows: ($ID$, discrete) specified as above, and ($SCH\_Id$, discrete) that acts as a foreign key of $Scheme\_TBL$ indicating the concerned scheme, ($Content$, text) contains the context of the associated information, and ($Type$, discrete $\in$ {P, C, CQ, CC}), such that P for premises, C for conclusion, CQ for critical question and CC for critical argumentation conclusion. For instance, the corresponding RADB formulation of the expert opinion scheme [12], shown in Section 2, is represented in $Fig.$ 5.

| | ID | SCH_ID | Type | Content |
|---|---|---|---|---|
| ⊞ **Scheme_Struct_TBL : Table** | | | | |
| + | 1 | 1 | P | Source E is an expert in the subject domain X containing proposition B. |
| + | 2 | 1 | P | E asserts that proposition B in domain X is true. |
| + | 3 | 1 | C | B may plausibly be taken to be true. |
| + | 4 | 8 | CC | Critical argumentation conclusion |
| + | 5 | 1 | CQ | Expertise Question: How credible is expert E as an expert source? |
| + | 6 | 1 | CQ | Field Question: Is E an expert in the field that the B is in? |
| + | 7 | 1 | CQ | Opinion Question: Does E's assertions imply B? |
| + | 8 | 1 | CQ | Trustworthiness Question: Is E reliable as source? |
| + | 9 | 1 | CQ | Consistency Question: Does B consistent with the assertions of other experts? |
| + | 10 | 1 | CQ | Backup Evidence Question: are there any evidences sustain B? |
| + | 11 | 5 | RA | RA-Node |
| + | 12 | 6 | CA | CA-Node |
| + | 13 | 7 | PA | PA-Node |
| + | 14 | 9 | C | a has property G |
| + | 15 | 9 | P | individual premise:a has property F |
| + | 16 | 9 | P | classification premise: for all x, if x has property F then x can be classified as ha |
| + | 17 | 9 | CQ | critical question1: what evidence is there that a definitely has property F? |
| + | 18 | 9 | CQ | critical question2: is the verbal classification in the classification premise based |

**Fig. 5.** The design and structure of Scheme_Struct_TBL table

### 3.1.3   The Data_TBL Table

The Data_TBL table, as seen in $Fig.$ 6, contains all users' transactions. The table gathers all the analysis done by the different users for specific argument contexts. It consists of ($ID$, discrete) attribute, defined as before, ($Stru\_Id$, discrete) that serves as foreign key of Scheme_Struct _TBL table referring to a specific part of the scheme details, ($Content$, text) attribute contains a portion of the analyzed context that fulfills the referred fixed part of the scheme details, and ($Type$, discrete $\in$ {-1, 0, 1}) attribute, which holds three values only, 1 for the supported node, -1 for rebuttal node, and 0 for undetermined value that denotes neither support nor rebuttal nodes. One of these values is to be given to the final conclusion of the analysis. Since we consider any argument network as a kind of directed root trees, ($Child\_Of$, discrete) attribute points to the parent of each node, whereas the root node has no parents (0 refers to no parent). ($level$, discrete) attribute refers to the level of each node in the tree, such that the value 0 indicates the root node of the argument. Finally, ($argumentation\_no$, text) attribute contains the identification name of the analyzed argument context.

### 3.2   Benefits of the RADB Structure

The described design and structure of the various RADB components emphasize a good foundation to gain valuable knowledge and extract useful information by i) querying the different argument component, ii) manipulating (add, update and delete) both schemes and arguments analyses in a flexible way, iii) mining the different analyses transactions using different mining techniques, which will be described in Section 4.

| | ID | Stru_ID | Content | Type | Child_of | level | argumentation_no |
|---|---|---|---|---|---|---|---|
| + | 1 | 3 | kyle mutch's tragic death was not an accident and he suffered in | 0 | 0 | 0 | argument_602 |
| + | 2 | 11 | RA-Node | 1 | 1 | 1 | argument_602 |
| + | 3 | 2 | Dr.James told the high court at Forfar the youngest death could r | 1 | 2 | 2 | argument_602 |
| + | 4 | 1 | Dr.James Grieveis a Pathologists who examine the baby shortly | 1 | 2 | 2 | argument_602 |
| + | 5 | 4 | Death, not an accident, court told | 1 | 2 | 2 | argument_602 |
| + | 6 | 11 | RA-Node | 1 | 5 | 3 | argument_602 |
| + | 7 | 6 | Field Question: Dr.James Grieve is an expert in pathology | 1 | 6 | 4 | argument_602 |
| + | 8 | 7 | Opinion Question: the baby is a child that is not walking, | 1 | 6 | 4 | argument_602 |
| + | 9 | 10 | Backup Evidence Question: the lecturer in forensic medicines | 1 | 6 | 4 | argument_602 |
| + | 10 | 12 | CA-Node | -1 | 5 | 3 | argument_602 |
| + | 11 | 9 | conflict from inconsistent testimony: the baby's stepfather denies | -1 | 10 | 4 | argument_602 |
| + | 12 | 3 | it is plausible true that sir mark was planning to leave the country | 0 | 0 | 0 | argument_533 |
| + | 14 | 11 | RA-Node | 1 | 12 | 1 | argument_533 |
| + | 15 | 4 | sir mark was planning to leave the country | 1 | 12 | 1 | argument_533 |
| + | 16 | 1 | Authorities are experts in south Africa wher sir mark live | 1 | 14 | 2 | argument_533 |
| + | 17 | 2 | Authorities said that Sir Mark was planning to leave the country \ | 1 | 14 | 2 | argument_533 |
| + | 18 | 7 | Opinion Question: prospecution authorities asserts that sir mark | 1 | 15 | 2 | argument_533 |
| + | 19 | 10 | Backup Evidence Question: sir mark's Planning to leave the cour | 1 | 15 | 2 | argument_533 |
| + | 20 | 6 | Field Question: Authorities are experts in south Africa where sir i | 1 | 15 | 2 | argument_533 |
| + | 21 | 5 | Expertise Question: Prosecution Authorities are credible as expe | 1 | 15 | 2 | argument_533 |

Record: 14 ◀ 21 ▶ ▶I ▶* of 74

**Fig. 6.** The design and structure of Data_TBL table (this is a table for Fig. 2)

### 3.2.1   Some SQL Queries Examples

The following examples illustrate the flexibility of the design of the RADB as a repository of structured arguments. They put forward different ways to query different argument components in order to extract various information, which in turn, could usefully be used by different end users.

*Example 3.* Starting with a simple example, if we want to retrieve all the details of the "expert opinion" scheme, we simply run the following SQL query:

- SELECT * FROM Scheme_Struct_TBL INNER JOIN Scheme_TBL ON Scheme_Struct_TBL.$SCH\_Id$ = Scheme_TBL.$ID$
  WHERE Scheme_TBL.$SCH\_Name$ LIKE 'Expert Opinion'.

Then the result is obtained as shown in *Fig. 7*.

*Example 4.* If we want to retrieve all the transactions that use the "Backup Evidence Question" in their analysis, the following SQL queries are consecutively applied, and the clumps from these queries are obtained as shown in *Fig. 8*.

1. id = "SELECT ID FROM Scheme_Struct_TBL WHERE Scheme_Struct_TBL.Content LIKE 'Backup Evidence Question%' ";
2. "SELECT * FROM Data_TBL WHERE Data_TBL.Stru_ID=id";

| | | ID | SCH_Name | |
|---|---|---|---|---|
| ▶ | - | | 1 Expert Opinion | |

| | | ID | Type | Content |
|---|---|---|---|---|
| | + | 1 | P | Source E is an expert in the subject domain X containing proposition B. |
| | + | 2 | P | E asserts that proposition B in domain X is true. |
| | + | 3 | C | B may plausibly be taken to be true. |
| | + | 5 | CQ | Expertise Question: How credible is expert E as an expert source? |
| | + | 6 | CQ | Field Question: Is E an expert in the field that the B is in? |
| | + | 7 | CQ | Opinion Question: Does E's assertions imply B? |
| | + | 8 | CQ | Trustworthiness Question: Is E reliable as source? |
| | + | 9 | CQ | Consistency Question: Does B consistent with the assertions of other experts? |
| | + | 10 | CQ | Backup Evidence Question: are there any evidences sustain B? |

**Fig. 7.** The SQL query result in Example 3

| | ID | SCH_ID | Type | Content | | | | |
|---|---|---|---|---|---|---|---|---|
| - | 10 | 1 | CQ | Backup Evidence Question: are there any evidences sustain B? | | | | |

| | | ID | Content | Type | Child_of | level | argumentation_no |
|---|---|---|---|---|---|---|---|
| | + | 9 | Backup Evidence Question: the lecturer in forensic medicines | 1 | 6 | 4 | argument_602 |
| | + | 19 | Backup Evidence Question: sir mark's Planning to leave the cour | 1 | 15 | 2 | argument_533 |
| | + | 23 | Backup Evidence Question: there were suitcases around the hoi | 1 | 22 | 4 | argument_533 |
| | + | 24 | Backup Evidence Question: he had disposed of some of the cars | 1 | 22 | 4 | argument_533 |
| | + | 25 | Backup Evidence Question: the house was in the market | 1 | 22 | 4 | argument_533 |
| | + | 39 | Backup Evidence Question:There has been a basic change in th | 1 | 37 | 2 | argument_800 |
| | + | 44 | Backup Evidence Question: Capital spending is rising, as are co | 1 | 43 | 4 | argument_800 |
| | + | 45 | Backup Evidence Question: Japanese banks are finally making r | 1 | 43 | 4 | argument_800 |
| | + | 52 | Backup Evidence Question: blood exchange is not needed becat | 1 | 50 | 2 | argument_810 |
| | + | 55 | Backup Evidence Question: oil accounts for between 65 and 95 ; | 1 | 54 | 4 | argument_810 |
| | + | 56 | Backup Evidence Question: oil makes up only about 7 percent o | 1 | 54 | 4 | argument_810 |
| | + | 63 | Backup Evidence Question: The correctness of these judgement | 1 | 61 | 2 | argument_831 |
| | + | 66 | Backup Evidence Question: stocks have been richly or inexpens | 1 | 63 | 3 | argument_831 |
| | + | 67 | Backup Evidence Question: The extremes of the Internet bubble, | 1 | 63 | 3 | argument_831 |

**Fig. 8.** The SQL queries result in Example 4

*Example 5.* Suppose we want to retrieve all the arguments that have the word "war" in their analyses' conclusion. Firstly, the schemes' structures are classified such that all different schemes' conclusions are gathered in one set "CRS". Secondly, the arguments' transactions or analysis are categorized such that each item in the CRS set has a corresponding set of transactions. Finally, the different sets of transactions are summarized such that the selected items are the groups that contains "war" word in their context. These steps can be expressed by the following two SQL queries, where the second query meets both the second and the last steps. The resulted clumps from these queries are shown in *Fig.* 9.

- CRS = "SELECT * FROM Scheme_Struct_TBL
  WHERE Scheme_Struct_TBL.Type LIKE 'C' ";
- For Each Item I ∈ CRS;
- "SELECT * FROM Data_TBL
  WHERE Data_TBL.Stru_ID = I.ID
  AND Data_TBL.Content LIKE "'% war %' ";

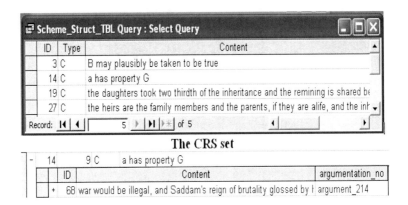

Fig. 9. The transaction final result

### 3.2.2    Data Manipulation

Argumentation can be defined as an intellectual process that depends on one's thoughts and beliefs and affected by the surrounding environment and culture.

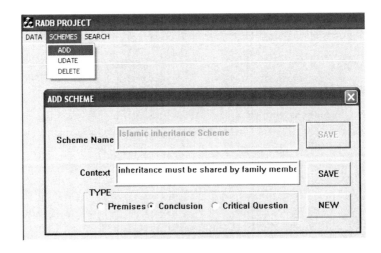

Fig. 10. The adding scheme tab

It is known that, cultural aspects are well-reflected in the style of mutual argument analysis, so appropriate schemes are obligatory to suit the different cultural influences. In response to this, an extendible repository is recognized. The developed RADB repository enjoys the extendibility feature, where it permits adding, deleting and updating both: schemes and arguments analyses. For example, new schemes, such as "Islamic Inheritance scheme" that divides the inheritance between different heirs based on Islamic regularities, can simply be added using the form shown in *Fig.* 10. Within the same form, different parts of the scheme can sequentially be inserted to the RADB repository using the ODBC connection through the object oriented (OOP) form. Then, the already existed schemes can be deleted as well as updated through the ODBC connection of the OOP forms that are shown in *Fig.* 11 and 12. These manipulations consider the different arguments analyses as well, where another different OOP forms are used.

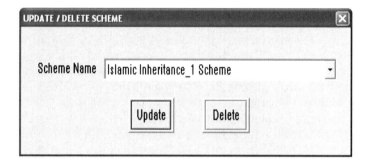

**Fig. 11.** The deleting scheme tab

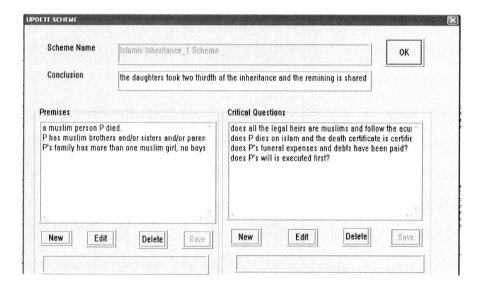

**Fig. 12.** The updating scheme tab

## 4    Mining Classifier Agent

So far we have described the retrieval methods of argument parts via some SQL statements to highlight the flexibility of the structured arguments formulation. However, those described methods retained some elementary tasks that are dedicated to arguments' parts only. The underlying RADB repository arguments can be more exploited for more general objectives. Some of these objectives are to:

- give the end user the freedom to navigate into the existing arguments to search for specific subjects,
- extricate information by classifying the RADB arguments and retrieving the most relevant arguments to the subject of search,
- discover hidden patterns and correlations between different pre-existing arguments,
- analyze any argument context based on a set of crucial hints that direct the analysis process.

To accomplish these objectives, different data mining techniques, which vary in functionality, were integrated together with the RADB repository. For example, the ApriorTid technique[4], the Substructure mining technique and the Rule Extraction mining technique [see subsections 4.1-4.3] that are used to classify the pre-existing arguments in order to retrieve the most relevant arguments, and discover hidden correlation between different argument parts.

Accordingly, a classifier agent that gathers and controls different mining techniques based on users' specification has been developed as discussed in the subsections below. The agent mines the RADB repository aiming to: (i) direct the search process towards hypotheses that are more relevant to users' queries; classifying the analogous arguments in different ways based on users' choice, seeking for the most relevant arguments to the users' subject of search. (ii) add flexibility to the retrieving process by providing different search techniques, and (iii) offer a myriad of arguments at users' fingertips.

### 4.1    AprioriTid Mining Classification

The AprioriTid algorithm [4] has been implemented and embedded to the classifier agent as "Priority Search" as seen in $Fig.$ 13. The Priority search aims to retrieve the most relevant arguments to the users' subject of search and queuing them based on the maximum support number, such that the first queued argument is the one that has more itemsets[3] related to the subject of search. Although the AprioriTid algorithm has originally been devised to discover all significant association rules between items in large database transactions, the agent employs its mechanism in the priority search to generate different combinations between different itemsets [4,3]. These combinations are then used to classify the retrieved contexts and queued them in a descending order based on its support number. As a response to the priority search purpose, an adapted version of the AprioriTid mining algorithm has been developed and applied. This adapted version, as seen in $Fig.$ 14, considers the single itemset (1-itemset) size as well as the maximum support number usage, rather than k-itemset for k≥2 and the minimum support number "minsup" mechanism.

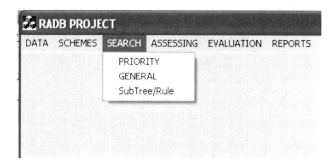

**Fig. 13.** The main window of our implemented system in Visual C++

*1) $L_1$ = {large 1-itemsets};*
*2) For each itemset $C_1 \in L_1$ repeat steps 5 and 6 for k=1;*
*3) For (k=2; $L_{k-1} \neq \Phi$; $K^{++}$) do begin*
*4)      $C_k$ = apriori-gen($L_{k-1}$);    //New candidates*
*5)      For all transactions t $\in D$ do begin*
*6)          $\bar{C}_k$=subset ($C_k$,t);  //candidates contained in t*
*7) end;*
*8) For all similar candidates c $\in U_k \bar{C}_k$ do*
*9)      c.count$^{++}$;  //the support number*
*10)Ans= {c $\in \bar{C}_k$ | descending ordered};*

**Fig. 14.** An enhanced version of AprioriTid

For more clarification, the priority search mines specific parts of the pre-existing arguments based on the users' search criteria. This search criteria enables the user to seek the premises, conclusions or the critical questions lying in the different arguments. For example, suppose the user queries the RADB searching for all information related to "Iraq war". Simply, he may write "the destructive war in Iraq" as the search statement and can choose the conclusion as the search criteria. In this case, the classifier agent receives the set of significant tokens {destructive, war, Iraq} from the parser module (for more explanation about parser functionality see [3]). This set is considered as the single size itemset (1-itemset) $C_1$={$w_1$, $w_2$, $w_3$} that contains the most crucial set of words in the search statement. Then, the agent uses the adapted version of the AprioriTid algorithm to generate the different super itemsets $C_{2\leq k\leq 3}$, which are the different combinations between different tokens. So, the generated super itemsets, as seen in $Fig.$15, will be the 2-itemset $C_2$={ $w_1w_2$, $w_1w_3$, $w_2w_3$ }, and the 3-itemset $C_3 = \{$ $w_1 w_2 w_3$ $\}$. Afterward, the different conclusions in the different arguments trees will be mined seeking for the most relevant set of arguments Ans$=\{d_1, d_2, ..., d_m \}$ such that $\forall\ d_i \in D\ \exists\ C_{k \in \{1,2,...,j\}} \subseteq d_i$ . Finally, the results will be queued in a descending order and exposed in a list, where the user can choose the argument name "Argument_314" from the list to expose the associated context and analysis as in $Fig.\ 21$.

$D = \{(1,\ Context_1,\ argument\_602),\ (2,\ Context_2,\ argument\_314),\ ......,etc.\}.$

| L₁ | |
|---|---|
| $C_1$ | $\overline{C}_1$ |
| $w_1$ | {1,2,7} |
| $w_2$ | {1,3,5} |
| $w_3$ | {1,2} |

| L₂ | |
|---|---|
| $C_2$ | $\overline{C}_2$ |
| {w₁ w₂} | {1,2} |
| {w₁ w₃} | {1,2} |
| {w₂ w₃} | {1} |

| L₃ | |
|---|---|
| $C_3$ | $\overline{C}_3$ |
| {w₁ w₂ w₃} | {1} |

Ans

| Argument ID | Support number |
|---|---|
| 1 | 7 |
| 2 | 4 |
| 7 | 1 |
| 3 | 1 |

| $C_k$ | Set of candidate k-itemsets |
|---|---|
| $\overline{C}_k$ | The ID transactions associated with each candidate |
| $L_K$ | Set of large k-itemsets. Each member of this set has two fields (i) $C_K$ (ii) $\overline{C}_K$ |
| D | The transactions "Data_TBL" query that contains three fields: (i) transaction ID (ii) the associated content/context (iii) the associated argument_no. |

**Fig. 15.** The adapted AprioriTid mechanism

## 4.2   Tree Substructure Mining

The substructure mining technique utilizes the tree structure lying in the RADB repository, and retrieves the most relevant arguments to the subject of search. This search mechanism is considered substantial for the structured repository and annotated as "General Search" (see $Fig.\ 13$). It uses the breadth first search technique [15][14] in order to encounter all nodes in the argument trees and retrieves the most relevant group. For example, suppose the user writes "the destructive war in Iraq" as a search statement. The revealed contexts, as shown in $Fig.\ 16$, will be ordered based on the nodes' cardinality. Which means that the first queued argument is the one that contains more nodes related to the subject of search.

With respect to $Fig.\ 17$, the breadth first search seeks each tree in our RADB forest, preserving the ancestor-descendant relation [7][14] by searching first the root $Fig.\ 17(a)$, then the children in the same level as in $Fig.\ 17(b)$ and so on as in $Fig.\ 17(c)$. Finally, if the user picks one of the resulted search arguments, the associated context and analysis are depicted as shown in $Fig.\ 21$.

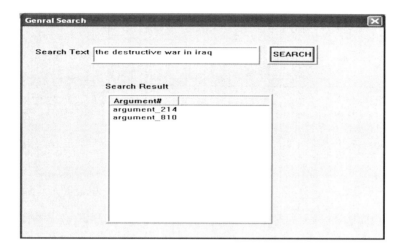

**Fig. 16.** The General search representation form

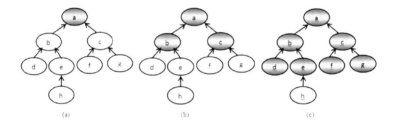

**Fig. 17.** The breadth first search

## 4.3   Rule Extraction Mining

Rule extraction mining is a search technique in which argument trees are encountered to discover all hidden patterns "embedded subtrees [7][14]" that coincide with the relation between some objects. These objects express a set of the most significant tokens of the user's subject of search. Precisely, suppose the user wants to report some information about the relation between the "USA war" and the "weapons of mass destruction". At the beginning, the user's search statements are reduced to the most significant set of tokens by the parser [1][3][2]. Then, the different argument trees, pre-existing in the RADB repository, are mined in order to fetch these different tokens. $Fig.$ 19(a) shows the analysis of an argument tree, where some enclosed nodes coincide with the user's search statements, while $Fig.$ 19(b) shows the revealed embedded subtree.

Finally, each resulted subtree is expressed in the form of a rule as shown in $Fig.$ 18, where "+" indicates that this node is a support to the final conclusion and its type is 1. Similarly, "-" is a rebuttal node to the final conclusion, and its type is -1 as specified in Section 3.1.3.

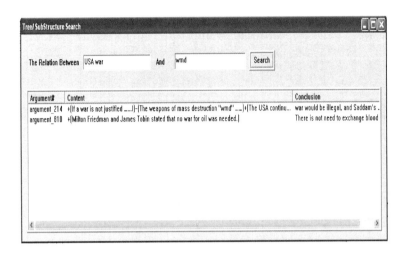

**Fig. 18.** The representation form of Rule Extraction search result

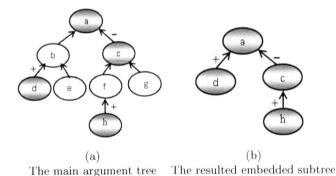

(a)                                (b)
The main argument tree    The resulted embedded subtree

**Fig. 19.** The Rule Extraction search

# 5  Use and Benefit of RADB and Argument Mining Agent in ITS

The RADB structure formulation and the mining classifier agent can be incorporated into an intelligent tutoring system (ITS) to construct an agent-based learning environment, which aims to teach argument analysis and construction. This section describes an architecture of an argument learning environment (ALES) as depicted in *Fig.* 20. The environment utilizes the RADB and the classifier agent in order to (i) guide the analysis process to refine the user's underlying classification, (ii) deepen the user's understanding of debating, decision making, and critical thinking. It consists of four main models: domain model, pedagogical model, student model, and the GUI model. The domain model utilizes the represented structured argument repository "RADB" (already discussed in Section 3.1), and enjoys the extendability feature as described in the subsection 3.2.2. The pedagogical model has three components (parser, classifier agent, teaching model) that co-operate together in order to achieve the required aims. The student model keeps track of the student performance and assists the pedagogical model in offering individualized teaching. Visual C++ was used to implement the graphical user interface "GUI" of the proposed environment. It provided a stable encoding application and satisfiable interface. Not only does ALES teach argument analysis, but also assesses the student and guides him through personalized feedback. Since the RADB design and structure have been discussed in Section 3.1. The next subsections will describe the pedagogical model, the student model and provide an illustrative example that explains the student-system interactions.

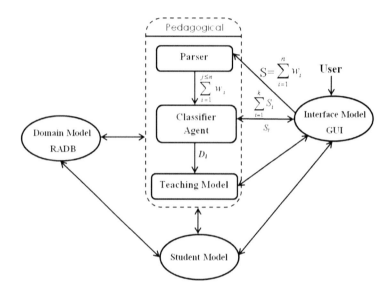

**Fig. 20.** ALES architecture

$S$ : The input statement S, which consists of a set of words $\{w_1, w_2, \ldots, w_n\}$.

$\sum_{i=1}^{j \leq n} w_i$ : The set of significant words existing in the search statement $S$.

$\sum_{i=1}^{k} S_i$ : The set of the retrieved arguments, which are most relevant to
the subject of search.

$S_i$    : The user's selected argument_i.

$D_i$    : The associated context and analysis of the selected argument $S_i$.

## 5.1  Pedagogical Model

The pedagogical model is responsible for reasoning about the student behavior according to the student model. It aims to: (i) retrieve the most relevant results "$\sum_{i=1}^{k} S_i$" to both the subject of search and the student's background, (ii) present the corresponding context and analysis "$D_i$" of the selected result $S_i$, (iii) guide the student analysis, during the assessing phase. The pedagogical model as seen in $Fig.$ 20 consists of three main components: the parser, the classifier agent, and the teaching model.

- The importance of the parser lies in reducing the set of tokens into a set of significant keywords, which in turn (i) improves the results of the classifier where combinations of unnecessary words vanish, (ii) reduces the number of iterations done by the classifier agent. For more details and discussion see [1,3,2].
- The classifier agent classifies the retrieved contexts depending on the student's specification, using general classification or rule extraction or priority based on specific search criteria such as premises (with/against) or conclusions. For more details about different retrieving techniques and the importance of the classifier agent see Section 4.
- The teaching model The teaching model monitors the student actions, guides the learning process and provides the appropriate feedback. The model starts its role when the classifier agent sends the document $D_i$ selected by the student. The teaching model checks, according to the current student model, whether the student is in the learning or the assessing phase. If the student is in the learning phase, the document is presented associated with the corresponding analysis as the shown in $Fig.$ 21. On the other hand, if the student is in the assessment phase, the student is able to do his own analysis, and the teaching model will guide him during analysis by providing personalized feedback whenever required. The feedback aims to guide the student and refine his analysis and intellectual skills. Two kinds of feedback are provided by the teaching model; partial argument negotiation and total argument negotiation.
    - **Case of partial argument negotiation:** In this case, the student starts analyzing the argument context in the form of a tree in which the root holds the final conclusion of the issue of discussion. The teaching pedagogy used in this case provides partial hints at each node of the analysis tree. They are results of comparing the student's current node analysis to the original one in the argument database. These hints are provided before allowing the student to proceed further in the analysis process; they aim to minimize the analysis error ratio, as much as possible, for the current analyzed node. Generally, the teaching model guides the student via the partial hints at each node till the error of the current

Argument 214

Sky News correspondent Jeff Reed rose after one of those military briefings in Doha to ask: "Can I ask the daily weapons of mass destruction "wmd" question? They haven't been deployed. They haven't been discovered. Is this war going to make history by being ended before you've found its cause?" After a significant pause, Maj Gen Victor Rennart replied, "That's a great question. We continue to look at sites around the country. If the wmd does not found, the very act of war would be illegal, and Saddam's reign of brutality glossed by his one truth: that Iraq possessed no wmd. Saddam was as bad a guy as one can get, but there is no law-national or international-that sanctions attacl on guys because you have good reason to believe they are bad, and could threaten you. A state's claim for self-defense does not justify military action against another state that has not attempted an attack on it."

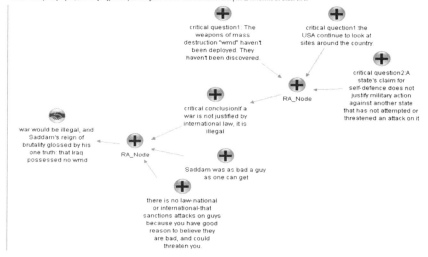

**Fig. 21.** The chosen context and the associated analysis

node is minimized to a specific ratio. After then, the student is able to move to the next analysis step (i.e., node).

- **Case of total argument negotiation:** The total argument negotiation is similar to the partial argument negotiation. However, the teaching pedagogy is different in that it provides hints only at the end of the analysis process. In other words, after the student builds the full analysis tree for the selected context, the system interprets and evaluates the student's analysis comparable to the pre-existing one and remarks the errors.

Generally, in the assessing phase, the teaching model presents the transcript of the chosen argument associated with an empty tree skeleton as seen in *Fig.* 22 and asks the student to start his own analysis. The student starts the analysis by copy and paste text passages from the transcript or freely enters text into the nodes. The teaching model traces each node text and divides it into set of significant tokens, then interprets and evaluates the errors ratios comparable to the pre-existing analysis underlying in the RABD. Finally the model provides the feedback, partially or totally, based on the student choice and records the student's errors for the current transcript, which in turn will be used, by the student model, to evaluate the performance and to follow the progress of the student.

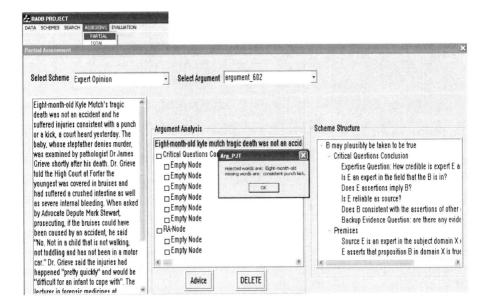

**Fig. 22.** The Partial Assessment Form

## 5.2   Student Model

The student model stores details about student's current problem-solving state and long term knowledge progress, that is essential for future student's performance evaluations. The model considers personal information, pre-test evaluation, and performance history. Personal information contains personal data as name, ID, password, ..., etc. The pre-test evaluation permanently assesses the student's argument analysis skills and follows the student progress through learning process. Finally, the performance history implicitly reflects how much the student has done and how well.

## 5.3   Student-System Interaction Example

This is an illustrative example of a complete run that shows the student-system interactions. The system interactions are written in normal font. The student's actions are in bold. My illustrations to some actions will capitalized.

THE SYSTEM ASKS THE USER TO LOG IN OR REGISTER IF IT IS HIS FIRST TIME.

System≫ please sign in or register as a new user.

**User≫ safia.**
**User≫******.**

System≫ Hello safia, do you have background about argumentation?

THE USER SHOULD CHOOSE AN ANSWER FROM AN EXISTED LIST. SUPPOSE THE USER HAVE NO IDEA ABOUT ARGUMENTATION OR ARGUMENT ANALYSIS.

User≫ **No.**

IN THIS CASE, THE SYSTEM GIVES THE USER A BRIEF INTRODUCTION ABOUT THE SYSTEM AND HOW HE CAN INTERACT WITH IT IN ORDER TO GAIN KNOWLEDGE ABOUT ARGUMENTATION.

System≫ OK. Let's see how you can use the different screens. You can navigate in our system searching for specific issue using different criteria through the search tab, then from the exposed arguments list you can pick any argument. This allows you to see the original context and the associated expert analysis. Next you can use the assessing tab for quiz arguments, you are asked to analyze it based on your selected scheme. This stage can provides you with two kinds of feedback "node/whole", this feedback will help and guide you during your analysis procedure.
System≫ Are you ready to start now?
USER≫ **Yes.**

THE MAIN FORM WILL APPEAR TO THE USER AS SHOWN IN $Fig.13$.

SUPPOSE THE USER WANTS TO SEARCH ABOUT ALL ARGUMENTS OR REPORTS RELATED TO "IRAQ WAR"

User≫ **presses the search tab**
User≫ **selects the General search tab.**
User≫ **writes in the search text "the destructive war in Iraq".**
User≫ **presses search button.**

THE SYSTEM RETRIEVES THE DIFFERENT RELATED ARGUMENTS BASED ON SUBSTRUCTURE MING TECHNIQUES.

System≫ the search results are the following list [argument_214, argument_810] as in 16.
User≫ **picks up argument_214 from the retrieved list to see the main context and the associated analysis.**
System≫ presents the associated context and analysis as in $Fig.21$.

DURING STUDENT NAVIGATION, THE STUDENT MODEL RECORDS
EACH ACCESSED ARGUMENT. AFTER THE USER FINISHES NAVIGAT-
ING, HE CAN move TO THE ASSESSING TAB IN ORDER TO START THE
LEARNING BY ASSESSING PHASE, WHICH PROVIDES THE ABILITY
TO ANALYZE SPECIFIC CONTEXT GUIDED BY FEEDBACK.

**User≫ presses the assessing tab.**
System≫ please select a specific scheme to be used in your analysis.
**User≫ "expert opinion scheme".**
System≫ select the kind of feedback "either node "partially" or whole "totally"
".

**User≫ node feedback.**

THE WHOLE ARGUMENTS, THAT USE THE "EXPERT OPINION
SCHEME" IN ITS ANALYSIS, WILL BE LISTED SUCH THAT THE PRI-
ORITY IS TO THE CONTEXTS THAT HAVE NOT BEEN ACCESSED YET
BY THE USER DURING NAVIGATIONS.

System≫ [argument_602, argument_1, argument_214].

**User≫ picks up one of the listed arguments, argument_602 as example.**

THE SYSTEM PRESENTS THE TRANSCRIPT OF THE CHOSEN
argument_602 ASSOCIATED WITH AN EMPTY TREE SKELETON AND
ASKS THE STUDENT TO START HIS OWN ANALYSIS.THE TRANSCRIPT
IS SHOWN AS IN EXAMPLE 2

SINCE THE USER CHOOSE THE NODE FEEDBACK THE SYSTEM WILL
SUPPLY ADVICE AT EACH NODE COMPARING THE USER ANALYSIS
WITH THE PRE-EXISTING ONE, AND WILL NOT GO STEP FURTHER
UNLESS THE USER ADJUST THE CURRENT NODE ANALYSIS BY RA-
TIO WITH THE EXPERT ANALYSIS FOR THE SAME NODE.
**User≫ starts to fill his first node waiting the system advices. Suppose
the student writes "final decision is the death was not accident".**

System≫ divides the user node statement in the set of significant tokens final,
decision, death, accident, and then compares this tokens with the expert analysis for the same node. then calculates and records the error ratios for that node.

System≫ shows out the following message "your analysis is partially correct
try to use the words Kyle Mutch, tragic, suffered, punch, in your node analysis, rather than the words final,decision,.... that have been used in the current
analysis".

**User≫ he will reanalyze the current node adding the context that contains the advised keywords.**

System≫ compares again the current context node with the pre-existing analysis and negotiate again, guiding the user, till he reaches to the correct analysis for this node.

**User≫ fills the other nodes.**

System≫ negotiates based on the pre-existing expert analysis guiding the user during his analyses.

AFTER THE USER FINISHES HIS ANALYSIS TO THE WHOLE CONTEXT, FILLING THE SUITABLE ANALYSIS FOR EACH NODE, THE SYSTEM WILL RECORD THE FIRST ANALYSIS RATIO FOR EACH NODE, THEN CALCULATE AND RECORD THE WHOLE ARGUMENT ANALYSIS RATIO FOR THAT ARGUMENT. THEN FOR EACH ASSESSMENT THE SYSTEM WILL RECORD THE CORRECTNESS RATIO TILL IT COMES TO BE MORE THAN OR EQUAL TO 90% IDENTICAL TO THE PRE-EXISTING ANALYSIS. THEN THE SYSTEM WILL ASK THE USER TO GO TO THE EVALUATION PHASE.

System≫ wow, you achieving well, it is better to accept the challenge and go to the evaluation phase.
**User≫ OK, and then press the tab evaluation.**

System≫ loads a context to be analyzed by the user; however the system shows a context that has not been accessed before by the user.

**User≫ starts to analyze without any help till press check analysis.**
System≫ compares each node of the user's analysis with the expert analysis and deduce a report for the wrong nodes, and record this analysis for future progress report.

# 6 Argument Mining-Driven Arguing Agents

In the previous sections, we have described some argument mining techniques from natural arguments organized in RADB, and their effective uses in ITS. In this section, we will consider how argument mining can enhance computational agents with an ability of argumentation that is well suited for agent self-deliberation, negotiation, communication, a decision aiding process, and so on in agent-oriented computing.

Computational argumentation has proven to be a useful mechanism for understanding and resolving several AI problems in the presence of inconsistent, uncertain and incomplete information. In particular, it has been considered as a best approach to dealing with those information in a networked distributed computing environment that is often open and changing. To this ends, logical models of argumentation have emerged as a promising paradigm for modeling agent reasoning and communication in light of interaction with other agents[6][16][19]. Arguing agents argues about issues concerned, through casting arguments and counter-arguments each other to reach an agreement. The formal argumentation frameworks have been mathematical apparatuses for mainly stipulating these items:

- What arguments are (well-formed argument and structure).
- What agent-interaction protocols are (dialectical process).
- What correct arguments are (semantics).

Thus, the formal argumentation frameworks tell both us and agents the final statuses of arguments such as 'justified', 'overruled', and 'defensive' or 'undecided' semantically and/or dialectically. In the argumentation process, agents are supposed to make arguments from their own knowledge bases provided by their originals.

The past works, unfortunately, lack a means of consulting other argument data bases in constructing arguments, nor a means of retrieving, reusing, processing (such as modifying, adjusting, etc.) preexisting arguments for an argument to be newly constructed. Put it differently, such an idea as argument construction from a large argument data base is missing. Humans' knowledge and wisdom produced usually have the form of arguments that are built up from more primitive knowledge of the form of facts and rules. Those repositories or treasuries of knowledge are now about to be organized to argument data bases or corpora that can be retrieved, stored, and reused online freely [20][9]. Recently, the online data on arguments of daily topics is rapidly growing due to Internet and Web technologies, such as in the form of blogs, SNS, and so on. Argument mining and argument discovery technology now turn out to play a particularly important role in supporting to create new ideas, opinions and thoughts by finding out meaningful arguments and reusing them from large-scale argument repositories.

## 6.1   A Way to Arguing Agents with Argument Mining Ability

For formal argumentation systems, the significance and importance of argument mining and argument discovery technology apply as well. If agents can consult argument data bases and have retrieving or mining methods proper to formal argumentation, they could make and put forward more fruitful and influential arguments to other parties, resulting in more successful results under more active argumentative dialogues. Thus, argument retrieval and mining are expected to strengthen arguing agent intelligence. In this subsection, we describe two possible scenarios for arguing agents to exploit argument discovery and mining, in order to enhance autonomous agents whose arguable abilities are based on formal logical models of argumentation.

(i) At the time of building up knowledge base: Agents usually are given their knowledge bases from their masters. However, it would be better for them to collect various forms of knowledge such as facts, rules and arguments from the outside argument DBs, which are relevant to the issues concerned and acceptable to them. The argument retrieval and mining techniques in the previous sections are applicable to formal argumentation systems as well.

(ii) At the time of arguing dynamically: During argumentation, agents might be aware of lack of knowledge and weakness of their arguments. Then, they immediately consult argument DBs and find out reasons or grounds supporting or augmenting their arguments. Then, the frequent substructure mining method [4] becomes most effective since frequently appearing substructures of arguments may be construed as commonly approved significant ideas and views among people. Interestingly, the frequent substructure mining might allow for extracting argument schemata not only like Walton' ones [26] but also like those argument patterns of broad sense that are deep-rooted in the culture.

In either case, our RADB and argument mining techniques are applicable and helpful for arguable agents. One major issue, however, occurs in these scenarios. It is how we can convert natural arguments to formal ones since argument sources around us consist of natural language sentences. As a first step to overcome this, we provided a preliminary work towards transforming natural arguments in Araucaria [20] to formal arguments in EALP/LMA [23], a formal argumentation framework. The overall flow of the semi-automated transformation consists of the following three processes:

(1) Analyze and diagrammatize natural arguments with Araucaria.
(2) Extract knowledge (rules and facts) from those diagrammatized arguments, and construct formal arguments in EALP semi-automatically for each agents participating in argumentation.
(3) Argue about an issue and decide the argument status with the dialectical proof theory of EALP/LMA.

It greatly relieves us from burden involved in knowledge base preparation for argumentation. In this manner, if analyzed and diagrammatized arguments have been exported to formal knowledge and argumentation bases that formal argumentation systems can deal with, agents could benefit a huge amount of knowledge for argumentation, and reciprocally argument mapping systems could gain a mechanism for deciding final argumentation statuses with the help of formal argumentation systems. Interested readers should refer to [24] for its details. This method can be applied to other argument mapping systems like Compendium, Rationale, etc. as well in a similar manner.

## 6.2   Knowledge and Argument Transformation

In this subsection, we describe some basic ideas on how to use facts, rules and sub-arguments in the formal argumentation frameworks, which have been found by the argument retrieval and mining techniques.

[**Argument transformation rules**] Agents can make their arguments better or more convincing by applying the following subtree replacement transformations: Rule replacement for information *refinement* and *diversification*, Fact replacement for information refinement and diversification, and Weak literal replacement for information *completion*.

Rule replacement allows agents to employ more persuasive or preferable rules in their arguments. Atom replacement allows agents to introduce more evidences to arguments. Weak literal replacement allows agents to reduce incomplete knowledge (belief) included in arguments. The defeasible rule $L_0 \Leftarrow L_1, ..., L_j, \sim L_k, ..., \sim L_n$ has the *assumptions* $\neg L_k$, ..., $\neg L_n$ [16][23]. This means that agents are allowed to submit arguments without any grounds for those weak literals as can be seen in the definition of arguments. However, we could reinforce arguments or make them better if the assumptions of weak literals were replaced by mined arguments with those assumptions as rule heads.

These transformations are not always accepted for arguments. We introduce the following acceptability conditions under which an agent can accept the mined subarguments: (i) The agent can neither undercut nor rebut any part of those subarguments . This acceptability condition is important since the agent should keep its knowledge base coherent or conflict-free, (ii) The agent replaces subtrees in arguments if the number of facts as evidences can be increased after the replacement, and (iii) let *arg* be a subargument to be replaced and *arg'* a candidate argument mined. Then, *arg'* is acceptable if the number of weak literals in *arg'* is less than or equal to the number of weak literals in *arg* to avoid increasing the incompleteness of weak literals.

Agents can learn rules (knowledge) that have been included in the subargument accepted. This may be said to be a sort of knowledge acquisition or discovery attained through argumentation. Three transformations are best illustrated by representing arguments in a tree form, as in Figure 23, where readers could see three kinds of replacements introduced above: the subtree with the node $b$ is transformed into a new subtree with the same node $b$ (Rule replacement), the leaf with the weak literal $\sim e$ is expanded to a new subtree with a strong literal $\neg e$ and the evidence $k$ (Weak literal replacement), the fact $d$ is further expanded to a new subtree (Fact replacement). As the results, the subtree with the node $c$ amounts to having been transformed into a bigger subtree that incorporates the above transformations.

It should be noted that the purposes of argument transformation is, in a sense, similar to those of proof transformation or program transformation that can be seen in logic or computer science. The typical ones in logic are the cut elimination in LK, proof normalization in ND and various deductive equivalences between proof architectures.

In addition to the argument transformation introduced above, there are other useful and versatile directions. They include: (iv) Argument transformation based on the concept of *similarity* (for example, an argument on the issue $p(a)$ is changed into the argument on the issue $p(b)$, using a certain similarity relation $a \sim b$), (v) Argument transformation based on the concept of *strengthening* (or

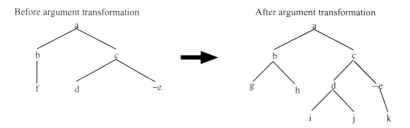

**Fig. 23.** Argument transformation

*specificity*) (for example, an argument on the issue $p$ is changed into the argument on the issue $q$, where $p \to q$), and (vi) Argument transformation based on the concept of *weakening* (for example, an argument on the issue $p$ is changed into the argument on the issue $q$, where $q \to p$). These three transformations are subject to such a condition that the transformed arguments are justified.

[**Knowledge transformation rules**] Agents may want to polish the knowledge bases beefed up by argument retrieval and mining techniques, for the future argumentation. For this, we provide two plausible rules: (vii) The reductant of rules and (viii) The rule abridgment. The reductant is a result obtained by reducing several rules with the same head to a single rule. The reductant is known to be a logical consequence from the several rules used. The rule abridgment, on the other hand, is its converse operation of the reductant. That is sort of a detachment and has a role of digesting complicated rules generated by the reductant. For example, $\{a \leftarrow b, c, d, e.\}$ is a reductant from $\{a \leftarrow b, c, .\}$ and $\{a \leftarrow d, e.\}$. Conversely, $\{a \leftarrow b, d.\}$ is an abridgment of the reductant, including $b$ from the first rule and $d$ from the second rule in their rule premises, where $b$ and $d$ are assumed to have most significant and relevant relationship with the conclusion $a$.

We have briefly described knowledge and argument transformation that can be used as a way to reorganize, improve and refine acquired knowledge and arguments. This can be also seen as a means for agents to grow through argumentation. Here, the term "grow" has a meaning similar to recent concepts such as those appearing in learning, evolutionary computing, genetic computing, emergent computing and so on, whose purpose is to realize not behavior-fixed computing entities but environment-sensitive ones.

## 7   Related Work and Discussion

Although a number of argument mark-up languages tools have been proposed to facilitate data preprocessing, using diagrams for argument representation, the integration between them is missing. This barrier limits the ability of the same user to access the same argument using different tools. The argument-markup language (AML, XML based language) behind the Araucaria, compendium, and reason!able systems [27][22][17][20] are examples of these trials. Indeed, Araucaria system provides many retrieving criteria, however an irrelevant result may be obtained when using more than one word in the search statement. Moreover, the user analyzes any document depending on his beliefs and understanding, and not based on a dedicated structure of different schemes or previous analysis. This kind of analysis encompasses an intractable knowledge, which undermines the mining process. These mining processes and the artificial intelligence influence together are needed to guide the user's understanding via the relation between scientific theories and evidences, and to refine his argument analysis ability.

Regarding the AI and education fields, many instructional systems have been developed to hone student's argumentation skills, for example SCHOLAR and WHY[14] systems. However, these systems were mainly designed to engage students in a Socratic dialog, which faces significant problems to develop a Socratic tutor[15]. Such as, knowledge representations, especially in the complex domains like legal reasoning, control or preprocessing, and manipulate the natural language. Later, as a response to these difficulties, a number of case-based models[11] and mining weblogs[8,16] have been proposed to tackle the mining and the artificial influence problem. The mining weblogs are considered as a classification problem for the legal or informal reasoning considering law. Though, it mines the textual data that is intractable to be processed. On the other hand, the case-based argumentation systems, such as the CATO[11], use the case based reasoning method in order to reify the argument structure through tools for analyzing, retrieving, and comparing cases in terms of factors.

Comparing CATO to our proposed application, both of them provides examples of specific issue to be studied by the different students, as well as evaluates students' arguments comparable to the pre-existing one. With respect to the search for arguments, both systems support students' search for the existing database, and retrieve the most relevant argument. However, CATO limits the students' search by a boolean combination of factors. On one hand, in the fulltext retrieval search, one can retrieve documents, by matching phrases, which is not relevant to the search subject. On the other hand, ALES provides different search criteria to tackle this problem, as seen in section 5, utilizing the provided search engine in order to: summon and provide a myriad of arguments at the student's fingertips, retrieve the most relevant results to the subject of search, and organize the retrieved result such that the most relevant is the first rowed.

Finally, I. Rahwan presents the ArgDf system [5,18], through which users can create, manipulate, and query arguments using different argumentation schemes. Comparing ArgDf system to our approach, both of them sustain creating new arguments based on existing argument schemes. In addition, the ArgDf system guides the user during the creation process based on the scheme structure only, the user relies on his efforts and his background to analyze the argument. However, in our approach, the user actions is monitored and guided not only by the scheme structure but also by crucial hints devolved through the appropriate feedback. Accordingly, the creation process is restricted by comparing the contrasting reconstruction of the user's analysis and the pre-existing one. Such restriction helps in refining the user's underlying classification.

In the ArgDf system, searching existing arguments is revealed by specifying text in the premises or the conclusion, as well as the type of relationship between them. Then the user is allowed to filter arguments based on a specific scheme. Whereas, in our approach, searching the existing arguments is not only done by specifying text in the premises or the conclusion but also by providing different strategies based on different mining techniques in order to: refine the learning environment by adding more flexible interoperability, guarantee the retrieval of the most convenient hypotheses relevant to the subject of search, facilitate the search process by providing a different search criteria. At last, ALES enjoys a certain advantage over ArgDf system, it can trace the users progress and produce representative reports about the learner analysis history, which in turn excavate the proper weakness points in the learners' analysis skills. Fig.24(a), as example, shows the analysis progress of the current student, spotting on the conclusion node analysis ratio for different arguments using different schemes. Looking deeply in this diagram we can conclude that this student cannot highlight the final conclusion of different context correctly, which means that the student cannot well understand the proposed contexts. Whereas $Fig.24(b)$ shows that the student total argument analysis skill starts to be improved after the fourth exam.

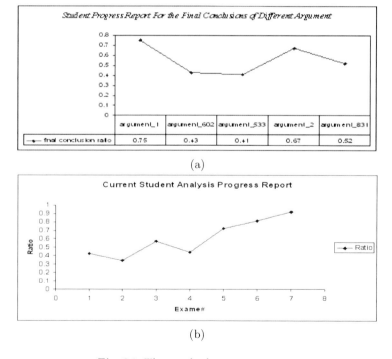

(a)

(b)

**Fig. 24.** The resulted progress reports

On the other hand, ALES handles special types of arguments, in which only
one scheme is used in the analysis process. So, if the context is much bigger and
needs more than one scheme in its analysis, ALES cannot be used.

## 8    Conclusion and Future Work

In this chapter, we have described an effective usage of argument mining for both
arguing agents and intelligent tutoring system. The mining techniques have been
implemented over the "RADB", which describes a novel approach of building a
highly structured argument repository aims to (i) summon and provide a myriad of arguments at the user's fingertips,(ii) facilitate the interoperability among
various agents/tools/humans. The novelty of the RADB approach lies in the
proposed structured knowledge base, that are used to answer users' queries related to the domain entities, rather than the diagrams forms that encoded by the
argument-markup language (AML, XML based language) and limits the argument interoperability between different argument mapping tools. Moreover, the
structured knowledge base (RADB) enjoys the extendable property, in which
adding of new schemes proper to different cultures and beliefs is applicable if
they are formulated in the form of Walton theory of argumentation.

In the latter usage, intelligent tutoring system, we have proposed an agent-
based educational environment (ALES), that utilizes both the RABD and the
mining classifier agent together, aiming to (i)guide the analysis process to refine the user's underlying classification, (ii)deepen the users' understanding of
negotiation, decision making, and critical thinking. The classifier agent has been
introduced as a managing tool that uses different mining techniques to enrich the
argument analysis, retrieval, and re-usage processes. The main purposes of the
mining agent are (i)retrieving the most relevant results to the subject of search,
(ii) supporting the fast interaction between the different mining techniques and
the existing arguments. ALES considers two phases, the learning and the assessment phases, and has four main components: domain model, pedagogical model,
student model, and the GUI model. These components co-operate together in order to verifying the stated targets. The system enjoys certain advantages when
compared to others, especially with respect to the search of pre-existing arguments. The results obtained are very promising, where highly relevant and
convenient arguments are obtained, especially when the search statement is in
this form "the destructive war in Iraq" or "USA war and the mass of destruction
weapons". It also can trace the users progress and produce representative reports
about the learner analysis history, which in turn excavate the proper weakness
points in the learners' analysis skills[see section7].

In the former usage, arguing agents, we have discussed how argument mining
and retrieval can strengthen arguing agent intelligence and can enhance computational agents with an ability of argumentation. The agents could make and
put forward more fruitful and influential arguments to other parties if they can
consult argument data bases and have retrieving or mining methods proper to
formal argumentation. They could benefit a huge amount of knowledge for argumentation, and reciprocally argument mapping systems could gain a mechanism

for deciding final argumentation statuses with the help of formal argumentation systems. Finally, we have also describe some basic ideas on how to use facts, rules and sub-arguments, founded by the argument retrieval and mining techniques, in the formal argumentation frameworks.

In the future, we intend to (i) continue in the learning environment implementation using the frequent tree mining techniques[7] in order to search for frequent patterns in different arguments, (ii) add more schemes together with related arguments that consider various cultures and beliefs, and (iii) integrate an NLP software to the classifier agent in order to aid in a new search criteria annotated as polarity classification that provides more flexibility to users' queries.

# References

1. Abbas, S., Sawamura, H.: Argument mining using highly structured argument repertoire. In: The First International Conference on Educational Data Mining (EDM), Montreal, Québec, Canada, pp. 202–210 (2008)
2. Abbas, S., Sawamura, H.: A first step towards argument mining and its use in arguing agents and its. In: Lovrek, I., Howlett, R.J., Jain, L.C. (eds.) KES 2008, Part I. LNCS (LNAI), vol. 5177, pp. 149–157. Springer, Heidelberg (2008)
3. Abbas, S., Sawamura, H.: Towards argument mining using relational argument database. In: The Second International Workshop on Juris-informatics (JURISIN), Asahikawa Convention Bureau, Hokkaido, Japan, pp. 22–31 (2008)
4. Agrawal, R., Srikant, R.: Fast algorithms for mining rules. In: The 20th VLDB Conference Santiago, Chile (1994)
5. Chesnevar, C., McGinnis, J., Modgil, S., Rahawan, I.: Towards an argument interchange format. In: The Knowledge Engineering Review, Cambridge University Press, Cambridge
6. Chesñevar, C.I., Maguitman, G., Loui, R.P.: Logical models of argument. ACM Computing Surveys 32, 337–383 (2000)
7. Chi, Y., Muntz, R.: Frequent subtree mining-an overview. Fundamenta Informaticae, 1001–1038 (2001)
8. Conrad, J., Schilder, F.: Opinion mining in legal blogs. In: ICAIL 2007, Palo Alto, California, USA, June 4-8, 2007, pp. 231–236 (2007)
9. Katzav, J., Reed, C., Rowe, G.: Argument research corpus. In: Huget, M.-P. (ed.) Communication in Multiagent Systems. LNCS, vol. 2650, pp. 269–283. Springer, Heidelberg (2003)
10. Macagno, F., Walton, D.: Argumentative reasoning patterns. In: Proceeding of 6th CMNA (Computational Models of Natural Argument) Workshop, ECAI (European Conference on Aetificial Intelligence), Rivadel Garda, Italy, Trento, Italy, University of Trento, pp. 48–51 (2006)
11. McBurney, P., Parsons, S.: Dialogue game protocols. In: Practical Applications in Language Corpora (2003)
12. Godden, M., Walton, D.: Argument from expert opinion as lgal evidence: Critical questions and admissibility criteria of expert testimony in the american legal system. In: R. Juris, vol. 19, pp. 261–286 (2006)
13. Moens, M., Boiy, E., Palau, R., Reed, C.: Automatic detection of arguments in legal texts. In: ICAIL 2007, Palo Alto, California, USA, June 4-8, 2007, pp. 225–230 (2007)

14. Zaki, M.: Efficiently mining frequent trees in a forest: Algorithms and applications. IEEE transactions on knowledge and data engineering 17, 1021–1035 (2005)
15. Nijssen, S., Kok, J.N.: A quickstart in frequent structure mining can make a difference. In: Tenth ACM SIGKDD international conference on Knowledge discovery and data mining, USA, pp. 647–652 (2004)
16. Prakken, H., Vreeswijk, G.: Logical systems for defeasible argumentation. In: Gabbay, D., Guenther, F. (eds.) Handbook of Philosophical Logic, pp. 219–318. Kluwer, Dordrecht (2002)
17. Rahawan, I., Sakeer, P.: Representing and querying arguments on semantic web. In: Dunne, P.E., Bench-Capon, T.J.M. (eds.) Computational Models of Argument, IOS Press, Amsterdam (2006)
18. Rahawan, I., Zablith, F., Reed, C.: The foundation for a world wide argument web. In: Artificial Intelligence Conference (AAAI), April 04 (2007) (published in the Artificial Intelligence Journal)
19. Reed, C., Norman, T.J. (eds.): Argumentation Machines. Kluwer Academinc Publishers, Dordrecht (2004)
20. Reed, C., Rowe, G.: Araucaria: Software for argument analysis, diagramming and representation. International Journal on Artificial Intelligence Tools 13, 983 (2004)
21. Reed, C., Rowe, G.: Translating toulmin diagrams: Theory neutrality in argument representation. Argumentation Journal 19, 267–286 (2005)
22. Rowe, G., Reed, C., Katzav, J.: Araucaria: Making up argument. In: European Conference on Computing and Philosophy (2003)
23. Takahashi, T., Sawamura, H.: A logic of multiple-valued argumentation. In: Proceedings of the third international joint conference on Autonomous Agents and Multi Agent Systems (AAMAS 2004), pp. 800–807. ACM, New York (2004)
24. Takahashi, Y., Sawamura, H., Zhang, J.: Transforming natural arguments in ARAUCARIA to formal arguments in LMA. In: Proc. of 2006 IEEE/WIC/ACM International Conference on Intelligent Agent Technology (IAT 2006), pp. 668–678. IEEE Computer Society Press, Los Alamitos (2006)
25. Walton, D.: Argument from appearance: a new argumentation scheme. Ligique et Analyse 195, 319–340 (2006)
26. Walton, D., Hansen, H.: Fundamentals of critical argumentation. Cambridge University Press, Cambridge, New York, Melbourne (2006)
27. Walton, D., Rowe, G., Macagno, F., Reed, C.: Araucaria as a tool for diagramming arguments in teaching and studying philosophy. Teaching Philosophy 29, 111–124 (2006)

Safia Abbas Mahmoud works as assistant lecturer in Department of Computer Science, Faculty of Computer and Information Sciences, University of Ain Shams, Cairo, Egypt. Her present concerns are the argument-based intelligent tutoring system, data mining and knowledge discovery.

Hajime Sawamura is associate professor at Niigata University since 1996. He received the B. E., M. E. and Doctor of Engineering degrees from Hokkaido University. During 1980-1996, he was with Institute for Social Information Science, Fujitsu Laboratories Ltd., where he was a senior research fellow of computational logic group. In 1990-1991, he was a visiting fellow of Austrian National University, and in 2002-2003 a visiting fellow of philosophy department, Victoria University of Wellington, New Zealand. His present main research concern is mathematical argumentation that is a true and realistic logic!

# Chapter 6

# Grouping and Anti-predator Behaviors for Multi-agent Systems Based on Reinforcement Learning Scheme

Koichiro Morihiro[1], Haruhiko Nishimura[2],
Teijiro Isokawa[3], and Nobuyuki Matsui[3]

[1] Hyogo University of Teacher Education, Hyogo 673-1494, Japan
mori@hyogo-u.ac.jp
[2] Graduate School of Applied Informatics, University of Hyogo,
Hyogo 650-0044, Japan
haru@ai.u-hyogo.ac.jp
[3] Graduate School of Engineering, University of Hyogo, Hyogo 671-2201, Japan
isokawa@eng.u-hyogo.ac.jp, matsui@eng.u-hyogo.ac.jp

**Abstract.** Several models have been proposed for describing grouping behavior such as bird flocking, terrestrial animal herding, and fish schooling. In these models, a fixed rule has been imposed on each individual a priori for its interactions in a reductive and rigid manner. We have proposed a new framework for self-organized grouping of agents by reinforcement learning. It is important to introduce a learning scheme for developing collective behavior in artificial autonomous distributed systems. This scheme can be expanded to cases in which predators are present. In this study we integrate grouping and anti-predator behaviors into our proposed scheme. The behavior of agents is demonstrated and evaluated in detail through computer simulations, and their grouping and anti-predator behaviors developed as a result of learning are shown to be diverse and robust by changing some parameters of the scheme.

## 1 Introduction

In various scenes in nature, the collective behavior of creature can often be seen. As its typical cases, bird flocking, land animal herding, and fish schooling are well-known. Many previous observations suggest that there are no leaders to control the behavior of the group; rather it emerges from the local interactions among individuals in the group[1,2]. Several models have been proposed for describing the flocking behavior. In these models, a fixed rule is given to each of individuals a priori for their interactions[3,4,5]. This reductive and rigid approach is plausible for modeling flocks of biological organisms, for they seem to inherit the ability of making a flock. However what is more, it will become important to introduce a learning scheme for making collective behavior. In a design of

D. Srinivasan & L.C. Jain (Eds.): Innovations in MASs and Applications – 1, SCI 310, pp. 149–182.
springerlink.com     © Springer-Verlag Berlin Heidelberg 2010

artificial autonomous distributed system, fixed interactive relationships among agents (individuals) lose the robustness against nonstationary environments. It is necessary for agents to be able to adjust their parameters of the ways of interactions. Some learning framework to form individual interaction will be of importance. In addition to securing the robustness of systems, this framework will give a possibility to design systems easier, because it determines the local interactions of agents adaptively as a certain function of the system.

The classes of learning paradigm have been roughly divided into two kinds; one is a supervised learning that needs external teacher data corresponding to input, and the other is an unsupervised learning to which the behavior of the system is automatically improved by its experience only. Reinforcement learning[6,7,8] characterizes its feature of the unsupervised learning introducing a process of trial and error called exploration to maximize the reward obtained from environment. Introducing appropriate relations between the agent behavior (action) and its reward, we could make a new scheme for flocking behavior emergence by reinforcement learning.

In this paper, we propose an adaptive scheme for self-organized making flock of agents[23,24,25,26,27,28]. Each of agents is trained in its perceptual internal space by Q-learning[9,10], which is a typical reinforcement learning algorithm. The behavior of agents is demonstrated through computer simulations. We further explore the characteristics of behavior under various situations.

## 2    Reinforcement Learning

Machine learning has been developed and used in various situations. Hence, it provides a computer system with the ability to learn. To acquire desired functions in a step-by-step manner, many learning algorithms and methods have been proposed for systems. Reinforcement learning [6,7,8] is one of such machine learning. In reinforcement learning, the system receives only an evaluative scalar feedback from its environment and not an external teacher data corresponding to the input is required as in supervised learning. Reinforcement learning was originally applied to the single agent problems and it has been applied extensively to multi-agent problems in recent studies [11]. The target level of its application has also been extended from an individual creature to a neuron in the brain [12]. At present, there are many studies in reinforcement learning like this [13], but the exploration factor in it has not been reflected enough.

In reinforcement learning, it is necessary to introduce a process of trial and error that is designed to maximize the rewards obtained from the environment. This trial and error process is called exploration. Because there is a trade-off between exploration and exploitation (avoiding bad rewards), balancing their usage is very important. This is known as the exploration-exploitation dilemma. The scheme of exploration is called a policy. There are many types of policies such as $\varepsilon$-greedy, softmax, weighted roulette, and so on. In these existing policies, exploration is performed by generating stochastic random numbers, according to the reference value and the provided criterion [7,14,15].

## 2.1   Q-Learning

Q-learning is considered to be the best-understood reinforcement learning algorithm [9,10]. It forms a Q-mapping from state-action pairs $(s, a)$ based on rewards $r$ obtained from interaction with the environment. Q-learning is defined as follows:

$$Q(s_{t+1}, a_{t+1}) = Q(s_t, a_t) + \alpha[r + \gamma \max_{a' \in A(s')} Q(s', a') - Q(s_t, a_t)] \qquad (1)$$

where $\alpha$ denotes the learning rate $(0 \leq \alpha \leq 1)$; and $\gamma$, the discount rate $(0 \leq \gamma \leq 1)$. Q-learning is employed for maximizing the sum of the rewards received. It attempts to learn the optimal policy by building a table of Q-values $Q(s, a)$ according to the above update equation. $Q(s, a)$ provides the estimated values of expected return by considering action $a$ in state $s$. Once these Q-values have been learned, the optimal action from any state is the one with the highest Q-value. In the original Q-learning algorithm, the greedy policy with pure exploitation is used. However, it is generally difficult to obtain satisfactory results by employing this policy. Therefore, in the present study, a policy that allows the adoption of a nonoptimal action (exploration) is introduced.

## 2.2   Exploration Policies

In reinforcement learning, many types of exploration policies have been proposed for the process of trial and error, such as $\varepsilon$-greedy, softmax, and weighted roulette. Further, exploration is performed by generating stochastic random numbers in these existing policies. We use $\varepsilon$-greedy as the exploration policy; this policy decides whether exploration or exploitation is required on the basis of a given threshold value $\varepsilon \in [0, 1]$. In $\varepsilon$-greedy, the action with the highest Q-value is selected as exploitation with probability $1 - \varepsilon$, and any other action is selected as exploration with probability $\varepsilon/(k - 1)$, where $k$ denotes the total number of actions in a given state.

In the reinforcement learning, many kinds of exploration policies have been proposed as a process of trial and error such as $\epsilon$-greedy, softmax, and weighted roulette action selection. In this research, I adopt softmax action selection, and the rule is given as follows:

$$p(a|s) = \frac{\exp\{Q(s, a)/T\}}{\sum_{a_i \in A} \exp\{Q(s, a_i)/T\}} \qquad (2)$$

where $T$ is a positive parameter called the temperature. High temperatures cause the actions to be all (nearly) equi-probable, and low temperatures cause a greater difference in selection probability for actions that differ in their value estimates.

## 2.3   Previous Models for Grouping Behavior of Multi-agents

In the design of an artificial autonomous distributed system, fixed interactive relationships among agents (individuals) lose their robustness in the case of nonstationary environments. It is necessary for agents to be able to adjust their manner of interaction. A learning framework for individual interactions is of importance. In addition to securing the robustness of systems, this framework could possibly make it easier to design systems because it adaptively determines the local interactions of agents as a function of the system.

The collective behavior of creatures can often be seen in nature. Bird flocking, land animal herding, and fish schooling are typical well-known cases. Many previous observations suggest that there are no leaders to control the behavior of a group; on the other hand, collective behavior emerges from the local interactions among individuals in the group and/or against predators [1,2,16]. Several models have been proposed for describing the grouping behavior and the anti-predator behavior [3,4,5,17,18,19,20,21,22]. In these models, a fixed rule is provided for each of the individuals a priori for their interactions. This reductive and rigid approach is suitable for modeling groups of biological organisms since they seem to inherit the ability of forming a group. However, it is important to introduce a learning scheme for causing collective behavior.

## 3    Model and Method for Grouping and Anti-predator Agent

### 3.1    Perceptual Internal Space for Each Agent

We employ a configuration where $N$ agents that can move in any direction are placed in a two-dimensional field. Learning of each agent (agent $i$) progresses asynchronously in the discrete time-step $t$ following timing $t_i = d_i t + o_i$ where $d_i$ and $o_i$ are integers proper to agent $i$ ($0 \leq o_i < d_i$). The agents act in discrete time $t$, and at each time-step $t_i$ an agent (agent $i$) finds other agent (agent $j$) among $N - 1$ agents and learns.

In the perceptual internal space, state $s_t$ of $Q(s_t, a_t)$ for agent $i$ is defined as $[R]$, which is the maximum integer not exceeding the Euclidean distance $R$ from agent $i$ to agent $j$. For action $a_t$ of $Q(s_t, a_t)$, four kinds of action patterns $(a_1, a_2, a_3, a_4)$ are considered as follows (also illustrated in Fig. 1).

$a_1$ : Attraction to agent $j$
$a_2$ : Parallel positive orientation to agent $j$
   $(\mathbf{m_a} \cdot (\mathbf{m_i} + \mathbf{m_j}) \geq 0)$
$a_3$ : Parallel negative orientation to agent $j$
   $(\mathbf{m_a} \cdot (\mathbf{m_i} + \mathbf{m_j}) < 0)$
$a_4$ : Repulsion to agent $j$

Here, $\mathbf{m_a}$ is the directional vector of $a_t$, and $\mathbf{m_i}$ and $\mathbf{m_j}$ are the velocity vectors of agents $i$ and $j$, respectively. If the velocities of agents are set to be one body length (1 BL), then $|\mathbf{m_a}| = |\mathbf{m_i}| = |\mathbf{m_j}| = 1\mathrm{BL}$. Agent $i$ moves in accordance with $\mathbf{m_i}$ in each time step, and $\mathbf{m_i}$ is updated by the expression

$$\mathbf{m_i} \leftarrow \frac{(1 - \kappa)\mathbf{m_i} + \kappa \mathbf{m_a}}{|(1 - \kappa)\mathbf{m_i} + \kappa \mathbf{m_a}|} \tag{3}$$

where $\kappa$ is a positive parameter ($0 \leq \kappa \leq 1$) called the inertia parameter.

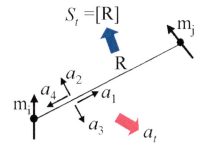

**Fig. 1.** Constitution of perceptual internal space for each agent

**Table 1.** Reward $r$ for the selected action $a_t$ in the state $s_t = [R]$ against the same kind of agent

| $s_t$ | $0 \sim R_1$ | | $R_1 \sim R_2$ | | $R_2 \sim R_3$ | | $R_3 \sim$ |
|---|---|---|---|---|---|---|---|
| $a_t$ | $a_4$ | $a_1 \sim a_3$ | $a_2$ | $a_1, a_3, a_4$ | $a_1$ | $a_2 \sim a_4$ | $a_1 \sim a_4$ |
| $r$ | 1 | -1 | 1 | -1 | 1 | -1 | 0 |

In this work, as we consider the same kind of agent and a predator as the perceptional objects, two sorts of the corresponding Q-value should be introduced.

### 3.2   Learning Mode against Agents of the Same Kind

In our proposed model, we prepare the reward $r$ for $(s_t, a_t)$ of each agent according to the distance $R$ from the perceived agent of same kind. The learning of the agents proceeds according to a positive or negative reward, as shown in Table 1, where $R_1$, $R_2$, and $R_3$ have the relationship of $R_1 < R_2 < R_3$. In case $0 < [R] \leq R_3$, agent $i$ can perceive another agent of same kind with the probability in proportion to $R^{-\beta}$, where $\beta$ is a positive parameter. This means that the smaller $R$ value is, the easier the agent at that position is selected. When $0 < [R] \leq R_1$, the agent gets the positive reward (+1) if it takes the repulsive action against the perceived agent ($a_4$); otherwise it gets the penalty (−1). In the cases of $R_1 < [R] \leq R_2$ and $R_2 < [R] \leq R_3$, the agent also gets the reward or penalty defined in Table 1 with respect to the actions. In case $[R] > R_3$, agent $i$ cannot perceive agent $j$, and then receives no reward and chooses an action from the four action patterns ($a_1, a_2, a_3, a_4$) randomly.

### 3.3   Learning Mode against Predators

When there is a predator within $R_3$, agent $i$ perceives the predator with the probability 1 and the above learning mode is switched to this mode. In this case,

**Table 2.** Reward $r$ for the selected action $a_t$ in the state $s_t = [R]$ against predator

| $s_t\ (=[R]\ )$ | $0 \sim R_3$ | | $R_3 \sim$ |
|---|---|---|---|
| $a_t$ | $a_4$ | $a_1 \sim a_3$ | $a_1 \sim a_4$ |
| $r$ | 1 | -1 | 0 |

the agent $i$ gets the positive reward $(+1)$ if it takes the repulsive action to evade the predator $(a_4)$; otherwise it gets the penalty $(-1)$ as defined in Table 2.

## 4 Grouping Behavior in No Predator Case

To demonstrate our proposed scheme via computer simulations, we consider the following experimental conditions: $\alpha = 0.1$ and $\gamma = 0.7$ in Eq. (1), $T = 0.5$ (under learning) in Eq. (2), and $\beta = 0.5$ for the distance dependence of $R^{-\beta}$. All the velocities of agents are set to one body-length (1BL). The total number of time-steps is set to 5000 in all simulations. Under these conditions, we check whether agents make a flock for the number of agents $N$, distance parameter $(R_1, R_2, R_3)$, and inertia parameter $\kappa$ in Eq. (3).

### 4.1 Typical Example: $N = 10$ with $(R_1, R_2, R_3) = (4, 20, 50)$

We simulated our model for the case in which $N = 10$, and $R_1 = 4$ (BL), $R_2 = 20$ (BL), and $R_3 = 50$ (BL). Figures 2(a), 2(b), and 2(c) show the trajectories in the range of 0–100 steps, 4900–5000 steps under learning, and 500–600 steps after learning, respectively. In Fig. 2(a), each agent changes its direction often without regard to the other agents' behavior; however, an agent maintains its direction for a long time-step with the others, as shown in Figs. 2(b) and 2(c). This indicates that the learning works well in flocks.

In order to quantitatively evaluate how the agents exhibit flocking behavior, we introduce a measure $|\mathbf{M}|$ of the uniformity in direction and a measure $E$ of the spread of agents:

$$|\mathbf{M}| = \frac{1}{N} \left| \sum_{i=1}^{N} \mathbf{m_i} \right| ,\tag{4}$$

$$E = \frac{1}{N} \sum_{i=1}^{N} \sqrt{(x_A^i - x_G)^2 + (y_A^i - y_G)^2} ,\tag{5}$$

where $(x_A^i, y_A^i)$ and $(x_G, y_G)$ are the two-dimensional coordinate of agent $i$ and the barycentric coordinate among the agents, respectively. The value of $|\mathbf{M}|$ approaches 1 when the directions of agents match to a greater excellent. The agents come close together when the value of $E$ decreases.

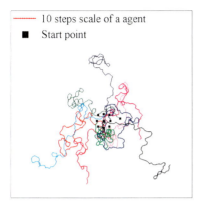

(a) Under learning, in the range of 0–100 steps

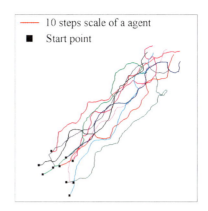

(b) Under learning, in the range of 4900–5000 steps

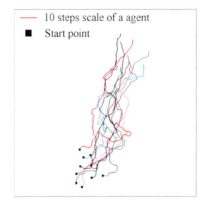

(c) After learning, in the range of 500–600 steps

**Fig. 2.** Trajectories of agents under and after learning for the case in which $N = 10$ and $(R_1, R_2, R_3) = (4, 20, 50)$

Figure 3 shows the time-step dependences of $|\mathbf{M}|$ and $E$ for this case. The transition of $|\mathbf{M}|$ evolves well except for the fluctuation caused by the exploration effect at every time-step. The value of $E$ takes a large value at the early stage of learning, and then, it decreases with fluctuations to around 8 as the learning proceeds. Further, we take the average of 100 events by repeating the above simulation with various random series in exploration. As a result, Fig. 4 is obtained in which the value of $|\mathbf{M}|$ increases up to around 0.9 and that of $E$ decreases to 8.

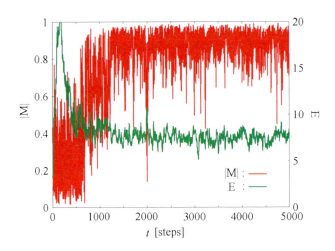

**Fig. 3.** Time-step dependence of |**M**| (Eq. (4)) and $E$ (Eq. (5)) for the case shown in Fig. 2

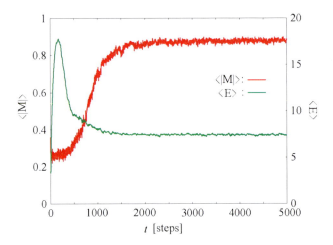

**Fig. 4.** Step-time dependence of the averaged |**M**| and $E$ for 100 events for the case shown in Fig. 3

## 4.2  Dependence on Number of Agents

We investigate the behaviors of the agents for various numbers of agents. The parameters $R_1$, $R_2$, and $R_3$ are the same as those described in Section 4.1. Figures 5(a)–5(e) show the trajectories of agents after learning, which correspond to the result shown in Fig. 2(c). The agents can make flocks in all these cases. We also calculated the values of $|\mathbf{M}|$ and $E$ for these cases and confirmed that their step-time evolutions exhibit a tendency that is very similar to that of the $N = 10$ case.

To clarify the $N$ dependence when making flocks, we show the values of $|\mathbf{M}|$ and $E$ at the end of the learning stage ($t = 5000$) with respect to $N$ in Fig. 6. These are the averaged values for 1000 events. $\langle |\mathbf{M}| \rangle$ does not decrease significantly with an increase in $N$, i.e., the uniformity of agents in a flock is maintained for a larger number of agents. It is observed that $\langle E \rangle$ for a smaller value of $N$ (except for $N = 2$) takes rather larger values. In this region of $N$, a flock tends to split into two flocks and each of them goes ahead in different direction; therefore the spread of agents increases. For $N = 2$, this situation rarely occurs unless the two agents are initially placed far away, because each of them can always catch the other.

## 4.3  Dependence on Distance Parameter $(R_1, R_2, R_3)$

In this section, we examine the cases in which each agent has different values of $R_1$, $R_2$, and $R_3$. Two settings of $(R_1, R_2, R_3)$ for $N = 10$, which are listed in Table 3, are examined. This results shown in Fig. 7 exhibit tendencies that are similar to those in the case of $(R_1, R_2, R_3) = (4, 20, 50)$ that is common for all agents, as described in Section 4.1. With regard to the learning performance, the agents in setting 2 exhibited a worse tendency as compared to those in setting 1 and the setting described in Section 4.1 in the early stage ($t < 1000$). This is because of the large values of $R_2$ and $R_3$ in setting 2. However, in the latter stage ($t > 3000$), flocking was attained well, as in the case of the other settings.

## 4.4  Effect of Inertia Parameter on Grouping

The effects of changing the inertia parameter $\kappa$ are investigated in this section. From the definition of updating the velocity vector of an agent $\mathbf{m_i}$ (Eq. (3)), an agent has stronger inertia (tendency to keep its own direction unchanged) when $(1 - \kappa)$ takes a larger value.

Figure 8 shows the $(1 - \kappa)$ dependences of $\langle |\mathbf{M}| \rangle$ and $\langle E \rangle$ at the end of the learning stage ($t = 5000$). The spread of agents $\langle E \rangle$ increases considerably and the directional uniformity of agents $\langle |\mathbf{M}| \rangle$ decreases when $(1 - \kappa)$ exceeds 0.5. This implies that two (or more) flocks of agents are formed due to the breakup of a flock. Because agents with strong inertia (large $(1 - \kappa)$) require some time-steps to change their directions according to other agents, they occasionally cannot keep track of other agents, and several agents get segregated. An example of the trajectories of agents when their actions have strong inertia are shown in Fig. 9,

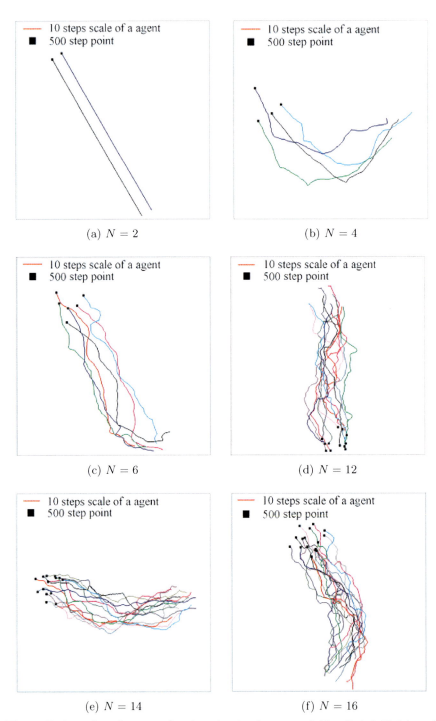

**Fig. 5.** Trajectories of agents after learning in the cases of $N = 2, 4, 6, 12, 14$, and $16$ with $(R_1, R_2, R_3) = (4, 20, 50)$

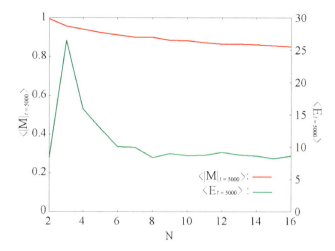

**Fig. 6.** $N$ dependences of $\langle|\mathbf{M}|\rangle$ and $\langle E\rangle$ at the end of the learning stage ($t = 5000$), averaged over 1000 events

**Table 3.** Two settings of $(R_1, R_2, R_3)$ for 10 agents

| Agent No. | Setting 1 | Setting 2 |
|---|---|---|
| 1 | (3,16,40) | (4,40,100) |
| 2 | (3,17,45) | (4,39,99) |
| 3 | (4,19,47) | (4,38,98) |
| 4 | (4,18,48) | (4,37,97) |
| 5 | (4,20,50) | (4,36,96) |
| 6 | (4,21,51) | (3,35,95) |
| 7 | (4,22,52) | (3,34,94) |
| 8 | (5,25,45) | (3,33,93) |
| 9 | (5,23,53) | (3,32,92) |
| 10 | (6,25,55) | (3,31,91) |

where agents are in the learning stage (0–1000 time-steps), and $(1 - \kappa) = 0.75$ is set for each agent. We observe that several agents depart from other agents and move far away.

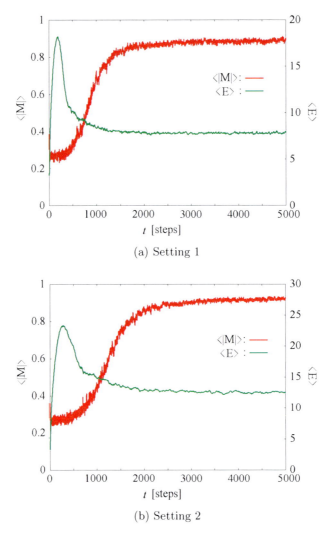

(a) Setting 1

(b) Setting 2

**Fig. 7.** Time-step dependence of the averaged $|\mathbf{M}|$ and $E$ in 100 events, where each agent has different value of $(R_1, R_2, R_3)$

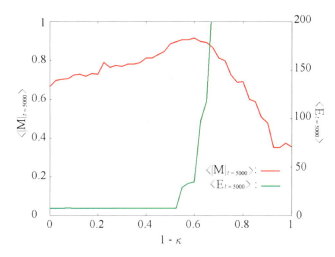

**Fig. 8.** $(1-\kappa)$ dependences of $\langle|\mathbf{M}|\rangle$ and $\langle E \rangle$ at the end of the learning stage ($t = 5000$), averaged for 100 events

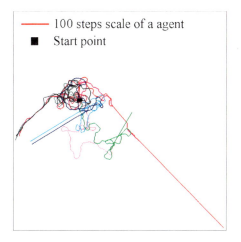

**Fig. 9.** Trajectories of agents in learning (0–1000 time-steps) with strong inertia ($(1 - \kappa) = 0.75$)

## 4.5   Asynchronous and Slow Pace Learning

The grouping behavior of biological organisms is a familiar phenomenon; however, the manner in which each agent learns synchronously cannot be observed. In addition, each agent possesses a certain individuality. Therefore, we consider the cases of slow pace learning and asynchronous learning. In the next simulation, the learning of each agent progresses asynchronously in discrete time-step $t$. We employ a configuration in which $N$ agents that can move in any direction are placed in a two-dimensional field. The learning of each agent (agent $i$) progresses asynchronously in the discrete time-step $t$ following timing $t_i = d_i t + o_i$, where $d_i$ and $o_i$ are integers proper to agent $i$ ($0 \leq o_i < d_i$). The agents act in discrete time $t$, and at each time-step $t_i$, an agent (agent $i$) finds another agent (agent $j$) among $N - 1$ agents and learns. In the next simulation, we consider a case in which there is only one slow learning agent in the group. That is, $d_{i=SLOW} > d_{i \neq SLOW}$ and all $d_{i \neq SLOW}$ are the same. We define the learning pace $d$ as $d = d_{i \neq SLOW} / d_{i=SLOW}$. We consider $o_i = i$ and $d_{i \neq SLOW} = N$ for consistency.

In order to evaluate how the slow pace learning agent affects the grouping behavior, we introduce another measure $\eta$ of the uniformity for the slow pace learning agent as follows.

$$|\eta| = \frac{1}{N-1} \sum_{i \neq SLOW} (\mathbf{m_i} \cdot \mathbf{m_{SLOW}}) , \tag{6}$$

If $\eta$ approaches $|\mathbf{M}|$, the direction of the slow pace learning agent and the group of other agents matches to a greater excellent.

We show the results for asynchronous learning and the slow pace learning agent. In 100 events without a slow pace learning agent, shown in Fig. 10, the averaged value of $\eta$ approaches $\langle|\mathbf{M}|\rangle$. In 100 events with one slow pace learning agent, shown in Fig. 11, the group of agents splits into two groups having different sizes in several cases and the measure $\langle \mathbf{E} \rangle$ increases. In almost all such cases, the slow pace learning agent is not included in the main group. Therefore $\langle|\mathbf{M}|\rangle$ remains large but $\eta$ decreases.

We show the dependence of $\langle \mathbf{E} \rangle$, $\langle|\mathbf{M}|\rangle$, and $\eta$ on the learning pace. Figure 12 shows the dependence of $\langle \mathbf{E} \rangle$, $\langle|\mathbf{M}|\rangle$, and $\eta$ on the learning pace in the case

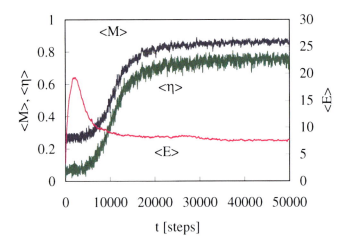

**Fig. 10.** Case of same pace learning agents

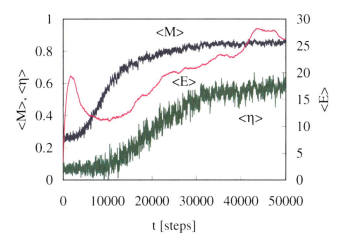

**Fig. 11.** Case of one slow pace learning agent

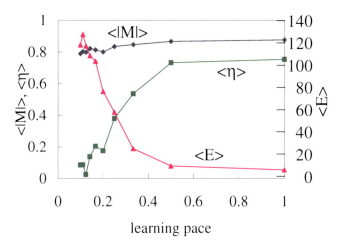

**Fig. 12.** Dependence of $\langle \mathbf{E} \rangle$, $\langle |\mathbf{M}| \rangle$, and $\eta$ on learning pace in the case of same pace learning agents

of same pace learning agents. $\langle |\mathbf{M}| \rangle$ and $\eta$ are sufficiently large and $\langle \mathbf{E} \rangle$ is sufficiently small when the learning pace approaches 1. When the learning pace decreases, $\langle |\mathbf{M}| \rangle$ maintains a good value but $\eta$ and $\langle \mathbf{E} \rangle$ worsen. This suggests that the slow pace learning agent is easily left when the learning pace decreases. Figure 13 shows the dependence on the learning pace in the case of one slow pace learning agent. Even if the agents perceive the slow learning pace agent by priority to include it in their group, the effect is not achieved. On the contrary, $\langle |\mathbf{M}| \rangle$ and $\eta$ maintain good values but $\langle \mathbf{E} \rangle$ worsens from the beginning. This suggests that a few of the agents leave the group in many cases.

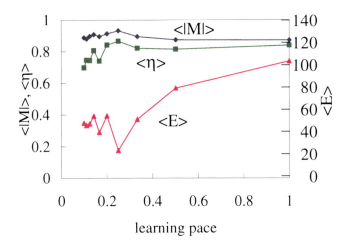

**Fig. 13.** Dependence of $\langle E \rangle$, $\langle |M| \rangle$, and $\eta$ on learning pace in the case of one slow pace learning agent

## 5     Grouping and Anti-predator Behavior

### 5.1     The Case Predator Appears Later

Each agent learns its flocking behavior without predator up to first 5000 time-steps and then a predator appears first. The predator appears behind the group of agents in the place where agents cannot perceive. Then, the predator approaches the center of the group from behind and passes straight. The predator appears in every 500 time-steps up to 10000 time-steps. The velocity of the predator is twice of the agent. Figure 14 shows the average of 100 events as like as in Fig. 4.

When the predator appears, the learning mode is changed. So, $\langle E \rangle$ takes a large value and $\langle |M| \rangle$ decreases down to near 0.3. This means that the agents do not make flocking behavior. When the predator disappears, the learning mode is changed back. $\langle E \rangle$ takes a small value and $\langle |M| \rangle$ increases up again to near 0.9 because of the re-flocking behavior of agents.

As shown in Fig. 14, the base line of $\langle E \rangle$ increases gradually. This is caused by appearing of a lone agent or splitting into two or more groups of agents. In 100 events in Fig. 14, there are 16 such events which have the maximum value of $E(t) > 40$. By eliminating these events from the average, the result becomes as in Fig. 15. The base line of $\langle E \rangle$ is kept to be almost constant.

Figure 16 is the magnification near the first appearance of the predator (5000 time-step) of Fig. 15. Also, Figure 17 is the magnification near the final appearance of the predator (9500 time-step) of Fig. 15. The peak of $\langle E \rangle$ grows up following the predator appearance. This suggests that each agent learns to escape from the predator gradually through the experience meeting it.

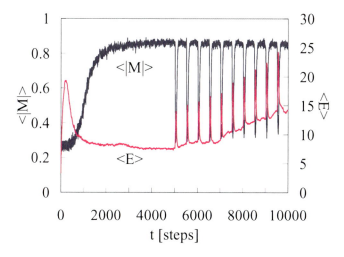

**Fig. 14.** The step-time dependence of the averaged |**M**| and $E$ in 100 events of the case predator appears later

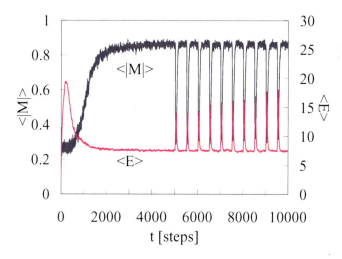

**Fig. 15.** The step-time dependence of the averaged |**M**| and $E$ in non-splitting 84 events corresponding to Fig. 14

To see the effect of these learning, we check the difference of the trajectories of agents in some learning stages. Stopping the learning of each agent corresponds to fixing the Q-value. Figure 18 shows the trajectories of the agents by fixing the Q-value at t=5000 before the first appearance of predator and by setting $T \rightarrow 0$ in Eq. (2) as the greedy behavioral policy. At this stage, the agents learn nothing about the predator, so they do not escape from the predator.

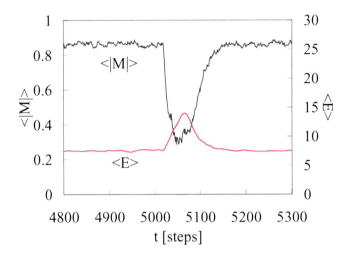

**Fig. 16.** The magnification near the first appearance of the predator (5000 time-step) in Fig. 15

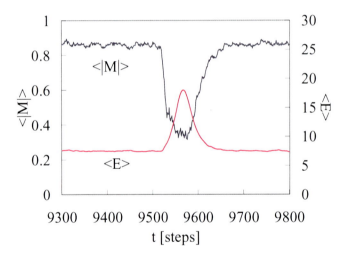

**Fig. 17.** The magnification near the final appearance of the predator (9500 time-step) in Fig. 15

Figure 19 shows the trajectories of agents including the range of the first learning stage against predator (5000–5099 steps). The agents in the first learning stage (5000–5099 steps) are panicked under learning.

Figure 20 shows the trajectories of the agents by the fixed Q(t=5500) after the first appearance of predator and the greedy behavioral policy. It seems that the agents learned a little from the predator, so they try to move away from the predator as they find it.

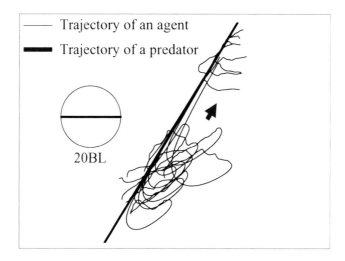

**Fig. 18.** The trajectories of agents of 100 steps against the predator in the fixed Q(t=5000) case

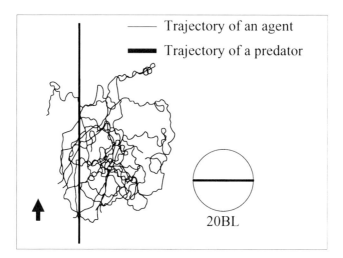

**Fig. 19.** The trajectories of agents in the range of 5000–5099 steps against the predator under learning

The trajectories of agents including the range of the 10th learning stage against predator (9450–9649 steps) are shown in Fig. 21. Compared with the panicked agents in Fig. 19, they learn well to escape from the predator. Due to some exploration for learning in the action of agents, the trajectories have a little fluctuation.

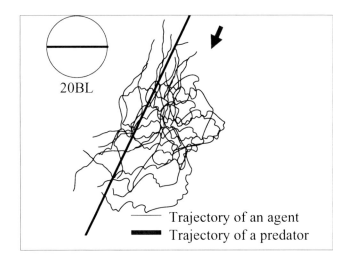

**Fig. 20.** The trajectories of agents of 100 steps against the predator in the fixed Q(t=5500) case

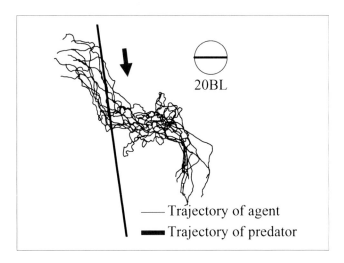

**Fig. 21.** The trajectories of agents in the range of 9450–9649 steps against the predator under learning

Figure 22 shows the trajectories of the agents by the fixed Q(t=10000) after the 10th appearance of predator and the greedy behavioral policy. Through 10 learning stages, they learned to evade well the predator compared with Figs. 18 and 20.

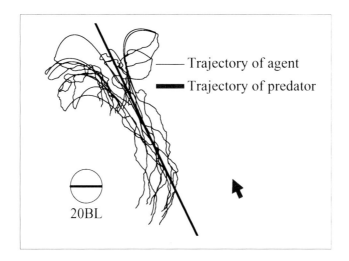

**Fig. 22.** The trajectories of agents of 200 steps against the predator in the fixed Q(t=10000) case

## 5.2 The Case Predator Appears from the Beginning

Figure 23 shows the case the predator appears from the beginning at the 1000 time-step. Before the agents learn the grouping behavior enough, they begin to learn the escape from the predator. In the examined 100 events, there are 5 events of the case of appearing of a lone agent or splitting into two or more groups of agents. Figure 24 shows the average of the remaining 95 events. In this case, similar relation as in Fig. 15 can be seen.

In the case predator appears from the beginning, the agents simultaneously learn the flocking behavior and the escape from the predator. Figure 25 shows the trajectories of agents in the range of 0–2200 steps under learning. In the insufficient learning stage, the trajectories of agents are crowded near the meeting point against the predator though the flocking behavior can be observed.

To see the effect of the learning, we check the difference of the trajectories of agents after learning. Stopping the learning of each agent corresponds to fixing the Q-value. Figure 26 shows the trajectories of the agents of 1000 steps against the predator in the fixed Q(t=5000) case. In this case, each agent uses fixed Q-value at t=5000 under learning and temperature parameter setting $T \to 0$ in Eq. (2) as the greedy behavioral policy. Through the learning stages, they learned to group together and evade well the predator compared with Fig. 25.

The magnifications of each 100 steps in Fig. 26 are shown in Fig. 27. After learning, there is no fluctuation by the exploration under learning. Finding the predator, the agents draw the shape like a (polarized) fountain to escape from it. This suggests that adaptive behaviors of agents including the escape from the predator emerge as a result of two mode learning.

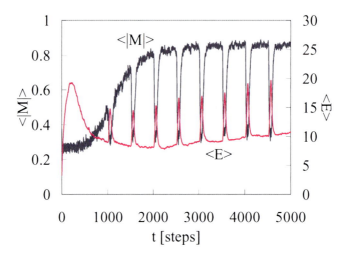

**Fig. 23.** The step-time dependence of the averaged |M| and $E$ in 100 events of the case predator appears from the beginning

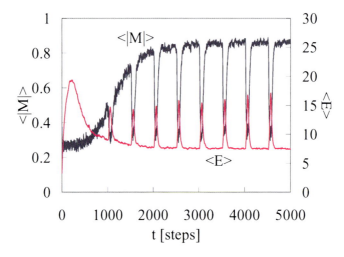

**Fig. 24.** The step-time dependence of the averaged |M| and $E$ in non-splitting 95 events corresponding to Fig. 23

Figure 28 shows the trajectories of the agents by the fixed $Q(t=5000)$ after the 8th appearance of the predator and greedy behavioral policy. Contrary to the insufficient learning stage, escaping behavior from the predator can be observed as well as flocking behavior.

Many kinds of anti-predator strategy are observed and recorded from a field study on predator-prey interactions, such as split, join vacuole, hourglass, ball,

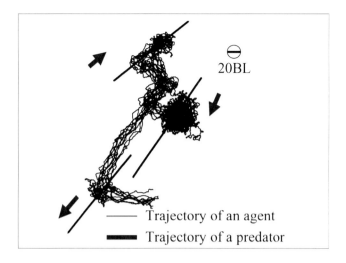

**Fig. 25.** The trajectories of agents in the range of 0–2200 steps against the predator under learning

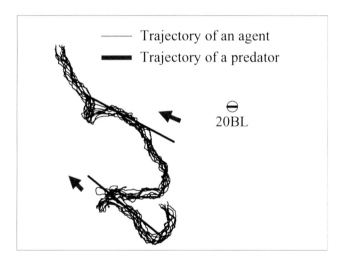

**Fig. 26.** The trajectories of agents of 1000 steps against the predator in the fixed Q(t=5000) case

bend, fountain and herd [16,20]. We show an anti-predator behavior of agents like vacuole observed in our simulation in Fig. 29. In this case, the speed of predator is changed to 0.5 BL (a half speed of agent) from 2 BL. Then, keeping their distance from the predator, the agents have surrounded the predator. Fig. 30 shows one more anti-predator behavior of agent like herd obtained by changing the speed of the predator to 1 BL (the same speed of agent) from 2 BL. Then,

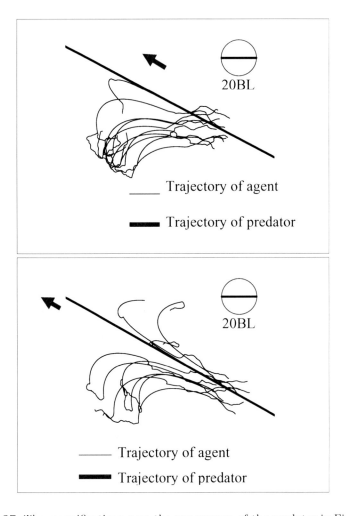

**Fig. 27.** The magnifications near the appearance of the predator in Fig. 26

keeping their distance from the predator, the agents advance as p party together with the predator.

### 5.3    Another Example: N = 30 with (R1;R2;R3) = (4;20;50) Case

Here, $\mathbf{m_a}$ is the directional vector of $a_t$, and $\mathbf{m_i}$ and $\mathbf{m_j}$ are the velocity vectors of agents $i$ and $j$, respectively. Agent $i$ moves in accordance with $\mathbf{m_i}$ in each time step, and $\mathbf{m_i}$ is updated by the expression

$$\mathbf{m_i} \leftarrow (1 - \kappa)\mathbf{m_i} + \kappa\mathbf{m_a} \quad , \qquad (7)$$

where $\kappa$ is a positive parameter $(0 \leq \kappa \leq 1)$ called the inertia parameter.

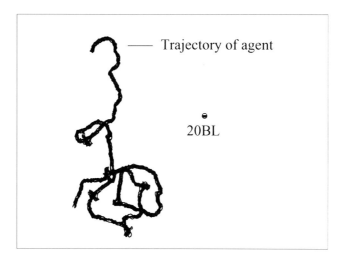

**Fig. 28.** The trajectories of agents of 5000 steps against the predator in the fixed $Q(t=5000)$ case

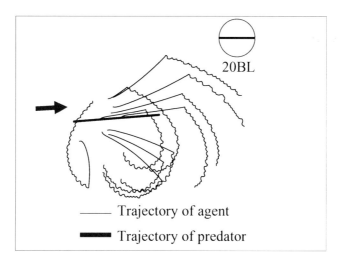

**Fig. 29.** Anti-predator behavior like vacuole in the case that the speed of predator is 0.5 BL as a half speed of agent

In the computer simulations, we have assumed the following experimental conditions: $\alpha = 0.1$, $\gamma = 0.7$ in Eq.(1), $T = 0.5$ (under learning) for the softmax action selection method, $\kappa = 0.4$ in Eq.(3), $\beta = 0.5$ for the distance dependence of $R^{-\beta}$, $d_i = 1$, and $o_i = 0$. The initial velocities of the same type of agents are set to one body length (1 BL). The velocity $|\mathbf{m_a}|$ which is the directional vector of $a_t$ is also set to one body length (1 BL). The velocity of the predator is set

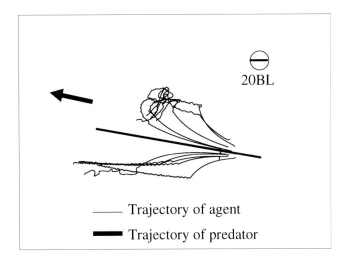

**Fig. 30.** Anti-predator behavior like herd in the case that the speed of predator is 1.0 BL as same speed of agent

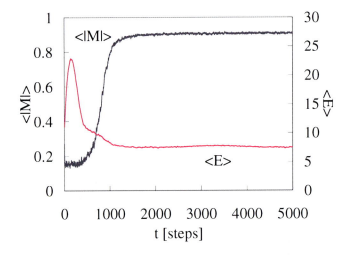

**Fig. 31.** Time step dependence of averaged $|\mathbf{M}|$ and $E$ in 100 events for no predator case

to two body lengths (2 BL). We have simulated our model for the number of agents $N = 30$ and $R_1 = 4$ (BL), $R_2 = 20$ (BL), and $R_3 = 50$ (BL).

In order to quantitatively evaluate how the agents develop grouping behavior, we use the measure $|\mathbf{M}|$ (Eq. (4)) of the uniformity in direction and the measure $E$ (Eq. (5)) of the spread of agents.

Figure 31 shows the time step dependences of averaged $|\mathbf{M}|$ and $E$ for no predator case. The transition of $\langle|\mathbf{M}|\rangle$ evolves good in every time step. The

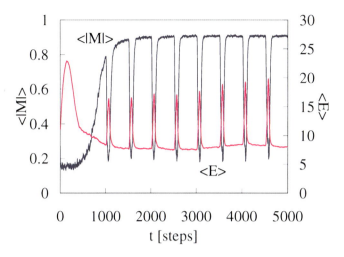

**Fig. 32.** Time step dependence of averaged $|\mathbf{M}|$ and $E$ in non-splitting 94 events for the case predator appears

value of $\langle E \rangle$ takes a large value at the early stage of learning, after which it decreases to a value around 8 as learning proceeds.

In the case predator appears, the predator appears behind the group of agents in the place where agents cannot perceive. Then, the predator approaches the center of the group from behind and passes straight. The predator appears in every 500th time step up to 5000 time steps. Figure 32 shows the average of non-splitting 94 events in 100 events. When the predator appears, the learning mode is changed. Hence, $\langle E \rangle$ takes a large value and $\langle |\mathbf{M}| \rangle$ decreases to around 0.2. This implies that the agents do not exhibit grouping behavior. When the predator disappears, the learning mode is reverted to the original mode. $\langle E \rangle$ takes a small value and $\langle |\mathbf{M}| \rangle$ increases again to around 0.9 because of the grouping behavior exhibited by the agents.

### 5.4   Effect of Inertia Parameter on Grouping in No Predator Case

From the definition of updating the velocity vector of an agent $\mathbf{m_i}$ (Eq.(3)), an agent has stronger inertia (tendency to keep its own direction unchanged) when $(1-\kappa)$ takes a larger value. Figure 33 shows $(1-\kappa)$ dependences of $\langle |\mathbf{M}| \rangle$ and $\langle E \rangle$ at the end of learning ($t = 5000$). The spread of agents $\langle E \rangle$ becomes considerably large and the directional uniformity of agents $\langle |\mathbf{M}| \rangle$ becomes lower when $(1 - \kappa)$ exceeds 0.6. This means that there grow two (or more) groups of agents, due to breakup of a group. Because agents with strong inertia (large $(1 - \kappa)$) need some time steps to change their directions according to other agents, they sometimes cannot keep track of other agents, and several agents get segregated. Figure 34 shows the case under the condition that the velocities $|\mathbf{m_i}|$ of all agents are fixed

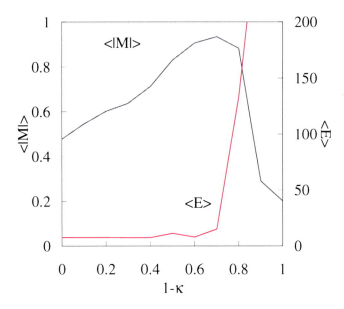

**Fig. 33.** $(1 - \kappa)$ dependences of averaged $|\mathbf{M}|$ and $E$ at $t = 5000$ in Eq. (3)

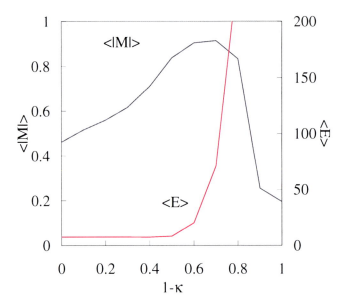

**Fig. 34.** $(1 - \kappa)$ dependences of averaged $|\mathbf{M}|$ and $E$ at $t = 5000$ in $|\mathbf{m_i}| = 1$

to 1 ($\mathbf{m_i} \leftarrow \frac{(1-\kappa)\mathbf{m_i}+\kappa\mathbf{m_a}}{|(1-\kappa)\mathbf{m_i}+\kappa\mathbf{m_a}|}$). Due to the restriction, the threshold $(1 - \kappa)$ for breakup of a group becomes 0.5 lower than 0.6 in Fig. 33.

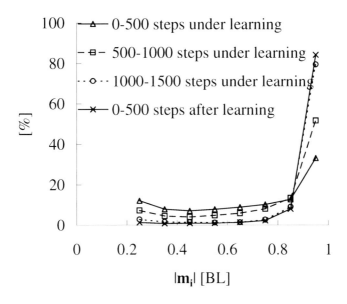

**Fig. 35.** The velocity distribution of the group in no predator case

## 5.5   Velocity Distribution of the Group

The agent changes the velocity $\mathbf{m}_i$ in each time step based on the definition for the velocity vector of an agent. Learning grouping and anti-predator behaviors, the agent also improves its speed for such behaviors. In order to confirm the influence of learning on the velocity of the agent, we check the velocity $|\mathbf{m}_i|$ of the agent under learning and after learning. Figure 35 shows the velocity $|\mathbf{m}_i|$ distribution of the group in no predator case for 500 steps intervals ($500steps \times 30agents$ data). In the distribution for 0-500 steps under learning, low speed holds more than 10% and high speed does not reach 40%. In the distribution for 500-1000 steps under learning, low speed decreases to 7% and high speed increases to 52%. At the stage of 1000-1500 steps under learning, the distribution becomes same as that for 500 steps after learning. The velocity $|\mathbf{m}_i|$ distributions of the group when a predator appears are shown in Fig. 36. A similar tendency to Fig. 35 is obtained for 100 steps intervals against the predator.

## 5.6   Trajectories of Agents and Predator

Figure 37 shows the trajectories of the agents in 1700 steps against the predator after learning. In this case, each agent uses fixed Q-value at t=5000 under learning and by setting $T \to 0$ in the softmax action selection method as the greedy behavioral policy. Through the learning stages, they have learned grouping and anti-predator behaviors. The magnification of 100 steps in Fig. 37 is shown in Fig. 38. On spotting the predator, the agents form a shape resembling a (polarized) fountain to escape from it. This suggests that the adaptive behaviors of

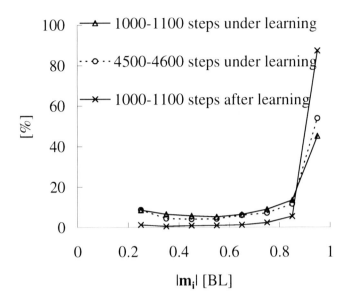

**Fig. 36.** The velocity distribution of the group in the case predator appears

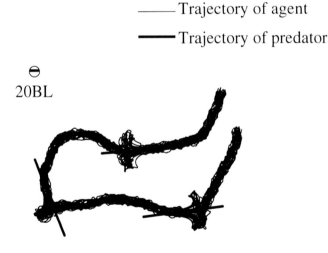

**Fig. 37.** Trajectories of agents in 1700 steps after learning

agents, including escaping from the predator, is developed as a result of the two learning modes. Many kinds of anti-predator strategy are observed and recorded from a field study on predator-prey interactions [16,20]. In our simulation, such anti-predator behaviors of agents like herd and vacuole are also observed. They are shown in Fig. 39 and Fig. 40, respectively.

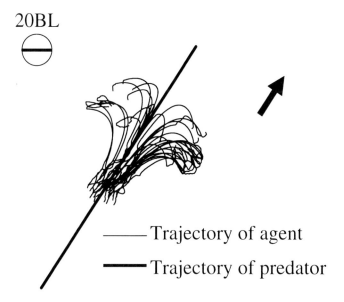

20BL

——— Trajectory of agent

▬▬▬ Trajectory of predator

**Fig. 38.** Magnification of Fig. 37 near appearance of predator

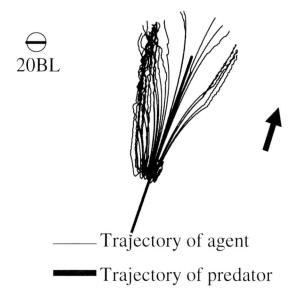

20BL

——— Trajectory of agent

▬▬▬ Trajectory of predator

**Fig. 39.** Anti-predator behavior like herd in the case that the speed of predator is 1.05 BL as same speed as agent

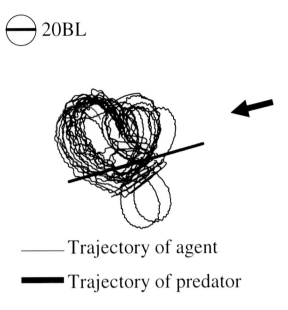

⊖ 20BL

—————— Trajectory of agent

━━━━━ Trajectory of predator

**Fig. 40.** Anti-predator behavior like vacuole in the case that the speed of predator is 0.5 BL as a half speed of agent

## 6    Conclusion

We have demonstrated a scheme for forming autonomous groups of agents by reinforcement Q-learning. In addition to the grouping behavior of agents, the anti-predator behavior exhibited while escaping from predators can be developed by learning. This indicates the adaptive flexibility of our proposed scheme. In order to confirm effectiveness of our scheme for various situations and patterns of escaping behavior, we have carried out further investigations. We are interested in the examination of the group that has complex and diverse learning conditions. We are carrying out a simulation on a group of agents in which learning progresses asynchronously and on a group that includes slow-learning agents.

## References

1. Shaw, E.: Schooling Fishes. American Scientist 66, 166–175 (1978)
2. Partridge, B.L.: The structure and function of fish schools. Scientific American 246, 90–99 (1982)
3. Aoki, I.: A Simulation Study on the Schooling Mechanism in Fish. Bulletin of the Japanese Society of Scientific Fisheries 48(8), 1081–1088 (1982)
4. Reynolds, C.W.: Flocks, Herds, and Schools: A Distributed Behavioral Model. Computer Graphics 21(4), 25–34 (1987)
5. Huth, A., Wissel, C.: The Simulation of the Movement of Fish Schools. Journal of Theoretical Biology 156, 365–385 (1992)

 6. Kaelbling, L.P., Littman, M.L., Moore, A.W.: Reinforcement Learning: A Survey. Journal of Artificial Intelligence Research 4, 237–285 (1996)
 7. Sutton, R.S., Barto, A.G.: Reinforcement Learning: An Introduction. MIT Press, Cambridge (1998)
 8. Glorennec, P.Y.: Reinforcement Learning: an Overview. In: Proceedings of ESIT 2000 - European Symposium on Intelligent Techniques, Aachen, Germany, pp. 17–35 (2000)
 9. Watkins, C.: Learning from delayed rewards, PhD Thesis, University of Cambridge, England (1989)
10. Watkins, C., Dayan, P.: Q-learning, Machine Learning, vol. 8, pp. 279–292 (1992)
11. Buşoniu, L., De Schutter, B., Babuška, R.: Learning and coordination in dynamic multiagent systems, Technical report 05-019, Delft Center for Systems and Control, Delft University of Technology, Delft, The Netherlands (2005)
12. Dietterich, T.G., Becker, S., Ghahramani, Z.: Advances in Neural Information Processing Systems 14. MIT Press, Cambridge (2002)
13. Mahadevan, S.: Reinforcement Learning Repository, University of Massachusetts, http://www-anw.cs.umass.edu/rlr/
14. Thrun, S.B.: Efficient Exploratio. In: Reinforcement Learning, Technical report CMU-CS-92-102, Carnegie Mellon University, Pittsburgh, PA (1992)
15. Thrun, S.B.: The role of exploration in learning control. In: White, A.D., Sofge, D.A. (eds.) Handbook of Intelligent Control: Neural, Fuzzy, and Adaptive Approaches, Van Nostrand Reinhold, New York (1992)
16. Pitcher, T.J., Wyche, C.J.: Predator avoidance behaviour of sand-eel schools: why schools seldom split. In: Noakes, D.L.G., Lindquist, B.G., Helfman, G.S., Ward, J.A. (eds.) Predators and Prey in Fishes, pp. 193–204. Junk, The Hague (1983)
17. Huth, A., Wissel, C.: The Simulation of the fish schools in comparison with experimental data. Ecological Modeling 75/76, 135–145 (1994)
18. Niwa, H.-S.: Self-organizing dynamic model of fish schooling. Journal of theoretical Biology 171, 123–136 (1994)
19. Shimoyama, N., Sugawara, K., Mizuguchi, T., Hayakawa, Y., Sano, M.: Collective Motion in a System of Motile Elements. Physical Review Letters 76, 3870–3873 (1996)
20. Vabo, R., Nottestad, L.: An individual based model of fish school reactions: predicting antipredator behaviour as observed in nature. Fisheries Oceanography 6, 155–171 (1997)
21. Inada, Y., Kawachi, K.: Order and Flexibility in the Motion of Fish Schools. Journal of theoretical Biology 214, 371–387 (2002)
22. Oboshi, T., Kato, S., Mutoh, A., Itoh, H.: A Simulation Study on the Form of Fish Schooling for Escape from Predator. Forma 18, 119–131 (2003)
23. Tomimasu, M., Morihiro, K., Nishimura, H., Isokawa, T., Matsui, N.: A Reinforcement Learning Scheme of Adaptive Flocking Behavior. In: Proc. of the 10th Int. Symp. on Artificial Life and Robotics (AROB), GS1-4, Oita, Japan (2005)
24. Morihiro, K., Isokawa, T., Nishimura, H., Tomimasu, M., Kamiura, N., Matsui, N.: Reinforcement Learning Scheme for Flocking Behavior Emergence. Journal of Advanced Computational Intelligence and Intelligent Informatics(JACIII) 11(2), 155–161 (2007)
25. Morihiro, K., Nishimura, H., Isokawa, T., Matsui, N.: Learning Grouping and Anti-Predator Behaviors for Multi-Agent Systems. In: Lovrek, I., Howlett, R.J., Jain, L.C. (eds.) KES 2008, Part II. LNCS (LNAI), vol. 5178, pp. 426–433. Springer, Heidelberg (2008)

26. Morihiro, K., Isokawa, T., Nishimura, H., Matsui, N.: Emergence of Flocking Behavior Based on Reinforcement Learning. In: Gabrys, B., Howlett, R.J., Jain, L.C. (eds.) KES 2006. LNCS (LNAI), vol. 4253, pp. 699–706. Springer, Heidelberg (2006)
27. Morihiro, K., Nishimura, H., Isokawa, T., Matsui, N.: Reinforcement Learning Scheme for Grouping and Anti-predator Behavior. In: Apolloni, B., Howlett, R.J., Jain, L. (eds.) KES 2007, Part III. LNCS (LNAI), vol. 4694, pp. 115–122. Springer, Heidelberg (2007)
28. Morihiro, K., Nishimura, H., Isokawa, T., Matsui, N.: Emergence of Grouping and Anti-Predator Behavior by Reinforcement Learning Scheme (submitted)

# Chapter 7

# Multi-agent Reinforcement Learning: An Overview[*]

Lucian Buşoniu[1], Robert Babuška[2], and Bart De Schutter[3]

[1] Center for Systems and Control, Delft University of Technology, The Netherlands
   i.l.busoniu@tudelft.nl
[2] Center for Systems and Control, Delft University of Technology, The Netherlands
   r.babuska@tudelft.nl
[3] Center for Systems and Control & Marine and Transport Technology Department,
   Delft University of Technology, The Netherlands
   b.deschutter@tudelft.nl

**Abstract.** Multi-agent systems can be used to address problems in a variety of domains, including robotics, distributed control, telecommunications, and economics. The complexity of many tasks arising in these domains makes them difficult to solve with preprogrammed agent behaviors. The agents must instead discover a solution on their own, using learning. A significant part of the research on multi-agent learning concerns reinforcement learning techniques. This chapter reviews a representative selection of multi-agent reinforcement learning (MARL) algorithms for fully cooperative, fully competitive, and more general (neither cooperative nor competitive) tasks. The benefits and challenges of MARL are described. A central challenge in the field is the formal statement of a multi-agent learning goal; this chapter reviews the learning goals proposed in the literature. The problem domains where MARL techniques have been applied are briefly discussed. Several MARL algorithms are applied to an illustrative example involving the coordinated transportation of an object by two cooperative robots. In an outlook for the MARL field, a set of important open issues are identified, and promising research directions to address these issues are outlined.

## 1 Introduction

A multi-agent system is a group of autonomous, interacting entities sharing a common environment, which they perceive with sensors and upon which they act with actuators [107, 135, 139]. Multi-agent systems are finding applications in a variety of domains including robotic teams, distributed control, resource management, collaborative decision support systems, data mining, etc. [4, 33, 88, 100, 115, 125]. They may arise as the most natural way of looking at a system, or may provide an alternative perspective on systems that are originally regarded as centralized. For instance, in robotic teams the control authority is naturally distributed among the robots [115]. In resource management, while

---

[*] Portions reprinted, with permission, from [20], 'A Comprehensive Survey of Multiagent Reinforcement Learning', by Lucian Buşoniu, Robert Babuška, and Bart De Schutter, IEEE Transactions on Systems, Man, and Cybernetics—Part C: Applications and Reviews, vol. 38, no. 2, March 2008, pages 156–172. © 2008 IEEE.

D. Srinivasan & L.C. Jain (Eds.): Innovations in MASs and Applications – 1, SCI 310, pp. 183–221.

the resources could be managed by a central authority, identifying each resource with an agent may provide a helpful, distributed perspective on the system [33].

Although the agents in a multi-agent system can be endowed with behaviors designed in advance, they often need to learn new behaviors online, such that the performance of the agent or of the whole multi-agent system gradually improves [106, 115]. This is usually because the complexity of the environment makes the *a priori* design of good agent behaviors difficult or even impossible. Moreover, in an environment that changes over time, a hardwired behavior may become unappropriate.

A reinforcement learning (RL) agent learns by interacting with its dynamic environment [58, 106, 120]. At each time step, the agent perceives the state of the environment and takes an action, which causes the environment to transit into a new state. A scalar reward signal evaluates the quality of each transition, and the agent has to maximize the cumulative reward along the course of interaction. The RL feedback (the reward) is less informative than in supervised learning, where the agent would be given the correct actions to take [27] (such information is unfortunately not always available). The RL feedback is, however, more informative than in unsupervised learning, where there is no explicit feedback on the performance [104]. Well-understood, provably convergent algorithms are available for solving the single-agent RL task. Together with the simplicity and generality of the setting, this makes RL attractive also for multi-agent learning.

This chapter provides a comprehensive overview of multi-agent reinforcement learning (MARL). We mainly focus on autonomous agents learning how to solve *dynamic* tasks online, using algorithms that originate in *temporal-difference RL*. We discuss the contributions of game theory to MARL, as well as important algorithms for static tasks.

We first outline the benefits and challenges of MARL. A central challenge in the field is the definition of an appropriate formal goal for the learning multi-agent system. We present the different learning goals proposed in the literature, which consider the stability of the agent's learning dynamics on the one hand, and its adaptation to the changing behavior of the other agents on the other hand. The core of the chapter consists of a detailed study of a representative selection of MARL algorithms, which allows us to identify the structure of the field and to provide insight into the state of the art. This study organizes the algorithms first by the type of task they address: fully cooperative, fully competitive, and mixed (neither cooperative nor competitive); and then by the type of learning goal they target: stability, adaptation, or a combination of both. Additionally, we briefly discuss the problem domains where MARL techniques have been applied, and we illustrate the behavior of several MARL algorithms in a simulation example involving the coordinated transportation of an object by two cooperative agents. In an outlook for the MARL field, we identify a set of important open issues and suggest promising directions to address these issues.

The remainder of this chapter is organized as follows. Section 2 introduces the necessary background in single-agent RL, multi-agent RL, and game theory. Section 3 reviews the main benefits and challenges of MARL, and Section 4 presents the MARL goals proposed in the literature. In Section 5, MARL algorithms are classified and reviewed in detail. Section 6 reviews several application domains of MARL, while Section 7 provides an example involving object transportation. Finally, Section 8 distills an outlook for the MARL field, Section 9 presents related work, and Section 10 concludes the chapter.

## 2   Background: Reinforcement Learning

In this section, the necessary background on single-agent and multi-agent RL is introduced. First, the single-agent task is defined and its solution is characterized. Then, the multi-agent task is defined. Static multi-agent tasks are introduced separately, together with necessary game-theoretic concepts. The discussion is restricted to discrete state and action spaces having a finite number of elements, as a large majority of MARL results is given for this setting.

### 2.1   The Single-Agent Case

The formal model of single-agent RL is the *Markov decision process.*

**Definition 1.** *A finite Markov decision process is a tuple* $\langle X, U, f, \rho \rangle$ *where $X$ is the finite set of environment states, $U$ is the finite set of agent actions, $f : X \times U \times X \to [0,1]$ is the state transition probability function, and $\rho : X \times U \times X \to \mathbb{R}$ is the reward function.*[1]

The state $x_k \in X$ describes the environment at each discrete time step $k$. The agent observes the state and takes an action $u_k \in U$. As a result, the environment changes its state to some $x_{k+1} \in X$ according to the transition probabilities given by $f$: the probability of ending up in $x_{k+1}$ after $u_k$ is executed in $x_k$ is $f(x_k, u_k, x_{k+1})$. The agent receives a scalar reward $r_{k+1} \in \mathbb{R}$, according to the reward function $\rho$: $r_{k+1} = \rho(x_k, u_k, x_{k+1})$. This reward evaluates the immediate effect of action $u_k$, i.e., the transition from $x_k$ to $x_{k+1}$. It says, however, nothing directly about the long-term effects of this action. We assume that the reward function is bounded.

For deterministic systems, the transition probability function $f$ is replaced by a simpler transition function, $\bar{f} : X \times U \to X$. It follows that the reward is completely determined by the current state and action: $r_{k+1} = \bar{\rho}(x_k, u_k)$, $\bar{\rho} : X \times U \to \mathbb{R}$. Some Markov decision processes have terminal states, i.e., states that once reached, can no longer be left; all the rewards received from a terminal state are 0. In such a case, the learning process is usually separated in distinct *trials* (episodes), which are trajectories starting from some initial state and ending in a terminal state.

The behavior of the agent is described by its policy, which specifies how the agent chooses its actions given the state. The policy may be either stochastic, $h : X \times U \to [0,1]$, or deterministic, $\bar{h} : X \to U$. A policy is called stationary if it does not change over time. The agent's goal is to find a policy that maximizes, from every state $x$, the expected discounted return:

---

[1] Throughout the chapter, the standard control-theoretic notation is used: $x$ for state, $X$ for state space, $u$ for control action, $U$ for action space, $f$ for environment (process) dynamics. We denote reward functions by $\rho$, to distinguish them from the instantaneous rewards $r$ and the returns $R$. We denote agent policies by $h$.

$$R^h(x) = E\left\{\sum_{k=0}^{\infty} \gamma^k r_{k+1} \,\Big|\, x_0 = x, h\right\} \tag{1}$$

where $\gamma \in [0,1)$ is the discount factor, and the expectation is taken over the probabilistic state transitions under the policy $h$. The return $R$ compactly represents the reward accumulated by the agent in the long run. Other possibilities of defining the return exist [58]. The discount factor $\gamma$ can be regarded as encoding an increasing uncertainty about rewards that will be received in the future, or as a means to bound the sum which otherwise might grow unbounded.

The task of the agent is therefore to maximize its long-term performance (return), while only receiving feedback about its immediate, one-step performance (reward). One way it can achieve this is by computing the optimal state-action value function ($Q$-function). The Q-function $Q^h : X \times U \rightarrow \mathbb{R}$ gives the expected return obtained by the policy $h$ from any state-action pair:

$$Q^h(x,u) = E\left\{\sum_{k=0}^{\infty} \gamma^k r_{k+1} \,\Big|\, x_0 = x, u_0 = u, h\right\}$$

The optimal Q-function is defined as $Q^*(x,u) = \max_h Q^h(x,u)$. It satisfies the Bellman optimality equation:

$$Q^*(x,u) = \sum_{x' \in X} f(x,u,x')\left[\rho(x,u,x') + \gamma \max_{u'} Q^*(x',u')\right] \quad \forall x \in X, u \in U \tag{2}$$

This equation states that the optimal value of taking $u$ in $x$ is the expected immediate reward plus the expected (discounted) optimal value attainable from the next state (the expectation is explicitly written as a sum since $X$ is finite).

Once $Q^*$ is available, an optimal policy (i.e., one that maximizes the return) can be computed by choosing in every state an action with the largest optimal Q-value:

$$\bar{h}^*(x) = \arg\max_u Q^*(x,u) \tag{3}$$

When multiple actions attain the largest Q-value, any of them can be chosen and the policy remains optimal. In such a case, here as well as in the sequel, the 'arg' operator is interpreted as returning only one of the equally good solutions. A policy that maximizes a Q-function in this way is said to be greedy in that Q-function. So, an optimal policy can be found by first determining $Q^*$ and then computing a greedy policy in $Q^*$.

A broad spectrum of single-agent RL algorithms exists, e.g., model-free methods based on the online estimation of value functions [6, 89, 118, 120, 137], model-based methods (typically called dynamic programming) [8, 96], and model-learning methods that estimate a model, and then learn using model-based techniques [79, 119]. The model comprises the transition probabilities and the reward function. Many MARL algorithms are derived from a model-free algorithm called $Q$-learning[2] [137], see e.g., [17, 42, 49, 67, 69, 70].

---

[2] Note that algorithm names are shown in italics throughout the chapter, e.g., $Q$-learning.

*Q-learning* [137] turns (2) into an iterative approximation procedure. *Q-learning* starts with an arbitrary Q-function, observes transitions $(x_k, u_k, x_{k+1}, r_{k+1})$, and after each transition updates the Q-function with:

$$Q_{k+1}(x_k, u_k) = Q_k(x_k, u_k) + \alpha_k \left[ r_{k+1} + \gamma \max_{u'} Q_k(x_{k+1}, u') - Q_k(x_k, u_k) \right] \qquad (4)$$

The term between square brackets is the temporal difference, i.e., the difference between the current estimate $Q_k(x_k, u_k)$ of the optimal Q-value of $(x_k, u_k)$ and the updated estimate $r_{k+1} + \gamma \max_{u'} Q_k(x_{k+1}, u')$. This new estimate is a sample of the right-hand side of the Bellman equation (2), applied to $Q_k$ in the state-action pair $(x_k, u_k)$. In this sample, $x'$ is replaced by the observed next state $x_{k+1}$, and $\rho(x_k, u_k, x')$ by the observed reward $r_{k+1}$. The learning rate $\alpha_k \in (0, 1]$ can be time-varying, and usually decreases with time.

The sequence $Q_k$ provably converges to $Q^*$ under the following conditions [53, 129, 137]:

- Explicit, distinct values of the Q-function are stored and updated for each state-action pair.
- The sum $\sum_{k=0}^{\infty} \alpha_k^2$ is finite, while $\sum_{k=0}^{\infty} \alpha_k$ is infinite.
- Asymptotically, all the state-action pairs are visited infinitely often.

The third requirement can be satisfied if, among others, the agent keeps trying all the actions in all the states with nonzero probabilities. This is called exploration, and can be done e.g., by choosing at each step a random action with probability $\varepsilon \in (0, 1)$, and a greedy action with probability $(1 - \varepsilon)$. The $\varepsilon$-greedy exploration procedure is obtained. The probability $\varepsilon$ is usually decreased over time. Another option is to use the Boltzmann exploration procedure, which in state $x$ selects action $u$ with probability:

$$h(x, u) = \frac{e^{Q(x,u)/\tau}}{\sum_{\bar{u}} e^{Q(x,\bar{u})/\tau}} \qquad (5)$$

where $\tau > 0$, the temperature, controls the randomness of the exploration. When $\tau \to 0$, (5) becomes equivalent with greedy action selection (3). When $\tau \to \infty$, action selection is purely random. For $\tau \in (0, \infty)$, higher-valued actions have a greater chance of being selected than lower-valued ones.

## 2.2   The Multi-agent Case

The generalization of the Markov decision process to the multi-agent case is the *stochastic game*.

**Definition 2.** *A stochastic game is a tuple* $\langle X, U_1, \ldots, U_n, f, \rho_1, \ldots, \rho_n \rangle$ *where n is the number of agents, X is the finite set of environment states,* $U_i$, $i = 1, \ldots, n$ *are the finite sets of actions available to the agents, yielding the joint action set* $\boldsymbol{U} = U_1 \times \cdots \times U_n$, $f : X \times \boldsymbol{U} \times X \to [0, 1]$ *is the state transition probability function, and* $\rho_i : X \times \boldsymbol{U} \times X \to \mathbb{R}$, $i = 1, \ldots, n$ *are the reward functions of the agents.*

We assume that the reward functions are bounded. In the multi-agent case, the state transitions are the result of the joint action of all the agents, $\boldsymbol{u}_k = [u_{1,k}^T, \ldots, u_{n,k}^T]^T$, $\boldsymbol{u}_k \in \boldsymbol{U}$, $u_{i,k} \in U_i$ (where T denotes vector transpose). The policies $h_i : X \times U_i \to [0,1]$ form together the joint policy $\boldsymbol{h}$. Because the rewards $r_{i,k+1}$ of the agents depend on the joint action, their returns depend on the joint policy:

$$R_i^{\boldsymbol{h}}(x) = \mathrm{E}\left\{ \sum_{k=0}^{\infty} \gamma^k r_{i,k+1} \,\middle|\, x_0 = x, \boldsymbol{h} \right\}$$

The Q-function of each agent depends on the joint action and on the joint policy, $Q_i^{\boldsymbol{h}} : X \times \boldsymbol{U} \to \mathbb{R}$, with $Q_i^{\boldsymbol{h}}(x, \boldsymbol{u}) = \mathrm{E}\left\{ \sum_{k=0}^{\infty} \gamma^k r_{i,k+1} \,|\, x_0 = x, \boldsymbol{u}_0 = \boldsymbol{u}, \boldsymbol{h} \right\}$.

In fully cooperative stochastic games, the reward functions are the same for all the agents: $\rho_1 = \cdots = \rho_n$. It follows that the returns are also the same, $R_1^{\boldsymbol{h}} = \cdots = R_1^{\boldsymbol{h}}$, and all the agents have the same goal: to maximize the common return. If $n = 2$ and $\rho_1 = -\rho_2$, the two agents have opposing goals, and the stochastic game is fully competitive.[3] Mixed games are stochastic games that are neither fully cooperative nor fully competitive.

### 2.3   Static, Repeated, and Stage Games

Many MARL algorithms are designed for static (stateless) games, or work in a stage-wise fashion, i.e., separately in the static games that arise in every state of the stochastic game. Next, we introduce some game-theoretic concepts regarding static games that are necessary to understand such algorithms [3, 39].

A *static (stateless) game* is a stochastic game with no state signal and no dynamics, i.e., $X = \emptyset$. A static game is described by a tuple $\langle U_1, \ldots, U_n, \rho_1, \ldots, \rho_n \rangle$, with the rewards depending only on the joint actions $\rho_i : \boldsymbol{U} \to \mathbb{R}$. When there are only two agents, the game is often called a bimatrix game, because the reward functions of each of the two agents can be represented as a $|U_1| \times |U_2|$ matrix with the rows corresponding to the actions of agent 1, and the columns to the actions of agent 2, where $|\cdot|$ denotes set cardinality. Fully competitive static games are also called zero-sum games, because the sum of the agents' reward matrices is a zero matrix. Mixed static games are also called general-sum games, because there is no constraint on the sum of the agents' rewards.

A *stage game* is the static game that arises in a certain state of a stochastic game. The reward functions of the stage game in state $x$ are the Q-functions of the stochastic game projected on the joint action space, when the state is fixed at $x$. In general, the agents visit the same state of a stochastic game multiple times, so the stage game is a *repeated* game. In game theory, a repeated game is a static game played repeatedly by the same agents. The main difference from a one-shot game is that the agents can use some of the game iterations to gather information about the other agents' behavior or about the reward functions, and make more informed decisions thereafter.

In a static or repeated game, the policy loses the state argument and transforms into a strategy $\sigma_i : U_i \to [0,1]$. An agent's strategy for the stage game arising in some state $x$

---

[3] Competition can also arise when more than two agents are involved. However, the literature on RL in fully-competitive games typically deals with the two-agent case only.

of the stochastic game is the projection of its policy $h_i$ on its action space $U_i$, when the state is fixed at $x$. MARL algorithms relying on the stage-wise approach learn strategies separately for every stage game. The agent's overall policy is then the aggregate of these strategies.

An important solution concept for static games is the Nash equilibrium. First, define the best response of agent $i$ to a vector of opponent strategies as the strategy $\sigma_i^*$ that achieves the maximum expected reward given these opponent strategies:

$$\mathrm{E}\left\{r_i \mid \sigma_1, \ldots, \sigma_i, \ldots, \sigma_n\right\} \leq \mathrm{E}\left\{r_i \mid \sigma_1, \ldots, \sigma_i^*, \ldots, \sigma_n\right\} \quad \forall \sigma_i \tag{6}$$

A Nash equilibrium is a joint strategy $[\sigma_1^*, \ldots, \sigma_n^*]^{\mathrm{T}}$ such that each individual strategy $\sigma_i^*$ is a best-response to the others (see e.g., [3]). The Nash equilibrium describes a *status quo*, from which no agent can benefit by changing its strategy as long as all the other agents keep their strategies constant. Any static game has at least one (possibly stochastic) Nash equilibrium; some static games have multiple Nash equilibria. Many MARL algorithms reviewed in the sequel strive to converge to Nash equilibria.

Stochastic strategies (and consequently, stochastic policies) are of a more immediate importance in MARL than in single-agent RL, because they are necessary to express certain solution concepts, such as the Nash equilibrium described above.

# 3  Benefits and Challenges in Multi-agent Reinforcement Learning

In addition to benefits owing to the distributed nature of the multi-agent solution, such as the speedup made possible by parallel computation, multiple RL agents can harness new benefits from sharing experience, e.g., by communication, teaching, or imitation. Conversely, besides challenges inherited from single-agent RL, including the curse of dimensionality and the exploration-exploitation tradeoff, several new challenges arise in MARL: the difficulty of specifying a learning goal, the nonstationarity of the learning problem, and the need for coordination.

## 3.1  Benefits of MARL

Experience sharing can help RL agents with similar tasks learn faster and reach better performance. For instance, the agents can exchange information using communication [123], skilled agents may serve as teachers for the learner [30], or the learner may watch and imitate the skilled agents [95].

A speed-up can be realized in MARL thanks to parallel computation, when the agents exploit the decentralized structure of the task. This direction has been investigated in e.g., [21, 38, 43, 61, 62].

When one or more agents fail in a multi-agent system, the remaining agents can take over some of their tasks. This implies that MARL is inherently robust. Furthermore, by design most multi-agent systems also allow the easy insertion of new agents into the system, leading to a high degree of scalability.

Existing MARL algorithms often require some additional preconditions to theoretically guarantee and to exploit the above benefits [67, 95]. Relaxing these conditions and further improving the performance of MARL algorithms in this context is an active field of study.

### 3.2  Challenges in MARL

The curse of dimensionality is caused by the exponential growth of the discrete state-action space in the number of state and action variables (dimensions). Because basic RL algorithms like *Q-learning* estimate values for each possible discrete state or state-action pair, this growth leads directly to an exponential increase of their computational complexity. The complexity of MARL is exponential also in the number of agents, because each agent adds its own variables to the joint state-action space. This makes the curse of dimensionality more severe in MARL than in single-agent RL.

Specifying a good MARL goal in the general stochastic game is a difficult challenge, because the agents' returns are correlated and cannot be maximized independently. Several types of MARL goals have been proposed in the literature, which consider stability of the agent's learning dynamics [50], adaptation to the changing behavior of the other agents [93], or both stability and adaptation [14, 16, 17, 26, 70]. A detailed analysis of this open problem is given in Section 4.

Nonstationarity arises in MARL because all the agents in the system are learning simultaneously. Each agent is therefore faced with a moving-target learning problem: the best policy changes as the other agents' policies change.

The exploration-exploitation trade-off requires online (single-agent as well as multi-agent) RL algorithms to strike a balance between the exploitation of the agent's current knowledge, and exploratory, information-gathering actions taken to improve that knowledge. For instance, the Boltzmann policy (5) is a simple way of trading off exploration with exploitation. The exploration procedure is crucial for the efficiency of RL algorithms. In MARL, further complications arise due to the presence of multiple agents. Agents explore to obtain information not only about the environment, but also about the other agents in order to adapt to their behavior. Too much exploration, however, can destabilize the other agents, thereby making the learning task more difficult for the exploring agent.

The need for coordination stems from the fact that the effect of any agent's action on the environment depends also on the actions taken by the other agents. Hence, the agents' choices of actions must be mutually consistent in order to achieve their intended effect. Coordination typically boils down to consistently breaking ties between equally good joint actions or strategies. Although coordination is typically required in cooperative settings, it may also be desirable for self-interested agents, e.g., if the lack of coordination negatively affects all the agents. Consider, as an example, that a number of countries have interconnected electricity networks, and each country's network is managed by an agent. Although each agent's primary goal is to optimize its own country's energy interests, the agents must still coordinate on the power flows between neighboring countries in order to achieve a meaningful solution [84].

## 4  Multi-agent Reinforcement Learning Goal

In fully cooperative stochastic games, the common return can be jointly maximized. In other cases, however, the agents' returns are typically different and correlated, and they cannot be maximized independently. Specifying a good general MARL goal is a difficult problem.

In this section, the learning goals proposed in the literature are reviewed. These goals incorporate the *stability* of the learning dynamics of the agent on the one hand, and the *adaptation* to the changing behavior of the other agents on the other hand. Stability essentially means the convergence to a stationary policy, whereas adaptation ensures that performance is maintained or improved as the other agents are changing their policies.

The goals typically formulate conditions for static games, in terms of strategies and rewards. Some of the goals can be extended to dynamic games by requiring that conditions are satisfied stage-wise for all the states of the dynamic game. In this case, the goals are formulated in terms of stage strategies and expected returns instead of strategies and rewards.

Convergence to equilibria is a basic stability requirement [42, 50]. It means the agents' strategies should eventually converge to a coordinated equilibrium. Nash equilibria are most frequently used. However, concerns have been raised regarding their usefulness [108]. For instance, one objection is that the link between stage-wise convergence to Nash equilibria and performance in the dynamic stochastic game is unclear.

In [16, 17], convergence is required for stability, and rationality is added as an adaptation criterion. For an algorithm to be convergent, the authors of [16, 17] require that the learner converges to a stationary strategy, given that the other agents use an algorithm from a predefined, targeted class of algorithms. Rationality is defined in [16, 17] as the requirement that the agent converges to a best response when the other agents remain stationary. Though convergence to a Nash equilibrium is not explicitly required, it arises naturally if all the agents in the system are rational and convergent.

An alternative to rationality is the concept of no-regret, which is defined in [14] as the requirement that the agent achieves a return that is at least as good as the return of any stationary strategy, and this holds for any set of strategies of the other agents. This requirement prevents the learner from 'being exploited' by the other agents. Note that for certain types of static games, no-regret learning algorithms converge to Nash equilibria [54, 143].

Targeted optimality/compatibility/safety are adaptation requirements expressed in the form of bounds on the average reward [93]. Targeted optimality demands an average reward, against a targeted set of algorithms, which is at least the average reward of a best-response. Compatibility prescribes an average reward level in self-play, i.e., when the other agents use the learner's algorithm. Safety demands a safety-level average reward against all other algorithms. An algorithm satisfying these requirements does not necessarily converge to a stationary strategy.

Other properties of (but not necessarily requirements on) MARL algorithms can also be related to stability and adaptation. For instance, opponent-independent learning is related to stability, whereas opponent-aware learning is related to adaptation [15, 70]. An opponent-independent algorithm converges to a strategy that is part of an equilibrium solution regardless of what the other agents are doing. An opponent-aware algorithm learns models of the other agents and reacts to them using some form of best-response. Prediction and rationality as defined in [26] are related to stability and adaptation, respectively. Prediction is the agent's capability to learn accurate models of the other agents. An agent is called rational in [26] if it maximizes its expected return given its models of the other agents.

Table 1 summarizes these requirements and properties of MARL algorithms. The names under which the authors refer to the stability and adaptation properties are given in the first two columns. Pointers to some relevant literature are provided in the last column.

**Table 1.** Stability and adaptation in MARL. Reproduced from [20], © 2008 IEEE.

| Stability property | Adaptation property | Some relevant work |
|---|---|---|
| convergence | rationality | [17, 31] |
| convergence | no-regret | [14] |
| — | targeted optimality, compatibility, safety | [93, 108] |
| opponent-independent | opponent-aware | [15, 70] |
| equilibrium learning | best-response learning | [13] |
| prediction | rationality | [26] |

**Remarks and Open Issues**

Stability of the learning process is needed, because the behavior of stable agents is more amenable to analysis and meaningful performance guarantees. Moreover, a stable agent reduces the nonstationarity in the learning problem of the other agents, making it easier to solve. Adaptation to the other agents is needed because their behavior is generally unpredictable. Therefore, a good MARL goal must include both components. Since 'perfect' stability and adaptation cannot be achieved simultaneously, an algorithm should guarantee bounds on both stability and adaptation measures. From a practical viewpoint, a realistic learning goal should also include bounds on the transient performance, in addition to the usual asymptotic requirements.

Convergence and rationality have been used in dynamic games in the stage-wise fashion already explained [16, 17]. No-regret has not been used in dynamic games, but it could be extended in a similar way. It is unclear how targeted optimality, compatibility, and safety could be extended.

## 5    Multi-agent Reinforcement Learning Algorithms

This section first provides a taxonomy of MARL algorithms, followed by a detailed review of a representative selection of algorithms.

MARL algorithms can be classified along several dimensions, among which some, such as the task type, stem from properties of multi-agent systems in general. Others, like the awareness of the other agents, are specific to learning multi-agent systems. The proposed classifications are illustrated using the set of algorithms selected for review. All these algorithms will be discussed separately in Sections 5.1–5.4.

The type of task considered by the learning algorithm leads to a corresponding classification of MARL techniques into those addressing fully cooperative, fully competitive, or mixed stochastic games. A significant number of algorithms are designed for static (stateless) tasks only. Figure 1 summarizes the breakdown of MARL algorithms by task type.

| Fully cooperative | |
|---|---|
| **Static** | **Dynamic** |
| *JAL* [29] <br> *FMQ* [59] | *Team-Q* [70] <br> *Distributed-Q* [67] <br> *OAL* [136] |

| Fully competitive |
|---|
| *Minimax-Q* [69] |

| Mixed | |
|---|---|
| **Static** | **Dynamic** |
| *Fictitious Play* [19] <br> *MetaStrategy* [93] <br> *IGA* [109] <br> *WoLF-IGA* [17] <br> *GIGA* [144] <br> *GIGA-WoLF* [14] <br> *AWESOME* [31] <br> *Hyper-Q* [124] | *Single-agent RL* [32, 75, 105] <br> *Nash-Q* [49] <br> *CE-Q* [42] <br> *Asymmetric-Q* [64] <br> *NSCP* [138] <br> *WoLF-PHC* [17] <br> *PD-WoLF* [5] <br> *EXORL* [116] |

**Fig. 1.** Breakdown of MARL algorithms by the type of task they address. Reproduced from [20], © 2008 IEEE.

The degree of awareness of other learning agents exhibited by MARL algorithms is strongly related to the learning goal that the agents aim for. Algorithms focused on stability (convergence) only are typically unaware and *independent* of the other learning agents. Algorithms that consider adaptation to the other agents clearly need to be *aware* to some extent of their behavior. If adaptation is taken to the extreme and stability concerns are disregarded, algorithms are only *tracking* the behavior of the other agents. The degree of agent awareness exhibited by the algorithms can be determined even if they do not explicitly target stability or adaptation goals. All agent-tracking algorithms and many agent-aware algorithms use some form of opponent modeling to keep track of the other agents' policies [25, 49, 133].

The field of origin of the algorithms is a taxonomy axis that shows the variety of research inspiration contributing to MARL. MARL can be regarded as a fusion of temporal-difference RL (especially *Q-learning*), game theory, and more general direct policy search techniques. Figure 2 presents the organization of the MARL algorithms considered by their field of origin.

Other classification criteria include the following:

- Homogeneity of the agents' learning algorithms: the algorithm only works if all the agents use it (homogeneous learning agents, e.g., *team-Q*, *Nash-Q*), or other agents can use other learning algorithms (heterogeneous learning agents, e.g., *AWESOME*, *WoLF-PHC*).
- Assumptions on the agent's prior knowledge about the task: a task model is available to the learning agent (model-based learning, e.g., *AWESOME*) or not (model-free learning, e.g., *team-Q*, *Nash-Q*, *WoLF-PHC*). The model consists of the transition function (unless the game is static) and of the reward functions of the agents.

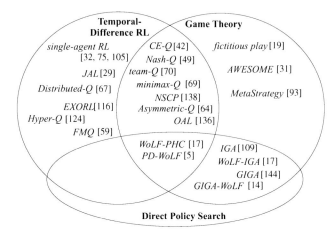

**Fig. 2.** MARL encompasses temporal-difference reinforcement learning, game theory and direct policy search techniques. Reproduced from [20], © 2008 IEEE.

- Assumptions on the agent's inputs. Typically the inputs are assumed to exactly represent the state of the environment. Differences appear in the agent's observations of other agents: it might need to observe the actions of the other agents (e.g., *team-Q*, *AWESOME*), their actions and rewards (e.g., *Nash-Q*), or neither (e.g., *WoLF-PHC*).

The remainder of this section discusses in detail the MARL algorithms selected for review. The algorithms are grouped first by the type of task they address, and then by the degree of agent awareness, as depicted in Table 2. So, algorithms for fully cooperative tasks are presented first, in Section 5.1. Explicit coordination techniques that can be applied to algorithms from any class are discussed separately in Section 5.2. Algorithms for fully competitive tasks are reviewed in Section 5.3. Finally, Section 5.4 presents algorithms for mixed tasks. Algorithms that are designed only for static tasks are discussed in separate paragraphs in the text. Simple examples are provided to illustrate several central issues that arise.

**Table 2.** Breakdown of MARL algorithms by task type and degree of agent awareness. Reproduced from [20], © 2008 IEEE.

| Task type → <br> ↓ Agent awareness | Cooperative | Competitive | Mixed |
|---|---|---|---|
| Independent | coordination-free | opponent-independent | agent-independent |
| Tracking | coordination-based | — | agent-tracking |
| Aware | indirect coordination | opponent-aware | agent-aware |

## 5.1   Fully Cooperative Tasks

In a fully cooperative stochastic game, the agents have the same reward function ($\rho_1 = \cdots = \rho_n$) and the learning goal is to maximize the common discounted return. If a centralized controller were available, the task would reduce to a Markov decision process, the action space of which would be the joint action space of the stochastic game. In this case, the goal could be achieved e.g., by learning the optimal joint-action values with *Q-learning*:

$$Q_{k+1}(x_k, \boldsymbol{u}_k) = Q_k(x_k, \boldsymbol{u}_k) + \alpha \left[ r_{k+1} + \gamma \max_{\boldsymbol{u}'} Q_k(x_{k+1}, \boldsymbol{u}') - Q_k(x_k, \boldsymbol{u}_k) \right] \qquad (7)$$

and then using a greedy policy. However, the agents are independent decision makers, and a coordination problem arises even if all the agents learn in parallel the common optimal Q-function using (7). It may seem that the agents could use greedy policies applied to $Q^*$ to maximize the common return:

$$\overline{h}_i^*(x) = \arg\max_{u_i} \max_{u_1,\dots,u_{i-1},u_{i+1},\dots,u_n} Q^*(x, \boldsymbol{u}) \qquad (8)$$

However, in certain states, multiple joint actions may be optimal. In the absence of additional coordination mechanisms, different agents may break these ties among multiple optimal joint actions in different ways, and the resulting joint action may be suboptimal.

*Example 1. The need for coordination.* Consider the situation illustrated in Figure 3: two mobile agents need to avoid an obstacle while maintaining formation (i.e., maintaining their relative positions). Each agent *i* has three available actions: go straight ($S_i$), left ($R_i$), or right ($L_i$).

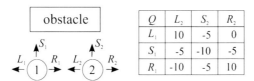

| $Q$ | $L_2$ | $S_2$ | $R_2$ |
|-----|-----|-----|-----|
| $L_1$ | 10 | -5 | 0 |
| $S_1$ | -5 | -10 | -5 |
| $R_1$ | -10 | -5 | 10 |

**Fig. 3.** Left: two mobile agents approaching an obstacle need to coordinate their action selection. Right: the common Q-values of the agents for the state depicted to the left. Reproduced from [20], © 2008 IEEE.

For a given state (position of the agents), the Q-function can be projected on the joint action space. For the state represented in Figure 3 (left), a possible projection is represented in the table on the right. This table describes a fully cooperative static (stage) game. The rows correspond to the actions of agent 1, and the columns to the actions of agent 2. If both agents go left, or both go right, the obstacle is avoided while maintaining the formation: $Q(L_1, L_2) = Q(R_1, R_2) = 10$. If agent 1 goes left, and agent 2 goes right, the formation is broken: $Q(L_1, R_2) = 0$. In all other cases, collisions occur and the Q-values are negative.

Note the tie between the two optimal joint actions: $(L_1, L_2)$ and $(R_1, R_2)$. Without a coordination mechanism, agent 1 might assume that agent 2 will take action $R_2$, and therefore it takes action $R_1$. Similarly, agent 2 might assume that agent 1 will take $L_1$, and consequently takes $L_2$. The resulting joint action $(R_1, L_2)$ is severely suboptimal, as the agents collide.

## Coordination-Free Methods

The *Team Q-learning* algorithm [70] avoids the coordination problem by assuming that the optimal joint actions are unique (which is not always the case). Then, if all the agents learn the common Q-function in parallel with (7), they can safely use (8) to select these optimal joint actions and maximize their return.

The *Distributed Q-learning* algorithm [67] solves the cooperative task without assuming coordination and with limited computation (its computational complexity is similar to that of single-agent *Q-learning*, see Section 5.4). However, the algorithm only works in deterministic problems with non-negative reward functions. Each agent $i$ maintains a local policy $\bar{h}_i(x)$, and a local Q-function $Q_i(x, u_i)$, depending only on its own action. The local Q-values are updated only when the update leads to an increase in the Q-value:

$$Q_{i,k+1}(x_k, u_{i,k}) = \max \left\{ Q_{i,k}(x_k, u_{i,k}), r_{k+1} + \gamma \max_{u_i'} Q_{i,k}(x_{k+1}, u_i') \right\} \qquad (9)$$

This ensures that the local Q-value always captures the maximum of the joint-action Q-values: $Q_{i,k}(x, u_i) = \max\limits_{u_1, \ldots, u_{i-1}, u_{i+1}, \ldots, u_n} Q_k(x, \boldsymbol{u})$ at all $k$, where $\boldsymbol{u} = [u_1, \ldots, u_n]^{\mathrm{T}}$ with $u_i$ fixed. The local policy is updated only if the update leads to an improvement in the Q-values:

$$\bar{h}_{i,k+1}(x_k) = \begin{cases} u_{i,k} & \text{if } \max_{\bar{u}_i} Q_{i,k+1}(x_k, \bar{u}_i) > \max_{\bar{u}_i} Q_{i,k}(x_k, \bar{u}_i) \\ \bar{h}_{i,k}(x_k) & \text{otherwise} \end{cases} \qquad (10)$$

This ensures that the joint policy $[\bar{h}_{1,k}, \ldots, \bar{h}_{n,k}]^{\mathrm{T}}$ is always optimal with respect to the global $Q_k$. Under the condition that $Q_{i,0} = 0 \; \forall i$, the local policies of the agents provably converge to an optimal joint policy.

## Coordination-Based Methods

Coordination graphs [43] simplify coordination when the global Q-function can be additively decomposed into local Q-functions that only depend on the actions of subsets of agents. For instance, in a stochastic game with 4 agents, the decomposition might be $Q(x, \boldsymbol{u}) = Q_1(x, u_1, u_2) + Q_2(x, u_1, u_3) + Q_3(x, u_3, u_4)$. The decomposition might be different for different states. The local Q-functions have smaller dimensions than the global Q-function. The maximization of the joint Q-value is done by solving simpler, local maximizations in terms of the local Q-functions, and aggregating their solutions. Under certain conditions, the coordinated selection of an optimal joint action is guaranteed [43, 61].

In general, all the coordination techniques described in Section 5.2 below can be applied to fully cooperative MARL tasks. For instance, a framework to explicitly reason about possibly costly communication is the communicative multi-agent team decision problem [97].

### Indirect Coordination Methods

Indirect coordination methods bias action selection toward actions that are likely to result in good rewards or returns. This steers the agents toward coordinated action selections. The likelihood of obtaining good rewards (returns) is evaluated using e.g., models of the other agents estimated by the learner, or statistics of the rewards observed in the past.

**Static tasks.** *Joint Action Learners (JAL)* learn joint-action values and empirical models of the other agents' strategies [29]. Agent $i$ learns models for all the other agents $j \neq i$, using:

$$\widehat{\sigma}_j^i(u_j) = \frac{C_j^i(u_j)}{\sum_{\bar{u}_j \in U_j} C_j^i(\bar{u}_j)} \tag{11}$$

where $\widehat{\sigma}_j^i$ is agent $i$'s model of agent $j$'s strategy and $C_j^i(u_j)$ counts the number of times agent $i$ observed agent $j$ taking action $u_j$. Note that agent $i$ has to observe the actions taken by the other agents. Several heuristics are proposed in [29] to increase the learner's Q-values for the actions with high likelihood of getting good rewards given the models.

The *Frequency Maximum Q-value (FMQ)* heuristic is based on the frequency with which actions yielded good rewards in the past [59]. Agent $i$ uses Boltzmann action selection (5), plugging in modified Q-values $\bar{Q}_i$ computed with the formula:

$$\bar{Q}_i(u_i) = Q_i(u_i) + v \frac{C_{\max}^i(u_i)}{C^i(u_i)} r_{\max}(u_i) \tag{12}$$

where $r_{\max}(u_i)$ is the maximum reward observed after taking action $u_i$, $C_{\max}^i(u_i)$ counts how many times this reward has been observed, $C^i(u_i)$ counts how many times $u_i$ has been taken, and $v$ is a weighting factor. Increasing the Q-values of actions that frequently produced good rewards in the past steers the agent toward coordination. Compared to single-agent *Q-learning*, the only additional computational demands of *FMQ* come from maintaining and using the counters. However, *FMQ* can fail in some problems with strongly stochastic rewards [59], and the weighting parameter $v$ must be tuned in a problem-specific fashion, which may be difficult to do.

**Dynamic tasks.** In *Optimal Adaptive Learning (OAL)*, virtual games are constructed on top of each stage game of the stochastic game [136]. In these virtual games, optimal joint actions are rewarded with 1, and the rest of the joint actions with 0. An algorithm is introduced that, by biasing the agent towards recently selected optimal actions, guarantees convergence to a coordinated optimal joint action for the virtual game, and therefore to a coordinated joint action for the original stage game. Thus, *OAL* provably converges to optimal joint policies in any fully cooperative stochastic game. This however comes at the cost of increased complexity: each agent estimates empirically a model of the stochastic game, virtual games for each stage game, models of the other agents, and an optimal value function for the stochastic game.

### Remarks and Open Issues

All the methods presented above rely on exact measurements of the state. Some of them also require exact measurements of the other agents' actions. This is most obvious for coordination-free methods: if at any point the perceptions of the agents differ, this may lead different agents to update their Q-functions differently, and the consistency of the Q-functions and policies can no longer be guaranteed.

The algorithms discussed above also suffer from the curse of dimensionality. *Distributed Q-learning* and *FMQ* are exceptions in the sense that they do not need to take into account the other agents' actions (but they only work in restricted settings: *Distributed Q-learning* only in deterministic tasks, and *FMQ* only in static tasks).

### 5.2   Explicit Coordination Mechanisms

A general approach to solving the coordination problem is to make sure that any ties are broken by all the agents in the same way, using explicit coordination or negotiation. Mechanisms for doing so based on social conventions, roles, and communication, are described next [135]. These mechanisms can be used for any type of task.

Both social conventions and roles restrict the action choices of the agents. An agent role restricts the set of actions available to that agent prior to action selection, as in e.g., [112]. This means that some or all of the ties in (8) are prevented. Social conventions encode *a priori* preferences toward certain joint actions, and help break ties during action selection. If properly designed, roles or social conventions eliminate ties completely. A simple social convention relies on a unique ordering of the agents and actions [11]. These two orderings must be known to all the agents. Combining them leads to a unique ordering of the joint actions, and coordination is ensured if in (8) the first joint action in this ordering is selected by all the agents.

Communication can be used to negotiate action choices, either alone or in combination with the above coordination techniques, as in [37, 135]. When combined with the above techniques, communication can relax their assumptions and simplify their application. For instance, in social conventions, if only an ordering between agents is known, they can select actions in turn, in that order, and broadcast their selection to the remaining agents. This is sufficient to ensure coordination.

Besides action choices, agents can also communicate various other types of information, including partial or complete Q-tables, state measurements, rewards, learning parameters, etc. For example, the requirements of complete and consistent perception among all the agents (discussed under Remarks in Section 5.1) can be relaxed by allowing agents to communicate interesting data (e.g., partial state measurements) instead of relying on direct measurement [123].

Learning coordination approaches have also been investigated, where the coordination mechanism is learned, rather than being hardwired into the agents. The agents learn social conventions in [11], role assignments in [81], and the structure of the coordination graph (see Section 5.1) together with the local Q-functions in [60].

*Example 2. Coordination using social conventions in a fully-cooperative task.* In Example 1 above (see Figure 3), suppose the agents are ordered such that agent 1 < agent 2 ($a < b$ means that $a$ precedes $b$ in the chosen ordering), and the actions of both agents are ordered in the following way: $L_i < R_i < S_i$, $i \in \{1,2\}$. To coordinate, the first agent in the ordering of the agents, agent 1, looks for an optimal joint action such that its action component is the first in the ordering of its actions: $(L_1, L_2)$. It then selects its component of this joint action, $L_1$. As agent 2 knows the orderings, it can infer this decision, and appropriately selects $L_2$ in response. If agent 2 would still face a tie (e.g., if $(L_1, L_2)$ and $(L_1, S_2)$ would both be optimal), it could break this tie by using the ordering of its own actions (which because $L_2 < S_2$ would also yield $(L_1, L_2)$).

If communication is available, only the ordering of the agents has to be known. Agent 1, the first in the ordering, chooses an action by breaking ties in some way between the optimal joint actions. Suppose it settles on $(R_1, R_2)$, and therefore selects $R_1$. It then communicates this selection to agent 2, which can then select an appropriate response, namely the action $R_2$.

### 5.3  Fully Competitive Tasks

In a fully competitive stochastic game (for two agents, when $\rho_1 = -\rho_2$), the minimax principle can be applied: maximize one's benefit under the worst-case assumption that the opponent will always endeavor to minimize it. This principle suggests using opponent-independent algorithms.

The *minimax-Q* algorithm [69, 70] employs the minimax principle to compute strategies and values for the stage games, and a temporal-difference rule similar to *Q-learning* to propagate the values across state transitions. The algorithm is given here for agent 1:

$$h_{1,k}(x_k, \cdot) = \arg \mathbf{m}_1(Q_k, x_k) \tag{13}$$

$$Q_{k+1}(x_k, u_{1,k}, u_{2,k}) = Q_k(x_k, u_{1,k}, u_{2,k}) + \alpha \left[ r_{k+1} + \gamma \mathbf{m}_1(Q_k, x_{k+1}) - Q_k(x_k, u_{1,k}, u_{2,k}) \right] \tag{14}$$

where $\mathbf{m}_1$ is the minimax return of agent 1:

$$\mathbf{m}_1(Q, x) = \max_{h_1(x, \cdot)} \min_{u_2} \sum_{u_1} h_1(x, u_1) Q(x, u_1, u_2) \tag{15}$$

The stochastic strategy of agent 1 in state $x$ at time $k$ is denoted by $h_{1,k}(x, \cdot)$, with the dot standing for the action argument. The optimization problem in (15) can be solved by linear programming [82]. The Q-table is not subscripted by the agent index, because the equations make the implicit assumption that $Q = Q_1 = -Q_2$; this follows from $\rho_1 = -\rho_2$.

*Minimax-Q* is truly opponent-independent, because even if the minimax optimization has multiple solutions (strategies), any of them will achieve at least the minimax return regardless of what the opponent is doing. However, if the opponent is suboptimal (i.e., does not always take the action that is the worst for the learner), and the learner has a model of the opponent's policy, it can actually do better than the minimax return (15). An opponent model can be learned using e.g., the $M^*$ algorithm described in [25], or a simple extension of (11) to multiple states:

$$\widehat{h}^i_j(x,u_j) = \frac{C^i_j(x,u_j)}{\sum_{\bar{u}_j \in U_j} C^i_j(x,\bar{u}_j)} \tag{16}$$

where $C^i_j(x,u_j)$ counts the number of times agent $i$ observed agent $j$ taking action $u_j$ in state $x$.

Such an algorithm then becomes opponent-aware. Even agent-aware algorithms for mixed tasks (see Section 5.4) can be used to exploit a suboptimal opponent. For instance, in [17] *WoLF-PHC* was used with promising results in a fully competitive task.

*Example 3. The minimax principle.* Consider the situation represented in the left part of Figure 4: agent 1 has to reach the goal in the middle while still avoiding capture by its opponent, agent 2. Agent 2 on the other hand, has to prevent agent 1 from reaching the goal, preferably by capturing it. The agents can only move to the left or to the right.

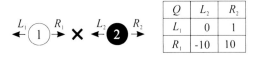

**Fig. 4.** Left: an agent ($\circ$) attempting to reach a goal ($\times$) while avoiding capture by another agent ($\bullet$). Right: the Q-values of agent 1 for the state depicted to the left ($Q_2 = -Q_1$). Reproduced from [20], © 2008 IEEE.

For this situation (state), a possible projection of agent 1's Q-function on the joint action space is given in the table on the right. This represents a zero-sum static game involving the two agents. If agent 1 moves left and agent 2 does likewise, agent 1 escapes capture, $Q_1(L_1,L_2) = 0$; furthermore, if at the same time agent 2 moves right, the chances of capture decrease, $Q_1(L_1,R_2) = 1$. If agent 1 moves right and agent 2 moves left, agent 1 is captured, $Q_1(R_1,L_2) = -10$; however, if agent 2 happens to move right, agent 1 achieves the goal, $Q_1(R_1,R_2) = 10$. As agent 2's interests are opposite to those of agent 1, the Q-function of agent 2 is $-Q_1$. For instance, when both agents move right, agent 1 reaches the goal and agent 2 is punished with a Q-value of $-10$.

The minimax solution for agent 1 in this case is to move left, because for $L_1$, regardless of what agent 2 is doing, it can expect a return of at least 0, as opposed to $-10$ for $R_1$. Indeed, if agent 2 plays well, it will move left to protect the goal. However, it might *not* play well and move right instead. If this is true and agent 1 can find it out (e.g., by learning a model of agent 2) it can take advantage of this knowledge by moving right and achieving the goal.

## 5.4   Mixed Tasks

In mixed stochastic games, no constraints are imposed on the reward functions of the agents. This model is most appropriate for self-interested (but not necessarily competing) agents. The influence of game-theoretic equilibrium concepts is the strongest in MARL algorithms for mixed stochastic games. When multiple equilibria exist in a particular state of a stochastic game, the equilibrium selection problem arises: the agents need to consistently pick their part of the same equilibrium.

A significant number of algorithms in this category are designed only for static tasks (i.e., repeated, general-sum games). Even in repeated games, the learning problem is still nonstationary due to the dynamic behavior of the agents playing the repeated game. This is why most of the methods in this category focus on adaptation to the other agents.

Besides agent-independent, agent-tracking, and agent-aware techniques, the application of single-agent RL methods to multi-agent learning is also presented here. That is because single-agent RL methods do not make any assumption on the type of task, and are therefore applicable to general stochastic games, although without guarantees of success.

### Single-Agent RL

Single-agent RL algorithms like *Q-learning* can be directly applied to the multi-agent case [105]. They learn Q-functions that only depend on the current agent's action, using the basic *Q-learning* update (4), and without being aware of the other agents. The nonstationarity of the MARL problem invalidates most of the single-agent RL theoretical guarantees. Despite its limitations, this approach has found a significant number of applications, mainly because of its simplicity [32, 73, 74, 75].

One important step forward in understanding how single-agent RL works in multi-agent tasks was made in [130]. The authors of [130] applied results in evolutionary game theory to analyze the learning dynamics of *Q-learning* with Boltzmann policies (5) in repeated games. It appeared that for certain parameter settings, *Q-learning* is able to converge to a coordinated equilibrium in particular games. In other cases, unfortunately, Q-learners exhibit non-stationary cyclic behavior.

### Agent-Independent Methods

Many algorithms that are independent of the other agents share a common structure based on *Q-learning*, where policies and state values are computed with game-theoretic solvers for the stage games arising in the states of the stochastic game [13, 42]. This structure is similar to that of (13)–(14); the difference is that for mixed games, solvers are usually different from minimax.

Denoting by $\{Q_{\cdot,k}(x,\cdot)\}$ the stage game arising in state $x$ and given by all the agents' Q-functions at time $k$, learning takes place according to:

$$h_{i,k}(x,\cdot) = \textbf{solve}_i\{Q_{\cdot,k}(x_k,\cdot)\} \tag{17}$$

$$Q_{i,k+1}(x_k,\boldsymbol{u}_k) = Q_{i,k}(x_k,\boldsymbol{u}_k) + \alpha\left[r_{i,k+1} + \gamma\cdot\textbf{eval}_i\{Q_{\cdot,k}(x_{k+1},\cdot)\} - Q_{i,k}(x_k,\boldsymbol{u}_k)\right] \tag{18}$$

where $\textbf{solve}_i$ returns agent $i$'s part of some type of equilibrium (a strategy), and $\textbf{eval}_i$ gives the agent's expected return given this equilibrium. The goal is to converge to an equilibrium in every state.

The updates use the Q-tables of all the agents. So, each agent needs to replicate the Q-tables of the other agents. It can do that by applying (18). This requires that all the agents use the same algorithm and can measure all the actions and rewards. Even under these assumptions, the updates (18) are only guaranteed to maintain identical results for all the agents if **solve** returns consistent equilibrium strategies for all the agents. This means the equilibrium selection problem arises when the solution of **solve** is not unique.

A particular instance of **solve** and **eval** for *Nash Q-learning* [49, 50] is:

$$\begin{cases} \mathbf{eval}_i \left\{ Q_{.,k}(x,\cdot) \right\} = V_i(x, \mathbf{NE} \left\{ Q_{.,k}(x,\cdot) \right\}) \\ \mathbf{solve}_i \left\{ Q_{.,k}(x,\cdot) \right\} = \mathbf{NE}_i \left\{ Q_{.,k}(x,\cdot) \right\} \end{cases} \tag{19}$$

where **NE** computes a Nash equilibrium (a set of strategies), $\mathbf{NE}_i$ is agent $i$'s strategy component of this equilibrium, and $V_i(x, \mathbf{NE} \left\{ Q_{.,k}(x,\cdot) \right\})$ is the expected return for agent $i$ from $x$ under this equilibrium. The algorithm provably converges to Nash equilibria for all the states if either: (a) every stage game encountered by the agents during learning has a Nash equilibrium under which the expected return of all the agents is maximal; or (b) every stage game has a Nash equilibrium that is a saddle point, i.e., not only does the learner not benefit from deviating from this equilibrium, but the other agents do benefit from this [12, 49]. This requirement is satisfied only in a small class of problems. In all other cases, some external mechanism for equilibrium selection is needed to guarantee convergence.

Instantiations for *correlated equilibrium Q-learning (CE-Q)* [42] or *asymmetric Q-learning* [64] can be performed in a similar fashion, by using correlated or Stackelberg (leader-follower) equilibria, respectively. For *asymmetric Q-learning*, the follower does not need to model the leader's Q-table; however, the leader must know how the follower chooses its actions.

*Example 4. The equilibrium selection problem.* Consider the situation illustrated in Figure 5, left: two cleaning robots (the agents) have arrived at a junction in a building, and each needs to decide which of the two wings of the building it will clean. It is inefficient if both agents clean the same wing, and both agents prefer to clean the left wing because it is smaller, and therefore requires less time and energy.

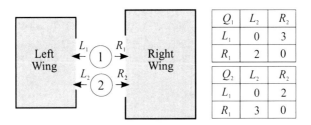

| $Q_1$ | $L_2$ | $R_2$ |
|-------|-------|-------|
| $L_1$ | 0 | 3 |
| $R_1$ | 2 | 0 |

| $Q_2$ | $L_2$ | $R_2$ |
|-------|-------|-------|
| $L_1$ | 0 | 2 |
| $R_1$ | 3 | 0 |

**Fig. 5.** Left: two cleaning robots negotiating their assignment to different wings of a building. Both robots prefer to clean the smaller left wing. Right: the Q-values of the two robots for the state depicted to the left. Reproduced from [20], © 2008 IEEE.

For this situation (state), possible projections of the agents' Q-functions on the joint action space are given in the tables on the right. These tables represent a general-sum static game involving the two agents. If both agents choose the same wing, they will not clean the building efficiently, $Q_1(L_1,L_2) = Q_1(R_1,R_2) = Q_2(L_1,L_2) = Q_2(R_1,R_2) = 0$. If agent 1 takes the (preferred) left wing and agent 2 the right wing, $Q_1(L_1,R_2) = 3$, and $Q_2(L_1,R_2) = 2$. If they choose the other way around, $Q_1(R_1,L_2) = 2$, and $Q_2(R_1,L_2) = 3$.

For these returns, there are two deterministic Nash equilibria[4]: $(L_1,R_2)$ and $(R_1,L_2)$. This is easy to see: if either agent unilaterally deviates from these joint actions, it can expect a (bad) return of 0. If the agents break the tie between these two equilibria independently, they might do so inconsistently and arrive at a suboptimal joint action. This is the equilibrium selection problem, corresponding to the coordination problem that arises in fully cooperative tasks. Its solution requires additional coordination mechanisms, e.g., social conventions.

## Agent-Tracking Methods

Agent-tracking algorithms estimate models of the other agents' strategies or policies (depending on whether static or dynamic games are considered) and act using some form of best-response to these models. Convergence to stationary strategies is not a requirement. Each agent is assumed capable to observe the other agents' actions.

**Static tasks.** In the *fictitious play* algorithm [19], agent $i$ acts at each iteration according to a best-response (6) to models $\widehat{\sigma}_1^i, \ldots, \widehat{\sigma}_{i-1}^i, \widehat{\sigma}_{i+1}^i, \ldots, \widehat{\sigma}_n^i$ learned with (11). Fictitious play converges to a Nash equilibrium in certain restricted classes of games, among which are fully cooperative, repeated games [29].

The *MetaStrategy* algorithm, introduced in [93], combines modified versions of fictitious play, minimax and a game-theoretic strategy called Bully [71] to achieve the targeted optimality, compatibility, and safety goals (see Section 4).

To compute best-responses, the *fictitious play* and *MetaStrategy* algorithms require a model of the static task, in the form of reward functions.

The *Hyper-Q* algorithm uses the other agents' models as a state vector and learns a Q-function $Q_i(\widehat{\sigma}_1, \ldots, \widehat{\sigma}_{i-1}, \widehat{\sigma}_{i+1}, \ldots, \widehat{\sigma}_n, u_i)$ with an update rule similar to *Q-learning* [124]. By learning values of strategies instead of only actions, *Hyper-Q* should be able to adapt better to nonstationary agents. One inherent difficulty is that the action selection probabilities in the models increase the dimensionality of the state space and therefore the severity of the curse of dimensionality. Additionally, the probabilities are continuous variables, which means that the classical, discrete-state *Q-learning* algorithm cannot be used. Approximate versions of *Q-learning* are required instead.

**Dynamic tasks.** The *Non-Stationary Converging Policies (NSCP)* algorithm [138] computes a best-response to the models and uses it to estimate state values. This algorithm is very similar to (13)–(14) and (17)–(18); this time, the stage game solver gives a best-response:

---

[4] There is also a stochastic (mixed) Nash equilibrium, where each agent goes left with probability 3/5. This is because the strategies $\sigma_1(L_1) = 3/5, \sigma_1(R_1) = 2/5$ and $\sigma_2(L_2) = 3/5, \sigma_2(R_2) = 2/5$ are best-responses to one another. The expected return of this equilibrium for both agents is 6/5, worse than for any of the two deterministic equilibria.

$$h_{i,k}(x_k, \cdot) = \arg \mathbf{br}_i(Q_{i,k}, x_k) \tag{20}$$

$$Q_{i,k+1}(x_k, \mathbf{u}_k) = Q_k(x_k, \mathbf{u}_k) + \alpha \left[ r_{i,k+1} + \gamma \mathbf{br}_i(Q_{i,k}, x_{k+1}) - Q_k(x_k, \mathbf{u}_k) \right] \tag{21}$$

where the best-response value operator **br** is implemented as:

$$\mathbf{br}_i(Q_i, x) = \max_{h_i(x, \cdot)} \sum_{u_1, \dots, u_n} h_i(x, u_i) \cdot Q_i(x, u_1, \dots, u_n) \prod_{j=1, j \neq i}^{n} \widehat{h}_j^i(x, u_j) \tag{22}$$

The empirical models $\widehat{h}_j^i$ are learned using (16). In the computation of **br**, the value of each joint action is weighted by the estimated probability of that action being selected, given the models of the other agents (the product term in (22)).

### Agent-Aware Methods

Agent-aware algorithms target convergence, as well as adaptation to the other agents. Some algorithms provably converge for particular types of tasks (mostly static), others use heuristics for which convergence is not guaranteed.

**Static tasks.** The algorithms presented here assume the availability of a model of the static task, in the form of reward functions. The *AWESOME* algorithm [31] uses fictitious play, but monitors the other agents and, when it concludes that they are nonstationary, switches from the best-response in fictitious play to a centrally precomputed Nash equilibrium (hence the name: Adapt When Everyone is Stationary, Otherwise Move to Equilibrium). In repeated games, *AWESOME* is provably rational and convergent [31] according to the definitions from [16, 17] given in Section 4.

Some methods in the area of direct policy search use gradient update rules that guarantee convergence in specific classes of static games: *Infinitesimal Gradient Ascent (IGA)* [109], *Win-or-Learn-Fast IGA (WoLF-IGA)* [17], *Generalized IGA (GIGA)* [144], and *GIGA-WoLF* [14]. For instance, *IGA* and *WoLF-IGA* work in two-agent, two-action games, and use similar gradient update rules:

$$\begin{cases} \alpha_{k+1} = \alpha_k + \delta_{1,k} \dfrac{\partial \mathrm{E} \{r_1 \mid \alpha, \beta\}}{\partial \alpha} \\ \beta_{k+1} = \beta_k + \delta_{2,k} \dfrac{\partial \mathrm{E} \{r_2 \mid \alpha, \beta\}}{\partial \beta} \end{cases} \tag{23}$$

The strategies of the agents are represented by the probability of selecting the first out of the two actions, denoted by $\alpha$ for agent 1 and by $\beta$ for agent 2. *IGA* uses constant gradient steps $\delta_{1,k} = \delta_{2,k} = \delta$. For an infinitesimal step size, i.e., when $\delta \to 0$, the average rewards achieved by the *IGA* policies converge to Nash rewards. In *WoLF-IGA*, $\delta_{i,k}$ switches between a smaller value when agent $i$ is winning, and a larger value when it is losing (hence the name, Win-or-Learn-Fast). *WoLF-IGA* is rational by the definition in Section 4, and convergent for an asymptotically infinitesimal step size [17] (i.e., if $\delta_{i,k} \to 0$ when $k \to \infty$).

**Dynamic tasks.** *Win-or-Learn-Fast Policy Hill-Climbing (WoLF-PHC)* [17] is a heuristic algorithm that updates Q-functions with the *Q-learning* rule (4), and policies with a WoLF rule inspired from (23):

$$h_{i,k+1}(x_k, u_i) = h_{i,k}(x_k, u_i) + \begin{cases} \sum_{\bar{u}_i \neq u_i} \delta_{i,k}^{\bar{u}_i} & \text{if } u_i = \arg\max_{u_i'} Q_{i,k+1}(x_k, u_i') \\ -\delta_{i,k}^{u_i} & \text{otherwise} \end{cases} \tag{24}$$

$$\text{where } \delta_{i,k}^{u_i} = \min\left\{ h_{i,k}(x_k, u_i), \frac{\delta_{i,k}}{|U_i| - 1} \right\} \tag{25}$$

$$\text{and } \delta_{i,k} = \begin{cases} \delta_{\text{win}} & \text{if winning} \\ \delta_{\text{lose}} & \text{if losing} \end{cases} \tag{26}$$

The probability decrements for the sub-optimal actions are bounded in (25) to ensure that all the probabilities remain non-negative, while the probability increment for the optimal action in (24) is chosen so that the probability distribution remains valid. The gradient step $\delta_{i,k}$ is larger when agent $i$ is losing than when it is winning: $\delta_{\text{lose}} > \delta_{\text{win}}$. For instance, in [17] $\delta_{\text{lose}}$ is 2 to 4 times larger than $\delta_{\text{win}}$. The rationale is that the agent should escape quickly from losing situations, while adapting cautiously when it is winning, in order to encourage convergence. The win/lose criterion in (26) is based on a comparison of an average policy with the current one in the original version of *WoLF-PHC*, and on the second-order difference of policy elements in *PD-WoLF* [5].

The *Extended Optimal Response (EXORL)* heuristic [116] applies a complementary idea in two-agent tasks: the policy update is biased in a way that minimizes the other agent's incentive to deviate from its current policy. Thus, convergence to a coordinated Nash equilibrium is encouraged.

### Remarks and Open Issues

Static, repeated games represent a limited set of applications. Algorithms for static games provide valuable theoretical insight; these algorithms should however be extended to dynamic stochastic games in order to become interesting for more general classes of applications (e.g., *WoLF-PHC* [17] is such an extension). Many algorithms for mixed stochastic games, especially agent-independent algorithms, are sensitive to imperfect observations.

Game theory induces a bias toward static, stage-wise solutions in the dynamic case, as seen e.g., in the agent-independent *Q-learning* template (17)–(18). However, the suitability of such state-wise solutions in the context of the dynamic task is not always clear [86, 108].

One important research direction is understanding the conditions under which single-agent RL works in mixed stochastic games, especially in light of the preference towards single-agent techniques in practice. This direction was pioneered by the analysis in [130].

## 6   Application Domains

MARL has been applied to a variety of problem domains, mostly in simulation but also in some real-life tasks. Simulated domains dominate for two reasons. The first reason it is easier to understand and to derive insight from results in simpler domains. The second reason is that scalability and robustness to imperfect observations are necessary in real-life tasks, and few MARL algorithms exhibit these properties. In real-life applications, more direct derivations of single-agent RL (see Section 5.4) are preferred [73, 74, 75, 113].

In this section, several representative application domains are reviewed: distributed control, multi-robot teams, trading agents, and resource management.

### 6.1 Distributed Control

In distributed control, a set of autonomous, interacting controllers act in parallel on the same process. Distributed control is a meta-application for cooperative multi-agent systems: any cooperative multi-agent system is a distributed control system where the agents are the controllers, and their environment is the controlled process. For instance, in cooperative robotic teams the control algorithms of the robots identify with the controllers, and the robots' environment together with their sensors and actuators identify with the process.

Particular distributed control domains where MARL is applied are process control [113], control of traffic signals [4, 141], and control of electrical power networks [100].

### 6.2 Robotic Teams

Robotic teams (also called multi-robot systems) are the most popular application domain of MARL, encountered under the broadest range of variations. This is mainly because robotic teams are a very natural domain for multi-agent systems, but also because many MARL researchers are active in the robotics field. The robots' environment is a real or simulated spatial domain, most often having two dimensions. Robots use MARL to acquire a wide spectrum of skills, ranging from basic behaviors like navigation to complex behaviors like playing soccer.

In navigation, each robot has to find its way to a fixed or changing goal position, while avoiding obstacles and harmful interference with other robots [17, 50].

Area sweeping involves navigation through the environment for one of several purposes: retrieval of objects, coverage of as much environment surface as possible, and exploration, for which the robots have to bring into sensor range as much of the environment surface as possible [73, 74, 75].

Multi-target observation is an extension of the exploration task, where the robots have to maintain a group of moving targets within sensor range [35, 128].

Pursuit involves the capture of moving targets by the robotic team. In a popular variant, several 'predator' robots have to capture a 'prey' robot by converging on it [52, 60].

Object transportation requires the relocation of a set of objects into a desired final configuration. The mass or size of some of the objects may exceed the transportation capabilities of one robot, thus requiring several robots to coordinate in order to bring about the objective [74]. Our example in Section 7 belongs to this category.

Robot soccer is a popular, complex test-bed for MARL, that requires most of the skills enumerated above [77, 114, 115, 131, 142]. For instance, intercepting the ball and leading it into the goal involve object retrieval and transportation skills, while the tactical placement of the players in the field is an advanced version of the coverage task.

## 6.3   Automated Trading

Software trading agents exchange goods on electronic markets on behalf of a company or a person, using mechanisms such as negotiations and auctions. For instance, the Trading Agent Competition is a simulated contest where the agents need to arrange travel packages by bidding for goods such as plane tickets and hotel bookings [140]. Multi-agent trading can also be applied to modeling electricity markets [117].

MARL approaches to automated trading typically involve *temporal-difference* [118] or *Q-learning* agents, using approximate representations of the Q-functions to handle the large state space [48, 68, 125]. In some cases, cooperative agents represent the interest of a single company or individual, and merely fulfill different functions in the trading process, such as buying and selling [68]. In other cases, self-interested agents interact in parallel with the market [48, 98, 125].

## 6.4   Resource Management

In resource management, the agents form a cooperative team, and they can be one of the following:

- Managers of resources, as in [33]. Each agent manages one resource, and the agents learn how to best service requests in order to optimize a given performance measure.
- Clients of resources, as in [102]. The agents learn how to best select resources such that a given performance measure is optimized.

A popular resource management domain is network routing [18, 28, 126]. Other examples include elevator scheduling [33] and load balancing [102]. Performance measures include average job processing times, minimum waiting time for resources, resource usage, and fairness in servicing clients.

## 6.5   Remarks

Though not an application domain *per se*, game-theoretic, stateless tasks are often used to test MARL approaches. Not only algorithms specifically designed for static games are tested on such tasks (e.g., *AWESOME* [31], *MetaStrategy* [93], *GIGA-WoLF* [14]), but also others that can, in principle, handle dynamic stochastic games (e.g., *EXORL* [116]).

As an avenue for future work, note that distributed control is poorly represented as a MARL application domain. This includes systems such as traffic, power, or sensor networks.

# 7    Example: Coordinated Multi-agent Object Transportation

In this section, several MARL algorithms are applied to an illustrative example.[5] The example, represented in Figure 6, is an abstraction of a task involving the coordinated transportation of an object by two agents. These agents (represented by numbered disks) travel on a two-dimensional discrete grid with $7 \times 6$ cells. The agents have to transport a target object (represented by a small rectangle) to the home base (delimited by a dashed black line) in minimum time, while avoiding obstacles (shown by gray blocks).

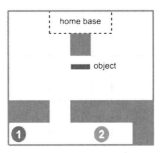

**Fig. 6.** The object transportation problem

The agents start from the positions in which they are shown in Figure 6, and can move at each time step by one cell to the left, right, up, or down; they can also stand still. If the target cell is not empty, the agent does not move; similarly, if both agents attempt to move to the same cell, neither agent moves. In order to transport the target, the agents first have to grasp it. When an agent reaches a cell immediately to the left or right of the target, it automatically grasps the target; once the target is grasped, it cannot be released. Only when the two agents have grasped either side the target, they can move it. The target only moves when both agents coordinately pull in the same direction. As soon as the target has reached the home base, the trial terminates, and the agents and target are reinitialized for a new trial.

The state variables describing each agent $i$ are its two position coordinates, $p_{i,X} \in \{1, 2, ..., 7\}$, $p_{i,Y} \in \{1, 2, ..., 6\}$, and a variable indicating if the agent is currently grasping the target, and if yes, to which side: $g_i \in \{\text{FREE}, \text{GRASPING-LEFT}, \text{GRASPING-RIGHT}\}$. Therefore, the complete state vector is $x = [p_{1,X}, p_{1,Y}, g_1, p_{2,X}, p_{2,Y}, g_2]^T$. The grasping variables are needed to ensure that the state vector has the Markov property (see Section 2). Each agent's action $u_i$ belongs to the set $U_i = \{\text{LEFT}, \text{RIGHT}, \text{UP}, \text{DOWN}, \text{STAND-STILL}\}$. So, the state space contains $|X| = (7 \cdot 6 \cdot 3)^2 = 15876$ elements, and the joint action space contains $5^2 = 25$ elements. Not all the states are valid, e.g., collisions prevent certain combinations from occurring.

---

[5] A MARL software package written by the authors in MATLAB was used for this example. This package can be downloaded at http://www.dcsc.tudelft.nl/~lbusoniu, and includes the coordinated object transportation example.

The task is fully cooperative, so both agents have the same reward function, which expresses the goal of grasping the target and transporting it to the home base:

$$r_{k+1} = \begin{cases} 1 & \text{if an agent has just grasped the target} \\ 10 & \text{if the target has reached the home base} \\ 0 & \text{in all other conditions} \end{cases}$$

The discount factor is $\gamma = 0.98$. The minimum-time character of the solution results from discounting: it is better to receive the positive rewards as early as possible, otherwise discounting will decrease their contribution to the return.

The agents face two coordination problems. The first problem is to decide which of them passes first through the narrow lower passage. The second problem is to decide whether they transport the target around the left or right side of the obstacle just below the home base.

We apply three algorithms to this problem: (i) single-agent *Q-learning*, which is a representative single-agent RL algorithm, (ii) *team Q-learning*, a representative MARL algorithm for fully cooperative tasks, and (iii) *WoLF-PHC*, a representative MARL algorithm for mixed tasks. An algorithm for competitive tasks (such as *minimax-Q*) is unlikely to perform well, because the object transportation problem is fully cooperative. In any given experiment, both agents use the same algorithm, so they are homogeneous. For all the algorithms, a constant learning rate $\alpha = 0.1$ is employed, together with an $\varepsilon$-greedy exploration procedure. The exploration probability $\varepsilon$ is initialized at 0.8, is constant within a trial, and decays exponentially with a factor 0.9 after every trial. For *WoLF-PHC*, the policy step sizes are $\delta_{\text{win}} = 0.1$ and $\delta_{\text{lose}} = 0.4$. Note that the Q-tables of *Q-learning* or *WoLF-PHC* agents contain $|X| \times 5 = 79380$ elements, because they only depend on that single agent's action, whereas the Q-tables of *team Q-learning* agents depend on the joint action and therefore contain $|X| \times 5^2 = 396900$ elements.

Figure 7 shows the mean learning performance of the three algorithms across 100 independent runs, together with 95% confidence intervals on this mean. Each graph shows the evolution of the number of steps taken to reach the home base, as the number of learning trials grows. The performance is measured while the agents are learning, and the effects of exploration are included. All the algorithms quickly converge to a good performance, usually after 20 to 30 trials. Remarkably, *Q-learning* performs very well, even though a Q-learner is unaware of the other agent except through its state variables. While *team Q-learning* and *WoLF-PHC* agents do take each other into account, they do not use explicit coordination. Instead, all three algorithms achieve an *implicit* form of coordination: the agents learn to prefer one of the equally good solutions by chance, and then ignore the other solutions. The fact that explicit coordination is not required can be verified e.g., by repeating the *team Q-learning* experiment after adding social conventions. Indeed, such an algorithm produces nearly identical results to those in Figure 7. Single-agent *Q-learning* is preferable in this problem, because it provides the same performance as the other two algorithms, but requires smaller Q-tables than *team Q-learning* and has simpler update formulas than *WoLF-PHC*.

Figure 8 shows a trajectory taken by the agents, after a representative run with *team Q-learning* (the trajectories followed by *Q-learning* and *WoLF-PHC* agents are similar). Agent 2 waits for agent 1 to go through the passage. Next, agent 1 grasps the target

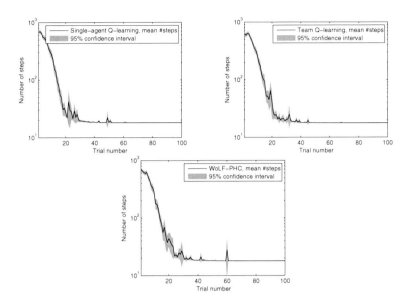

**Fig. 7.** Learning performance of single-agent *Q-learning*, *team Q-learning*, and *WoLF-PHC*. Note the logarithmic scale of the vertical axes.

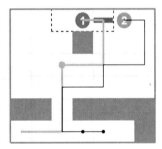

**Fig. 8.** A solution obtained with *team Q-learning*. Agent 1 travels along the thick gray line, and agent 2 along the thin black line. Dots mark the states where the agents stand still; the size of each dot is proportional with the time spent standing still in that cell.

waiting for agent 2, and after agent 2 also arrives, they transport the target to the home base around the right of the obstacle.

## 8   Outlook

In this section, we discuss some general open issues in MARL, concerning the suitability of MARL algorithms in practice, the choice of the multi-agent learning goal, and the study of the joint environment and learning dynamics.

## 8.1    Practical MARL

MARL algorithms are typically applied to small problems only, such as static games and small grid-worlds (like the grid-world of Section 7). As a consequence, it is unclear whether these algorithms scale up to realistic multi-agent problems, where the state and action spaces are large or even continuous. Few algorithms are able to deal with incomplete, uncertain observations. This situation can be explained by noting that scalability and dealing with imperfect observations are also open problems in single-agent RL. Nevertheless, improving the suitability of MARL to problems of practical interest is an essential research step. Below, we describe several directions in which this research can proceed, and point to some pioneering work done along these directions.

*Scalability* is a central concern for MARL. Most algorithms require explicit tabular storage of the agents' Q-functions and possibly of their policies. This means they only work in problems with a relatively small number of discrete states and actions. When the state and action spaces contain a large or infinite number of elements (e.g., when they are continuous), the tabular storage of the Q-function becomes impractical or impossible. Instead, the Q-function must be represented approximately. Approximate MARL algorithms have been proposed e.g., for discrete, large state-action spaces [1], for continuous states and discrete actions [22, 52], and for continuous states and actions [35, 122]. Unfortunately, many of these algorithms only work in restricted classes of problems and are heuristic in nature. Significant advances in approximate MARL can be made by putting to use the extensive results on single-agent approximate RL, which are outlined next.

While offline, batch algorithms have been very successful in single-agent approximate RL [34, 66, 80, 83, 85, 99], they are impractical in MARL, because an offline solution can become inappropriate as the behavior of the other agents changes. Instead, online approximate RL algorithms are required, such as approximate variants of *Q-learning*. *Q-learning* has been combined with Q-function approximators relying on, e.g., basis-function representations [110, 121], neural networks and self-organizing maps [127], and fuzzy rule bases [41, 47, 56]. Approximate *Q-learning* is provably convergent if a linearly parameterized approximator is employed, and under the restrictive condition that the agent's policy is kept fixed while learning [76]. Linearly parameterized approximators include basis-function representations, as well as certain classes of neural networks and fuzzy rule bases. An alternative to using *Q-learning* is to develop online, incremental versions of the successful offline, batch algorithms, see e.g., [57]. Another promising category of algorithms consists of *actor-critic* techniques, many of which are specialized for continuous state and action spaces [7, 10, 63, 90]. The reader interested in approximate RL is referred to [9, 23, 92], to Chapter 8 of [120], and to Chapter 6 of [8].

A complementary avenue for improving scalability is the discovery and exploitation of decentralized, modular structure in the multi-agent task [21, 38, 43].

*Incomplete, uncertain* state measurements could be handled with techniques related to partially observable Markov decision processes [72], as in [44, 51].

Providing *domain knowledge* to the agents can greatly help them to learn solutions of realistic tasks. In contrast, the large size of the state-action space and the delays in receiving informative rewards mean that MARL without any prior knowledge is very slow. Domain knowledge can be supplied in several forms. Informative reward functions, also rewarding promising behaviors rather than just the achievement of the goal, can be provided to the agents [74, 75]. Humans or skilled agents can teach unskilled agents how to solve the task [94]. Shaping is a technique whereby the learning process starts by presenting the agents with simpler tasks, and progressively moves toward complex ones [24]. Preprogrammed reflex behaviors can be built into the agents [74, 75]. Knowledge about the task structure can be used to decompose it in subtasks, and learn a modular solution with e.g., hierarchical RL [40]. If approximate solutions are used, a good way to incorporate domain knowledge is to structure the approximator in a way that ensures high accuracy in important regions of the state-action space, e.g., close to the goal. Last, but not least, if a (possibly incomplete) task model is available, this model can be used with model-based RL algorithms to initialize Q-functions to reasonable, rather than arbitrary, values.

### 8.2   Learning Goal

Defining a suitable MARL goal for general, dynamic stochastic games is a difficult open problem. MARL goals are typically formulated in terms of static games. Their extension to dynamic tasks, as discussed in Section 4, is not always clear or even possible.

Stability of the learning process is needed, because the behavior of stable agents is more amenable to analysis and meaningful performance guarantees. Adaptation to the other agents is needed because their behavior is generally unpredictable. Therefore, a good multi-agent learning goal must include both components. This means that MARL algorithms should not be totally independent of the other agents, nor just track their behavior without concern for convergence.

Moreover, from a practical viewpoint, a realistic learning goal should include bounds on the transient performance, in addition to the usual asymptotic requirements. Examples of such bounds include maximum time constraints for reaching a desired performance level, or a lower bound on the instantaneous performance. Some steps in this direction have been taken in [14, 93].

### 8.3   Joint Environment and Learning Dynamics

So far, game-theory-based analysis has only been applied to the learning dynamics of the agents [130, 132, 134], while the dynamics of the environment have not been explicitly considered. Tools developed in the area of robust control can play an important role in the analysis of the learning process as a whole, comprising the interacting environment and learning dynamics. In addition, this framework can incorporate prior knowledge about bounds on imperfect observations, such as noise-corrupted variables; and can help study the robustness of MARL algorithms against uncertainty in the other agents' dynamics.

## 9   Related Work

Besides its heritage relationship with single-agent RL, the MARL field has strong connections with game theory, evolutionary computation, and more generally with the direct optimization of agent policies. These relationships are described next.

Game theory [3] and especially the theory of learning in games [39] make an essential contribution to MARL. In this chapter, we have reviewed relevant game-theoretic algorithms for static and repeated games, and we have investigated the contribution of game theory to MARL algorithms for dynamic tasks. Other authors have investigated more closely the relationship between game theory and MARL. For instance, Bowling and Veloso [17] have discussed several MARL algorithms, showing that these algorithms combine temporal-difference RL with game-theoretic solvers for the static games that arise in each state of the dynamic stochastic game. Shoham et al. [108] have critically examined the focus of MARL research on game-theoretic equilibria, using a selection of MARL algorithms to illustrate their arguments.

Rather than estimating value functions and using them to derive policies, it is also possible to directly explore the space of agent behaviors using, e.g., nonlinear optimization techniques. Evolutionary multi-agent learning is a prominent example of such an approach. Evolutionary computation applies principles of biological evolution to the search for solutions (agent behaviors) of the given task [2, 55]. Populations of candidate behaviors are maintained. Candidates are evaluated using a fitness function related to the return, and selected for breeding or mutation on the basis of their fitness. Panait and Luke [86] have offered a comprehensive survey of evolutionary learning and MARL for cooperative agent teams. For the interested reader, examples of co-evolution techniques, where the behaviors of the agents evolve in parallel, can be found in [36, 87, 91]. Complementary, team learning techniques, where the entire set of agent behaviors is discovered by a single evolution process, can be found e.g., in [45, 78, 101]. Besides evolutionary computation, other approaches to the direct optimization of agent behaviors are gradient search [65], probabilistic hill-climbing [46], as well as more general behavior modification heuristics [103]. Because direct behavior optimization cannot readily benefit from the structure of the RL task, we did not focus on it in this chapter. Instead, we have mainly discussed the contribution of direct behavior optimization to MARL algorithms based on temporal-difference RL.

Evolutionary game theory sits at the intersection of evolutionary learning and game theory [111]. Tuyls and Nowé [132] have investigated the relationship between MARL and evolutionary game theory, focusing on static tasks.

## 10  Conclusions

Multi-agent reinforcement learning (MARL) is a young, but active and rapidly expanding field of research. MARL aims to provide an array of algorithms that enable multiple agents to learn the solution of difficult tasks, using limited or no prior knowledge. To this end, MARL integrates results from single-agent RL, game theory, and direct policy search.

This chapter has provided an extensive overview of MARL. First, we have presented the main benefits and challenges of MARL, as well as the different viewpoints on defining the MARL learning goal. Then, we have discussed in detail a representative set of MARL algorithms for fully cooperative, fully competitive, and mixed tasks. We have focused on autonomous agents that learn how to solve dynamic tasks online, with algorithms originating in temporal-difference RL, but we have also investigated techniques for static tasks. Additionally, we have reviewed some representative problem domains where MARL techniques have been applied, and we have illustrated the behavior of several MARL algorithms in a simulation example involving multi-agent object transportation.

Many avenues for MARL are open at this point, and many research opportunities present themselves. We have provided an outlook synthesizing these open issues and opportunities. In particular, approximate RL is needed to apply MARL to realistic problems, and control theory can help analyzing the learning dynamics and assessing the robustness to uncertainty in the observations or in the other agents' behavior. In our view, significant progress in the field of multi-agent learning can be achieved by a more intensive cross-fertilization between the fields of machine learning, game theory, and control theory.

*Acknowledgement.* This work was financially supported by the BSIK project "Interactive Collaborative Information Systems" (grant no. BSIK03024).

# References

1. Abul, O., Polat, F., Alhajj, R.: Multiagent reinforcement learning using function approximation. IEEE Transactions on Systems, Man, and Cybernetics—Part C: Applications and Reviews 4(4), 485–497 (2000)
2. Bäck, T.: Evolutionary Algorithms in Theory and Practice: Evolution Strategies, Evolutionary Programming, Genetic Algorithms. Oxford University Press, Oxford (1996)
3. Başar, T., Olsder, G.J.: Dynamic Noncooperative Game Theory, 2nd edn. Society for Industrial and Applied Mathematics, SIAM (1999)
4. Bakker, B., Steingrover, M., Schouten, R., Nijhuis, E., Kester, L.: Cooperative multi-agent reinforcement learning of traffic lights. In: Workshop on Cooperative Multi-Agent Learning, 16th European Conference on Machine Learning (ECML-2005), Porto, Portugal (2005)
5. Banerjee, B., Peng, J.: Adaptive policy gradient in multiagent learning. In: Proceedings 2nd International Joint Conference on Autonomous Agents and Multiagent Systems (AAMAS 2003), Melbourne, Australia, pp. 686–692 (2003)
6. Barto, A.G., Sutton, R.S., Anderson, C.W.: Neuronlike adaptive elements that can solve difficult learning control problems. IEEE Transactions on Systems, Man, and Cybernetics 13(5), 833–846 (1983)
7. Berenji, H.R., Vengerov, D.: A convergent actor-critic-based FRL algorithm with application to power management of wireless transmitters. IEEE Transactions on Fuzzy Systems 11(4), 478–485 (2003)
8. Bertsekas, D.P.: Dynamic Programming and Optimal Control, 3rd edn., vol. 2. Athena Scientific (2007)
9. Bertsekas, D.P., Tsitsiklis, J.N.: Neuro-Dynamic Programming. Athena Scientific (1996)
10. Borkar, V.: An actor-critic algorithm for constrained Markov decision processes. Systems & Control Letters 54(3), 207–213 (2005)
11. Boutilier, C.: Planning, learning and coordination in multiagent decision processes. In: Proceedings 6th Conference on Theoretical Aspects of Rationality and Knowledge (TARK-1996), pp. 195–210. De Zeeuwse Stromen, The Netherlands (1996)
12. Bowling, M.: Convergence problems of general-sum multiagent reinforcement learning. In: Proceedings 17th International Conference on Machine Learning (ICML-2000), Stanford University, US, pp. 89–94 (2000)

13. Bowling, M.: Multiagent learning in the presence of agents with limitations. Ph.D. thesis, Computer Science Dept., Carnegie Mellon University, Pittsburgh, US (2003)
14. Bowling, M.: Convergence and no-regret in multiagent learning. In: Saul, L.K., Weiss, Y., Bottou, L. (eds.) Advances in Neural Information Processing Systems 17, pp. 209–216. MIT Press, Cambridge (2005)
15. Bowling, M., Veloso, M.: An analysis of stochastic game theory for multiagent reinforcement learning. Tech. rep., Computer Science Dept., Carnegie Mellon University, Pittsburgh, US (2000), http://www.cs.ualberta.ca/~bowling/papers/00tr.pdf
16. Bowling, M., Veloso, M.: Rational and convergent learning in stochastic games. In: Proceedings 17th International Conference on Artificial Intelligence (IJCAI-2001), San Francisco, US, pp. 1021–1026 (2001)
17. Bowling, M., Veloso, M.: Multiagent learning using a variable learning rate. Artificial Intelligence 136(2), 215–250 (2002)
18. Boyan, J.A., Littman, M.L.: Packet routing in dynamically changing networks: A reinforcement learning approach. In: Moody, J. (ed.) Advances in Neural Information Processing Systems 6, pp. 671–678. Morgan Kaufmann, San Francisco (1994)
19. Brown, G.W.: Iterative solutions of games by fictitious play. In: Koopmans, T.C. (ed.) Activitiy Analysis of Production and Allocation, ch. XXIV, pp. 374–376. Wiley, Chichester (1951)
20. Buşoniu, L., Babuška, R., De Schutter, B.: A comprehensive survey of multi-agent reinforcement learning. IEEE Transactions on Systems, Man, and Cybernetics. Part C: Applications and Reviews 38(2), 156–172 (2008)
21. Buşoniu, L., De Schutter, B., Babuška, R.: Multiagent reinforcement learning with adaptive state focus. In: Proceedings 17th Belgian-Dutch Conference on Artificial Intelligence (BNAIC-2005), Brussels, Belgium, pp. 35–42 (2005)
22. Buşoniu, L., De Schutter, B., Babuška, R.: Decentralized reinforcement learning control of a robotic manipulator. In: Proceedings 9th International Conference of Control, Automation, Robotics, and Vision (ICARCV-2006), Singapore, pp. 1347–1352 (2006)
23. Buşoniu, L., De Schutter, B., Babuška, R.: Approximate dynamic programming and reinforcement learning. In: Babuška, R., Groen, F.C.A. (eds.) Interactive Collaborative Information Systems. Studies in Computational Intelligence, vol. 281, pp. 3–44. Springer, Heidelberg (2010)
24. Buffet, O., Dutech, A., Charpillet, F.: Shaping multi-agent systems with gradient reinforcement learning. Autonomous Agents and Multi-Agent Systems 15(2), 197–220 (2007)
25. Carmel, D., Markovitch, S.: Opponent modeling in multi-agent systems. In: Weiß, G., Sen, S. (eds.) Adaptation and Learning in Multi-Agent Systems, ch. 3, pp. 40–52. Springer, Heidelberg (1996)
26. Chalkiadakis, G.: Multiagent reinforcement learning: Stochastic games with multiple learning players. Tech. rep., Dept. of Computer Science, University of Toronto, Canada (2003), http://www.cs.toronto.edu/~gehalk/DepthReport/DepthReport.ps
27. Cherkassky, V., Mulier, F.: Learning from Data: Concepts, Theory, And Methods. Wiley, Chichester (1998)
28. Choi, S.P.M., Yeung, D.Y.: Predictive Q-routing: A memory-based reinforcement learning approach to adaptive traffic control. In: Touretzky, D.S., Mozer, M., Hasselmo, M.E. (eds.) Advances in Neural Information Processing Systems 8, pp. 945–951. MIT Press, Cambridge (1995)
29. Claus, C., Boutilier, C.: The dynamics of reinforcement learning in cooperative multiagent systems. In: Proceedings 15th National Conference on Artificial Intelligence and 10th Conference on Innovative Applications of Artificial Intelligence (AAAI/IAAI-1998), Madison, US, pp. 746–752 (1998)

30. Clouse, J.: Learning from an automated training agent. In: Working Notes Workshop on Agents that Learn from Other Agents, 12th International Conference on Machine Learning (ICML-1995), Tahoe City, US (1995)
31. Conitzer, V., Sandholm, T.: AWESOME: A general multiagent learning algorithm that converges in self-play and learns a best response against stationary opponents. In: Proceedings 20th International Conference on Machine Learning (ICML-2003), Washington, US, pp. 83–90 (2003)
32. Crites, R.H., Barto, A.G.: Improving elevator performance using reinforcement learning. In: Touretzky, D.S., Mozer, M.C., Hasselmo, M.E. (eds.) Advances in Neural Information Processing Systems 8, pp. 1017–1023. MIT Press, Cambridge (1996)
33. Crites, R.H., Barto, A.G.: Elevator group control using multiple reinforcement learning agents. Machine Learning 33(2–3), 235–262 (1998)
34. Ernst, D., Geurts, P., Wehenkel, L.: Tree-based batch mode reinforcement learning. Journal of Machine Learning Research 6, 503–556 (2005)
35. Fernández, F., Parker, L.E.: Learning in large cooperative multi-robot systems. International Journal of Robotics and Automation, Special Issue on Computational Intelligence Techniques in Cooperative Robots 16(4), 217–226 (2001)
36. Ficici, S.G., Pollack, J.B.: A game-theoretic approach to the simple coevolutionary algorithm. In: Deb, K., Rudolph, G., Lutton, E., Merelo, J.J., Schoenauer, M., Schwefel, H.-P., Yao, X. (eds.) PPSN 2000. LNCS, vol. 1917, pp. 467–476. Springer, Heidelberg (2000)
37. Fischer, F., Rovatsos, M., Weiss, G.: Hierarchical reinforcement learning in communication-mediated multiagent coordination. In: Proceedings 3rd International Joint Conference on Autonomous Agents and Multiagent Systems (AAMAS-2004), New York, US, pp. 1334–1335 (2004)
38. Fitch, R., Hengst, B., Suc, D., Calbert, G., Scholz, J.B.: Structural abstraction experiments in reinforcement learning. In: Zhang, S., Jarvis, R.A. (eds.) AI 2005. LNCS (LNAI), vol. 3809, pp. 164–175. Springer, Heidelberg (2005)
39. Fudenberg, D., Levine, D.K.: The Theory of Learning in Games. MIT Press, Cambridge (1998)
40. Ghavamzadeh, M., Mahadevan, S., Makar, R.: Hierarchical multi-agent reinforcement learning. Autonomous Agents and Multi-Agent Systems 13(2), 197–229 (2006)
41. Glorennec, P.Y.: Reinforcement learning: An overview. In: Proceedings European Symposium on Intelligent Techniques (ESIT-2000), Aachen, Germany, pp. 17–35 (2000)
42. Greenwald, A., Hall, K.: Correlated-Q learning. In: Proceedings 20th International Conference on Machine Learning (ICML-2003), Washington, US, pp. 242–249 (2003)
43. Guestrin, C., Lagoudakis, M.G., Parr, R.: Coordinated reinforcement learning. In: Proceedings 19th International Conference on Machine Learning (ICML-2002), Sydney, Australia, pp. 227–234 (2002)
44. Hansen, E.A., Bernstein, D.S., Zilberstein, S.: Dynamic programming for partially observable stochastic games. In: Proceedings 19th National Conference on Artificial Intelligence (AAAI-2004), San Jose, US, pp. 709–715 (2004)
45. Haynes, T., Wainwright, R., Sen, S., Schoenefeld, D.: Strongly typed genetic programming in evolving cooperation strategies. In: Proceedings 6th International Conference on Genetic Algorithms (ICGA-1995), Pittsburgh, US, pp. 271–278 (1995)
46. Ho, F., Kamel, M.: Learning coordination strategies for cooperative multiagent systems. Machine Learning 33(2–3), 155–177 (1998)
47. Horiuchi, T., Fujino, A., Katai, O., Sawaragi, T.: Fuzzy interpolation-based Q-learning with continuous states and actions. In: Proceedings 5th IEEE International Conference on Fuzzy Systems (FUZZ-IEEE-1996), New Orleans, US, pp. 594–600 (1996)

48. Hsu, W.T., Soo, V.W.: Market performance of adaptive trading agents in synchronous double auctions. In: Yuan, S.-T., Yokoo, M. (eds.) PRIMA 2001. LNCS (LNAI), vol. 2132, pp. 108–121. Springer, Heidelberg (2001)

49. Hu, J., Wellman, M.P.: Multiagent reinforcement learning: Theoretical framework and an algorithm. In: Proceedings 15th International Conference on Machine Learning (ICML-1998), Madison, US, pp. 242–250 (1998)

50. Hu, J., Wellman, M.P.: Nash Q-learning for general-sum stochastic games. Journal of Machine Learning Research 4, 1039–1069 (2003)

51. Ishii, S., Fujita, H., Mitsutake, M., Yamazaki, T., Matsuda, J., Matsuno, Y.: A reinforcement learning scheme for a partially-observable multi-agent game. Machine Learning 59(1–2), 31–54 (2005)

52. Ishiwaka, Y., Sato, T., Kakazu, Y.: An approach to the pursuit problem on a heterogeneous multiagent system using reinforcement learning. Robotics and Autonomous Systems 43(4), 245–256 (2003)

53. Jaakkola, T., Jordan, M.I., Singh, S.P.: On the convergence of stochastic iterative dynamic programming algorithms. Neural Computation 6(6), 1185–1201 (1994)

54. Jafari, A., Greenwald, A.R., Gondek, D., Ercal, G.: On no-regret learning, fictitious play, and Nash equilibrium. In: Proceedings 18th International Conference on Machine Learning (ICML-2001), pp. 226–233. Williams College, Williamstown, US (2001)

55. Jong, K.D.: Evolutionary Computation: A Unified Approach. MIT Press, Cambridge (2005)

56. Jouffe, L.: Fuzzy inference system learning by reinforcement methods. IEEE Transactions on Systems, Man, and Cybernetics—Part C: Applications and Reviews 28(3), 338–355 (1998)

57. Jung, T., Polani, D.: Kernelizing LSPE($\lambda$). In: Proceedings 2007 IEEE Symposium on Approximate Dynamic Programming and Reinforcement Learning (ADPRL-2007), Honolulu, US, pp. 338–345 (2007)

58. Kaelbling, L.P., Littman, M.L., Moore, A.W.: Reinforcement learning: A survey. Journal of Artificial Intelligence Research 4, 237–285 (1996)

59. Kapetanakis, S., Kudenko, D.: Reinforcement learning of coordination in cooperative multi-agent systems. In: Proceedings 18th National Conference on Artificial Intelligence and 14th Conference on Innovative Applications of Artificial Intelligence (AAAI/IAAI-2002), Menlo Park, US, pp. 326–331 (2002)

60. Kok, J.R., 't Hoen, P.J., Bakker, B., Vlassis, N.: Utile coordination: Learning interdependencies among cooperative agents. In: Proceedings IEEE Symposium on Computational Intelligence and Games (CIG 2005), Colchester, United Kingdom, pp. 29–36 (2005)

61. Kok, J.R., Spaan, M.T.J., Vlassis, N.: Non-communicative multi-robot coordination in dynamic environment. Robotics and Autonomous Systems 50(2–3), 99–114 (2005)

62. Kok, J.R., Vlassis, N.: Sparse cooperative Q-learning. In: Proceedings 21st International Conference on Machine Learning (ICML-2004), Banff, Canada, pp. 481–488 (2004)

63. Konda, V.R., Tsitsiklis, J.N.: On actor-critic algorithms. SIAM Journal on Control and Optimization 42(4), 1143–1166 (2003)

64. Könönen, V.: Asymmetric multiagent reinforcement learning. In: Proceedings IEEE/WIC International Conference on Intelligent Agent Technology (IAT-2003), Halifax, Canada, pp. 336–342 (2003)

65. Könönen, V.: Gradient based method for symmetric and asymmetric multiagent reinforcement learning. In: Liu, J., Cheung, Y.-m., Yin, H. (eds.) IDEAL 2003. LNCS, vol. 2690, pp. 68–75. Springer, Heidelberg (2003)

66. Lagoudakis, M.G., Parr, R.: Least-squares policy iteration. Journal of Machine Learning Research 4, 1107–1149 (2003)

67. Lauer, M., Riedmiller, M.: An algorithm for distributed reinforcement learning in cooperative multi-agent systems. In: Proceedings 17th International Conference on Machine Learning (ICML-2000), Stanford University, US, pp. 535–542 (2000)

68. Lee, J.-W., Jang Min, O.: A multi-agent Q-learning framework for optimizing stock trading systems. In: Hameurlain, A., Cicchetti, R., Traunmüller, R. (eds.) DEXA 2002. LNCS, vol. 2453, pp. 153–162. Springer, Heidelberg (2002)

69. Littman, M.L.: Markov games as a framework for multi-agent reinforcement learning. In: Proceedings 11th International Conference on Machine Learning (ICML-1994), New Brunswick, US, pp. 157–163 (1994)

70. Littman, M.L.: Value-function reinforcement learning in Markov games. Journal of Cognitive Systems Research 2(1), 55–66 (2001)

71. Littman, M.L., Stone, P.: Implicit negotiation in repeated games. In: Meyer, J.-J.C., Tambe, M. (eds.) ATAL 2001. LNCS (LNAI), vol. 2333, pp. 96–105. Springer, Heidelberg (2002)

72. Lovejoy, W.S.: Computationally feasible bounds for partially observed Markov decision processes. Operations Research 39(1), 162–175 (1991)

73. Matarić, M.J.: Reward functions for accelerated learning. In: Proceedings 11th International Conference on Machine Learning (ICML-1994), New Brunswick, US, pp. 181–189 (1994)

74. Matarić, M.J.: Learning in multi-robot systems. In: Weiß, G., Sen, S. (eds.) Adaptation and Learning in Multi-Agent Systems, ch. 10, pp. 152–163. Springer, Heidelberg (1996)

75. Matarić, M.J.: Reinforcement learning in the multi-robot domain. Autonomous Robots 4(1), 73–83 (1997)

76. Melo, F.S., Meyn, S.P., Ribeiro, M.I.: An analysis of reinforcement learning with function approximation. In: Proceedings 25th International Conference on Machine Learning (ICML-2008), Helsinki, Finland, pp. 664–671 (2008)

77. Merke, A., Riedmiller, M.A.: Karlsruhe brainstormers - A reinforcement learning approach to robotic soccer. In: Birk, A., Coradeschi, S., Tadokoro, S. (eds.) RoboCup 2001. LNCS (LNAI), vol. 2377, pp. 435–440. Springer, Heidelberg (2002)

78. Miconi, T.: When evolving populations is better than coevolving individuals: The blind mice problem. In: Proceedings 18th International Joint Conference on Artificial Intelligence (IJCAI 2003), Acapulco, Mexico, pp. 647–652 (2003)

79. Moore, A.W., Atkeson, C.G.: Prioritized sweeping: Reinforcement learning with less data and less time. Machine Learning 13, 103–130 (1993)

80. Munos, R., Szepesvári, C.: Finite time bounds for fitted value iteration. Journal of Machine Learning Research 9, 815–857 (2008)

81. Nagendra Prasad, M.V., Lesser, V.R., Lander, S.E.: Learning organizational roles for negotiated search in a multiagent system. International Journal of Human-Computer Studies 48(1), 51–67 (1998)

82. Nash, S., Sofer, A.: Linear and Nonlinear Programming. McGraw-Hill, New York (1996)

83. Nedić, A., Bertsekas, D.P.: Least-squares policy evaluation algorithms with linear function approximation. Discrete Event Dynamic Systems: Theory and Applications 13(1–2), 79–110 (2003)

84. Negenborn, R.R., De Schutter, B., Hellendoorn, H.: Multi-agent model predictive control for transportation networks: Serial versus parallel schemes. Engineering Applications of Artificial Intelligence 21(3), 353–366 (2008)

85. Ormoneit, D., Sen, S.: Kernel-based reinforcement learning. Machine Learning 49(2–3), 161–178 (2002)

86. Panait, L., Luke, S.: Cooperative multi-agent learning: The state of the art. Autonomous Agents and Multi-Agent Systems 11(3), 387–434 (2005)

87. Panait, L., Wiegand, R.P., Luke, S.: Improving coevolutionary search for optimal multiagent behaviors. In: Proceedings 18th International Joint Conference on Artificial Intelligence (IJCAI-2003), Acapulco, Mexico, pp. 653–660 (2003)

88. Parunak, H.V.D.: Industrial and practical applications of DAI. In: Weiss, G. (ed.) Multi–Agent Systems: A Modern Approach to Distributed Artificial Intelligence, ch. 9, pp. 377–412. MIT Press, Cambridge (1999)
89. Peng, J., Williams, R.J.: Incremental multi-step Q-learning. Machine Learning 22(1–3), 283–290 (1996)
90. Peters, J., Schaal, S.: Natural actor-critic. Neurocomputing 71(7–9), 1180–1190 (2008)
91. Potter, M.A., Jong, K.A.D.: A cooperative coevolutionary approach to function optimization. In: Davidor, Y., Männer, R., Schwefel, H.-P. (eds.) PPSN 1994. LNCS, vol. 866, pp. 249–257. Springer, Heidelberg (1994)
92. Powell, W.B.: Approximate Dynamic Programming: Solving the Curses of Dimensionality. Wiley, Chichester (2007)
93. Powers, R., Shoham, Y.: New criteria and a new algorithm for learning in multi-agent systems. In: Saul, L.K., Weiss, Y., Bottou, L. (eds.) Advances in Neural Information Processing Systems 17, pp. 1089–1096. MIT Press, Cambridge (2005)
94. Price, B., Boutilier, C.: Implicit imitation in multiagent reinforcement learning. In: Proceedings 16th International Conference on Machine Learning (ICML-1999), Bled, Slovenia, pp. 325–334 (1999)
95. Price, B., Boutilier, C.: Accelerating reinforcement learning through implicit imitation. Journal of Artificial Intelligence Research 19, 569–629 (2003)
96. Puterman, M.L.: Markov Decision Processes—Discrete Stochastic Dynamic Programming. Wiley, Chichester (1994)
97. Pynadath, D.V., Tambe, M.: The communicative multiagent team decision problem: Analyzing teamwork theories and models. Journal of Artificial Intelligence Research 16, 389–423 (2002)
98. Raju, C., Narahari, Y., Ravikumar, K.: Reinforcement learning applications in dynamic pricing of retail markets. In: Proceedings 2003 IEEE International Conference on E-Commerce (CEC-2003), Newport Beach, US, pp. 339–346 (2003)
99. Riedmiller, M.: Neural fitted Q iteration - first experiences with a data efficient neural reinforcement learning method. In: Gama, J., Camacho, R., Brazdil, P.B., Jorge, A.M., Torgo, L. (eds.) ECML 2005. LNCS (LNAI), vol. 3720, pp. 317–328. Springer, Heidelberg (2005)
100. Riedmiller, M.A., Moore, A.W., Schneider, J.G.: Reinforcement learning for cooperating and communicating reactive agents in electrical power grids. In: Hannebauer, M., Wendler, J., Pagello, E. (eds.) Balancing Reactivity and Social Deliberation in Multi-Agent Systems, pp. 137–149. Springer, Heidelberg (2000)
101. Salustowicz, R., Wiering, M., Schmidhuber, J.: Learning team strategies: Soccer case studies. Machine Learning 33(2–3), 263–282 (1998)
102. Schaerf, A., Shoham, Y., Tennenholtz, M.: Adaptive load balancing: A study in multi-agent learning. Journal of Artificial Intelligence Research 2, 475–500 (1995)
103. Schmidhuber, J.: A general method for incremental self-improvement and multi-agent learning. In: Yao, X. (ed.) Evolutionary Computation: Theory and Applications, ch. 3, pp. 81–123. World Scientific, Singapore (1999)
104. Sejnowski, T.J., Hinton, G.E. (eds.): Unsupervised Learning: Foundations of Neural Computation. MIT Press, Cambridge (1999)
105. Sen, S., Sekaran, M., Hale, J.: Learning to coordinate without sharing information. In: Proceedings 12th National Conference on Artificial Intelligence (AAAI-1994), Seattle, US, pp. 426–431 (1994)
106. Sen, S., Weiss, G.: Learning in multiagent systems. In: Weiss, G. (ed.) Multiagent Systems: A Modern Approach to Distributed Artificial Intelligence, ch. 6, pp. 259–298. MIT Press, Cambridge (1999)
107. Shoham, Y., Leyton-Brown, K.: Multiagent Systems: Algorithmic, Game Theoretic and Logical Foundations. Cambridge University Press, Cambridge (2008)

108. Shoham, Y., Powers, R., Grenager, T.: If multi-agent learning is the answer, what is the question? Artificial Intelligence 171(7), 365–377 (2007)
109. Singh, S., Kearns, M., Mansour, Y.: Nash convergence of gradient dynamics in general-sum games. In: Proceedings 16th Conference on Uncertainty in Artificial Intelligence (UAI 2000), San Francisco, US, pp. 541–548 (2000)
110. Singh, S.P., Jaakkola, T., Jordan, M.I.: Reinforcement learning with soft state aggregation. In: Tesauro, G., Touretzky, D.S., Leen, T.K. (eds.) Advances in Neural Information Processing Systems 7, pp. 361–368. MIT Press, Cambridge (1995)
111. Smith, J.M.: Evolution and the Theory of Games. Cambridge University Press, Cambridge (1982)
112. Spaan, M.T.J., Vlassis, N., Groen, F.C.A.: High level coordination of agents based on multiagent Markov decision processes with roles. In: Workshop on Cooperative Robotics, 2002 IEEE/RSJ International Conference on Intelligent Robots and Systems (IROS-2002), Lausanne, Switzerland, pp. 66–73 (2002)
113. Stephan, V., Debes, K., Gross, H.M., Wintrich, F., Wintrich, H.: A reinforcement learning based neural multi-agent-system for control of a combustion process. In: Proceedings IEEE-INNS-ENNS International Joint Conference on Neural Networks (IJCNN-2000), Como, Italy, pp. 6217–6222 (2000)
114. Stone, P., Veloso, M.: Team-partitioned, opaque-transition reinforcement learning. In: Proceedings 3rd International Conference on Autonomous Agents (Agents-1999), Seattle, US, pp. 206–212 (1999)
115. Stone, P., Veloso, M.: Multiagent systems: A survey from the machine learning perspective. Autonomous Robots 8(3), 345–383 (2000)
116. Suematsu, N., Hayashi, A.: A multiagent reinforcement learning algorithm using extended optimal response. In: Proceedings 1st International Joint Conference on Autonomous Agents and Multiagent Systems (AAMAS-2002), Bologna, Italy, pp. 370–377 (2002)
117. Sueyoshi, T., Tadiparthi, G.R.: An agent-based decision support system for wholesale electricity markets. Decision Support Systems 44, 425–446 (2008)
118. Sutton, R.S.: Learning to predict by the method of temporal differences. Machine Learning 3, 9–44 (1988)
119. Sutton, R.S.: Integrated architectures for learning, planning, and reacting based on approximating dynamic programming. In: Proceedings 7th International Conference on Machine Learning (ICML-1990), Austin, US, pp. 216–224 (1990)
120. Sutton, R.S., Barto, A.G.: Reinforcement Learning: An Introduction. MIT Press, Cambridge (1998)
121. Szepesvári, C., Smart, W.D.: Interpolation-based Q-learning. In: Proceedings 21st International Conference on Machine Learning (ICML-2004), Bannf, Canada, pp. 791–798 (2004)
122. Tamakoshi, H., Ishii, S.: Multiagent reinforcement learning applied to a chase problem in a continuous world. Artificial Life and Robotics 5(4), 202–206 (2001)
123. Tan, M.: Multi-agent reinforcement learning: Independent vs. cooperative agents. In: Proceedings 10th International Conference on Machine Learning (ICML 1993), Amherst, US, pp. 330–337 (1993)
124. Tesauro, G.: Extending Q-learning to general adaptive multi-agent systems. In: Thrun, S., Saul, L.K., Schölkopf, B. (eds.) Advances in Neural Information Processing Systems 16, MIT Press, Cambridge (2004)
125. Tesauro, G., Kephart, J.O.: Pricing in agent economies using multi-agent Q-learning. Autonomous Agents and Multi-Agent Systems 5(3), 289–304 (2002)
126. Tillotson, P., Wu, Q., Hughes, P.: Multi-agent learning for routing control within an Internet environment. Engineering Applications of Artificial Intelligence 17(2), 179–185 (2004)
127. Touzet, C.F.: Neural reinforcement learning for behaviour synthesis. Robotics and Autonomous Systems 22(3–4), 251–281 (1997)

128. Touzet, C.F.: Robot awareness in cooperative mobile robot learning. Autonomous Robots 8(1), 87–97 (2000)
129. Tsitsiklis, J.N.: Asynchronous stochastic approximation and Q-learning. Machine Learning 16(1), 185–202 (1994)
130. Tuyls, K., 't Hoen, P.J., Vanschoenwinkel, B.: An evolutionary dynamical analysis of multi-agent learning in iterated games. Autonomous Agents and Multi-Agent Systems 12(1), 115–153 (2006)
131. Tuyls, K., Maes, S., Manderick, B.: Q-learning in simulated robotic soccer – large state spaces and incomplete information. In: Proceedings 2002 International Conference on Machine Learning and Applications (ICMLA-2002), Las Vegas, US, pp. 226–232 (2002)
132. Tuyls, K., Nowé, A.: Evolutionary game theory and multi-agent reinforcement learning. The Knowledge Engineering Review 20(1), 63–90 (2005)
133. Uther, W.T., Veloso, M.: Adversarial reinforcement learning. Tech. rep., School of Computer Science, Carnegie Mellon University, Pittsburgh, US (1997),
    http://www.cs.cmu.edu/afs/cs/user/will/www/papers/Uther97a.ps
134. Vidal, J.M.: Learning in multiagent systems: An introduction from a game-theoretic perspective. In: Alonso, E., Kudenko, D., Kazakov, D. (eds.) AAMAS 2000 and AAMAS 2002. LNCS (LNAI), vol. 2636, pp. 202–215. Springer, Heidelberg (2003)
135. Vlassis, N.: A Concise Introduction to Multiagent Systems and Distributed Artificial Intelligence. Synthesis Lectures in Artificial Intelligence and Machine Learning. Morgan & Claypool Publishers (2007)
136. Wang, X., Sandholm, T.: Reinforcement learning to play an optimal Nash equilibrium in team Markov games. In: Becker, S., Thrun, S., Obermayer, K. (eds.) Advances in Neural Information Processing Systems 15, pp. 1571–1578. MIT Press, Cambridge (2003)
137. Watkins, C.J.C.H., Dayan, P.: Q-learning. Machine Learning 8, 279–292 (1992)
138. Weinberg, M., Rosenschein, J.S.: Best-response multiagent learning in non-stationary environments. In: Proceedings 3rd International Joint Conference on Autonomous Agents and Multiagent Systems (AAMAS-2004), New York, US, pp. 506–513 (2004)
139. Weiss, G. (ed.): Multiagent Systems: A Modern Approach to Distributed Artificial Intelligence. MIT Press, Cambridge (1999)
140. Wellman, M.P., Greenwald, A.R., Stone, P., Wurman, P.R.: The 2001 Trading Agent Competition. Electronic Markets 13(1) (2003)
141. Wiering, M.: Multi-agent reinforcement learning for traffic light control. In: Proceedings 17th International Conference on Machine Learning (ICML-2000), pp. 1151–1158. Stanford University, US (2000)
142. Wiering, M., Salustowicz, R., Schmidhuber, J.: Reinforcement learning soccer teams with incomplete world models. Autonomous Robots 7(1), 77–88 (1999)
143. Zapechelnyuk, A.: Limit behavior of no-regret dynamics. Discussion Papers 21, Kyiv School of Economics, Kyiv, Ucraine (2009)
144. Zinkevich, M.: Online convex programming and generalized infinitesimal gradient ascent. In: Proceedings 20th International Conference on Machine Learning (ICML-2003), Washington, US, pp. 928–936 (2003)

# Chapter 8

# Multi-Agent Technology for Fault Tolerant and Flexible Control

Pavel Tichý[1] and Raymond J. Staron[2]

[1] Rockwell Automation Research Center, Pekarska 10a,
15500 Prague 5, Czech Republic
[2] Rockwell Automation, 1 Allen-Bradley Drive,
Mayfield Hts., OH 44124-6118, USA
{ptichy,rjstaron}@ra.rockwell.com

## 1 Summary

One of the main characteristics of multi-agent systems (MAS) is fault tolerance. When an agent is unavailable for some reason, another agent with similar capabilities can theoretically compensate for this loss. Many key aspects of fault tolerance in MAS are described in this chapter including social knowledge, physical distribution, agent development, and validation. Therefore, the focus is not only on a fault tolerant agent platform with necessary services (e.g., fault tolerant social knowledge), but also on the design that can significantly reduce mistakes in agent programming and validation that can discover faults that manifest as failures during the testing phase.

Another aspect of MAS is flexibility or agile control. For example, the ability to locate agents based on an agent name or based on a capability that is needed to accomplish some task results in a more flexible design of control systems. Any new agent in the system is therefore immediately considered for planning and similarly, agent removal will not disrupt the overall functionality of the system if there are agents still present that offer similar capabilities. Nevertheless, the same flexibility has to be present also in a visualization system to ensure that any structural change in MAS is reflected in a user interface. The same applies also to simulation that is used to test the system before it is deployed to control real hardware.

In summary, it is not enough to apply MAS to immediately obtain fault tolerance and flexibility. The MAS has to be designed with all aspects of fault tolerance in mind and flexibility is achieved only when the application allows changes to be made on the fly and the MAS is designed accordingly. The goal of this chapter is to summarize all these aspects and present relevant solutions.

## 2 Introduction

An ability to tolerate failures, either foreseen or unexpected, is usually referred to as fault tolerance or robustness. For example, [30] declares that it is the ability of MAS to show predefined qualitative behavior in the presence of unaccounted types of events and technical disturbances. Fault tolerance is one of the key aspects of MAS

D. Srinivasan & L.C. Jain (Eds.): Innovations in MASs and Applications – 1, SCI 310, pp. 223–246.
springerlink.com © Springer-Verlag Berlin Heidelberg 2010

[46] that classical control is unable to satisfy for large-scale systems. For example, system components can fail in a variety of ways, and the control system (CS) must respond to all such possible situations correctly. A classical CS must expect all possible combinations of failures, a number that can have exponential complexity (see [22] for definitions of mistake, fault, failure, etc. from the area of software testing).

Since a MAS consists of autonomous agents that are able to dynamically exchange information, any combination of failures can be resolved without preprogramming all possible combinations. Nevertheless, blindly applying an ordinary MAS to increase fault tolerance can lead to the opposite result. A fault tolerant MAS must satisfy the following characteristics:

- *Reliable communication.* The system has to guarantee that no messages are lost or duplicated. Every message must be delivered, or else the sender must be notified if a message cannot be delivered. MAS can be designed to gracefully degrade, but keep functioning to some extent, when there is a problem with communication and reliable communication helps in this respect (e.g., there is a huge difference when an agent is immediately notified about undelivered message and when it has to wait for its communication response timeout).

- *Fault tolerant agent platform.* Since the agent platform is an environment where agents live, the possible failure of some agent should affect neither its own environment nor the other agents in the system.

- *Fault tolerant social knowledge.* Knowledge about other agents [26] is a fundamental part of a MAS. Thus this knowledge has to be present or dynamically obtained in a fault tolerant manner. For example, having only one directory facilitator that manages this knowledge is a single point of failure in the system.

- *Physical distribution.* It should be possible to physically distribute the agents of a MAS and allow communication among them. This distribution increases fault tolerance in the case of a hardware failure, power failure, etc.

- *Fault tolerant agent architecture.* An agent should be internally designed with respect to possible internal and external failures, e.g., when an agent is disconnected from other agents, it should still work correctly (possibly with decreased efficiency).

- *Fault tolerant negotiation.* Communication protocols and interaction among agents have to ensure fault tolerance. For example, according to a given protocol an agent expects an answer from another agent. The receiving agent must know how long to wait for the answer and how to proceed if an answer is not received in time.

There are also several approaches to decrease the number of failures in a MAS. The most important ones are the following:

- *Minimize agent programming mistakes.* Agents are currently programmed by humans and thus the presence of mistakes in this process is unavoidable. Using a development environment (DE) that is based on libraries and provides automatic generation of code can decrease the number of faults in a MAS. The user can program the library once and deploy the agents to multiple projects.

- *Ability to validate.* Since a MAS can have emergent behavior [14], it is important to be able to efficiently validate overall behavior to ensure its correct functionality.

Fault tolerance becomes more and more important in the design of MAS frameworks, e.g., the DARX framework [25] focuses on selectable agent replication within MAS and corresponding name service enhancements. Mobile agents represent another point of view for fault tolerance in MAS design [35].

We are focusing not only on a fault tolerant agent platform with necessary services (e.g., social knowledge), but also on the design that can significantly reduce mistakes in agent programming, and validation that can discover faults that manifest to failures during the testing phase. All of these aspects are important especially for industrial applications where the system is required to be as operational as possible to minimize production loss. The same applies for mission critical systems designed, for example, for space flights.

The chapter is organized as follows. Section 3 describes traditional fault tolerance techniques that can be directly applied to MAS followed by classification of agent and MAS failures in Section 4. Section 5 presents the main characteristics of agent platform and their effect on fault tolerance. Social knowledge in MAS is another area where fault tolerance is very important and general techniques that can be applied here are presented in Section 6. Development Environment that is based on libraries and minimizes agent programming mistakes is presented in Section 7. How to program agents to increase robustness of an agent and the whole MAS is shown in Section 8. Section 9 is dedicated to validating agents that are used to observe behavior of the whole system and to discover potential problems and Section 10 shows examples of industrial applications from the chilled water and material handling domains.

# 3 Traditional Fault Tolerance Techniques

The traditional fault tolerance techniques have not been developed to be specifically used in MAS. The development of these techniques is targeted to database systems, application servers, resource managers, and distributed systems, but these techniques can be used in a much broader range of software or hardware systems. The summary of these techniques is described for instance in [24]. One of the first fault tolerant approaches is SWIF that is targeted to synchronization among processors and memory [45]. We will focus only on general techniques that can be directly applied to MAS.

## 3.1 Warm and Hot Backup

Assume that the system has one primary component (for example an agent or hardware equipment) in an active state and one secondary component in an inactive state. In the active state, a component can be contacted by other components and can respond to their requests. While in the inactive state, a component can not be contacted.

In the *warm backup* technique, the initial goal of the secondary is only to observe the primary and wait for a failure of the primary. When the failure occurs, the secondary becomes active and starts recovering to the last known state of the primary. After the recovery process has been completed the secondary becomes the primary component.

In the *hot backup* technique, the initial goal of the secondary component is not only to observe the primary and wait for a failure of the primary, but also to monitor inputs and outputs of the primary. When the failure of the primary component occurs, the secondary component can immediately take over as primary without the recovery phase.

### 3.2  N-Version Programming

$N$-version programming is defined as an independent generation of $N \geq 2$ functionally equivalent programs, called 'versions', from the same initial specification [5].

By the 'independent generation of programs' we mean that the programming efforts are carried out by $N$ individuals or groups that do not interact with each other relative to the programming process. Whenever possible, different algorithms, programming languages, and translators are used in each effort.

All $N$ versions are running simultaneously and a voting mechanism is used to compute the result. Either exact voting, where all versions have the same number of ballots, or inexact voting, where each version has a specific number of ballots, can be used as the voting mechanism.

A fault tolerant pipeline technique [29] is an example of $N$-version programming in MAS. In this technique a fault tolerant computation is achieved by multiple computations of the same problem running simultaneously. Several agents are used for the same phase of the computation and each partial result is obtained by the voting process.

Both warm and hot backup techniques and $N$-version programming can be used throughout the development of MAS to increase fault tolerance of the whole system. For example, there can be $N$ agents with the same capabilities programmed in different languages and/or by different algorithms. In this case when one agent fails then other agents with the same capabilities can still remain functional.

Nevertheless, applying only these techniques is not sufficient in MAS. For example, assume that an agent can compute and execute only one solution to some task. Now, if this solution can not execute for some reason the agent must compute an alternative solution that is possibly worse or less efficient, but still better than no solution or the same solution again which is already known to fail. The backup techniques or $N$-version programming alone cannot help in this case.

## 4  Classification of Failures in MAS

Different types of failures can occur in agents and in MAS and each one has an impact on the robustness and survivability of the system architecture or on the particular implementation of the system. For example, [6] classifies basic failures such as timing failure, omission failure, and response failure. Nevertheless, more detailed classification can be found, for example, in [43] where possible types of agent failures and MAS failures are defined and briefly summarized as follows.

## 4.1  Agent Failure Types

Failures of an agent can be classified by their observability (see Fig. 1). *Unobservable* agent failure can be defined as a failure of the agent that cannot be proven to have occurred by all possible observers of inputs and outputs of the agent. *Observable* agent failure can be defined as a failure that an observer of inputs and outputs of the agent can prove has occurred.

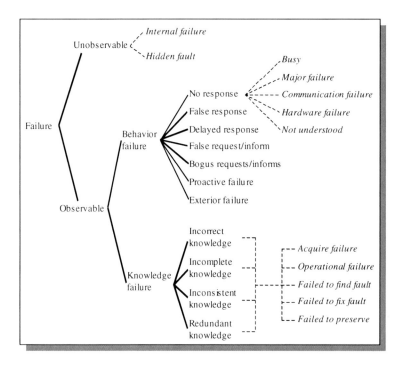

**Fig. 1.** Categorization of agent failures

Unobservable failures can have two possible reasons that are again unobservable:

- *Internal failure* - a failure in an internal processing of an agent that is not observable from outside the agent. The agent has been able to either recover from the failure or use an alternative way to avoid the manifestation of the failure.
- *Hidden fault* - a fault in the implementation of an agent that has not manifested yet. After the fault gets triggered, the fault transforms into another type of failure, possibly an internal failure.

The observable failures can be further classified as. *Behavior failure* manifests when an agent does not behave as expected according to a given specification. *Knowledge failure* appears when an agent deals with or provides incorrect, irrelevant, or old information according to a given specification for a particular situation.

The behavior failure manifests itself as one of the following types of failures:

- *No response* – an agent fails to respond to a particular request in a given timeframe for the response.
- *False response* – an agent responds to a particular request in a different way than noted in the given specification of the communication protocol.
- *Delayed response* – an agent responds to a particular request in a given timeframe for response but slower than noted in a given specification.
- *False request/inform* – an agent sends a request or an inform that is not correct according to the given specification.
- *Bogus requests/informs* – an agent sends numerous requests or informs that can be considered as fussing.
- *Proactive failure* – an agent either fails to take an action or chooses an incorrect action, according to a given specification, in a given situation, and in a current internal state.
- *Exterior failure* – an agent fails to behave correctly, according to a given specification, other than in communication among agents. For instance, an agent agreed upon some action and failed to finish this action in a given timeframe due to an external failure. Another example is that the agent agreed to change its location and failed to do it.

Furthermore, see [43] for *reasons* of these failures that do not have to be necessarily observable.

Knowledge failures can also be further classified using the classification of faults in knowledge bases [4]. We omit the intractability knowledge failure since it is not observable and, according to the presented classification of agent failures, it is either a reason why a knowledge failure occurred or a type of unobservable internal failure. Thus we classify knowledge failures as:

- *Incorrect knowledge* – some part of knowledge contains information that is not correct.
- *Incomplete knowledge* – some part of knowledge that should be present is missing. For instance, the message of type request-for-search does not contain the subject that should be searched for.
- *Inconsistent knowledge* – knowledge or a part of knowledge that is complete and correct per se, but in a broader view and/or in the current situation it is not correct.
- *Redundant knowledge* – knowledge or a part of knowledge that can be found or extracted elsewhere. The redundant knowledge can be, for instance, in a message that is used to register an agent and this message contains the same capabilities of the agent twice.

Again, see [43] for *reasons* of these failures that do not have to be necessarily observable.

## 4.2  Multi Agent System Failure Types

All of the agent failure categories provide a framework to classify failures of a single agent. If we consider the whole MAS, we cannot simply assume that failures of the MAS consist just of failures of a single agent. The MAS can be designed either by applying the connectionist model or using the concept of a hierarchical collective system [14].

The *connectionist model* is based on the fact that the creator of the system designs the behavior of components (agents) instead of designing the behavior of the whole system. As the components are connected together, they create an emergent behavior that is possibly much more complex than if we sum up the behavior of the components. As the system creator does not design the behavior of the whole system but only the behavior of system components, this approach can lead to mistakes that cannot be considered as a fault in any particular component. In this case the fault that manifests in a failure can be located in the emergent behavior.

The concept of a *hierarchical collective system* is based on a top-down design of the system. The system creator designs a rough large task that is divided into smaller and finer subtasks. By using this concept, the system creator has a possibility to identify the appropriate fault when a failure occurs. In another words, if there is no emergent behavior in the system, then fault in this behavior simply cannot exist.

Agents are not running in isolation, but are a part of a MAS architecture. This architecture provides time and space to run the agents and a communication system, usually with given languages, protocols, and ontology definitions, that allows the agents to cooperate. These elements not only provide the possibility to create agents (components), but also represent the glue that connects them together.

Based on these remarks, we can now define possible types of MAS failures.

- *Agent failure* – any failure of an agent implicitly means a failure within the MAS. This failure does not necessarily mean the failure of the whole MAS.
- *Agent environment failure* – the agent environment (platform) is not functioning properly. For instance, an agent tries to create a timer, but the creation failed for some reason.
- *Incorrect emergent behavior* – the agents do not behave as expected although the behavior of each particular agent is according to a specification for a given situation and given internal states. This usually means that the designer of the specification did not consider a given situation, internal states, or a combination of these.
- *Missing capability* – an agent that is running within a MAS fails in the search for a capability provider. This failure occurs when the system is missing an agent with the desired capability and agents are dependent on this capability.
- *Missing knowledge* – an agent is unable to obtain some particular knowledge using all available requests at its disposal. This failure occurs when the MAS is not designed correctly to store or to provide this knowledge.
- *Hardware failure* – any failure of hardware other than the hardware on which agents are running, for instance, the failure of communication media.

- *Real-time response failure* – a MAS fails to provide correct outputs in a given timeframe, occurring when the real time reaction of a MAS on a particular set of inputs is slower or faster than expected.
- *Robustness failure* – a MAS fails to provide the desired or expected fault-tolerant behavior after some other type of failure occurs.

Note that in all of these MAS failure types, except for the agent failure and the robustness failure, no particular agent failed in the system at run-time.

## 5   Fault Tolerant Agent Platform

An agent platform is an environment where agents live, similar to the universe for human beings. To specify the main characteristics and requirements of an agent platform, we focus first on agents and their characteristics to make sure that we cover all their requirements. Consequently the agent architecture is organized mainly according to the following characteristics:

- *Autonomy* – Each agent makes its own decisions and is responsible for carrying out its decisions toward successful completion.
- *Communication* – Agents share a common language. They communicate directly via message passing or indirectly by other interaction means.
- *Collaboration and Coordination* – Agents combine their capabilities with simple rules of interaction into collaboration groups (clusters) to adapt and respond to events and goals. Or, they at least coordinate their actions with others to avoid collisions and deadlocks or to improve their efficiency.
- *Negotiation* – Agents use common protocols and communication mechanisms for interaction purposes, defined for example in [12].
- *Fault tolerance* – Agents possess the capability to detect equipment failures and to prevent failures from propagating, and the ability to process multiple solutions (plans) to overcome situations where the best solution is not applicable or the best plan fails to execute.
- *Pro-action* – Agents periodically or asynchronously propose strategies to enhance the system performance or to prevent the system from entering harmful states.
- *Organization* – Agents have the ability to create and maintain organizational structure in which each agent has its role and position. See [13] for major types of organizational structures.
- *Learning* – Agents have the ability to adapt their behavior and knowledge of another agent to new facts to improve efficiency and performance. This adaptation can be achieved in a centralized manner, but usually is better implemented in a distributed manner.
- *Diagnostics* – Agents can monitor and analyze itself or other agents to detect and identify failures. Again, a distributed implementation is usually better than a centralized one.

- *Planning* – Agents can construct a plan of future actions [10] including distributed plan decomposition, distributed resource allocation with resolution of deadlocks, and plan execution.

To support agent autonomy, the agent platform provides agents with their own thread of execution (or interpretation), i.e., an agent has execution time available and can manage how this time is used efficiently, similar to a human being who is deciding how to use his time to live according to his preferences. The agent platform has to ensure that agents have threads of execution as independent as possible to make sure that a failure or deadlock in one agent does not affect others. Any resource that is accessed by more than one agent has to be synchronized in a fault tolerant way to avoid deadlocks, e.g., when an agent has a resource locked and terminates for some reason then the resource has to be unlocked.

An agent platform has to ensure communication that is reliable. In addition, an agent platform should provide multiple communication protocols, languages, and communication media to increase fault tolerance. For example, when communication over a wire is malfunctioning then the agent platform can switch to wireless. An agent registers multiple addresses where it can be reached, e.g., one wire-based and the other wireless.

There can be even more advanced techniques applied to increase robustness, such as *transactional conversations* proposed in [30], where agent conversations are treated as distributed transactions.

An agent platform can guide cooperation among agents via *platform agents*, such as the Agent Management System (AMS), Directory Facilitator (DF), and Agent Communication Channel (ACC) agents defined in the Foundation for Intelligent Physical Agents (FIPA) [12].

- The Agent Management System (AMS) – provides the authentication of resident agents and the control of registration services ('white pages').
- The Directory Facilitator (DF) – offers agent capabilities registration and lookup services ('yellow pages'). The behavior of the DF agent is similar to the service matchmaking mechanism. One of the earliest matchmaker systems is the ABSI (Agent-Based Software Interoperability) facilitator [39]. The service matchmaking approach is also used in the coordination within the coalition formation methodology [38] and in the acquaintance model [32] during an initial search for possible cooperators. The matchmaker then steps aside and is not involved in any further negotiation process.
- The Agent Communication Channel (ACC) – provides message routing (bridging) within or over the borders of an agent platform. The behavior of the ACC agent is similar to the embassy agent [16].

An agent platform need not contain all mentioned platform agents. But as long as these agents are offered by an agent platform it has to ensure that they are always present and redundant. We further focus on this task in next section.

An *ontology* [31] can be used to ensure a good level of understanding among agents that can be part of a heterogeneous MAS. The use of a common ontology can increase the possibility for cooperation and thus indirectly increase fault tolerance.

An agent platform should allow distributing the intelligence of the system among multiple execution units, e.g., personal computers and PLCs (programmable logical controllers) for industrial purposes. The agents are loosely coupled, but they are cohesive and adaptable.

Moreover, an agent platform can support *mobility*, i.e., the possibility to move an agent from one execution unit to another while running. An agent can replicate (clone) itself and move to a different location to increase fault tolerance and/or processing power. How to efficiently improve fault tolerance by agent replication without increasing system complexity is elaborated, for example, in [11].

There are numerous agent platforms available and it is beyond the scope of this chapter to provide a survey of them. Some of them even offer an Integrated Development Environment (IDE) for development. One of the most popular and widely used platforms is the FIPA-compliant Java Agent DEvelopment Framework (JADE) [19]. AGlobe [2] is a platform that has built-in support for communication inaccessibility and is known for its fast communication. JACK [18] supports the Belief-Desire-Intention (BDI) architecture. A list of platforms can be found for example at [1], unfortunately not further updated.

For PLCs it is usually required to extend firmware to support MAS. For example, Rockwell Automation Inc. developed the Autonomous Cooperative System (ACS) [27] to run in its Logix family of controllers (see Fig. 2). With these extensions, component-level intelligence is possible and the physical devices can be converted into intelligent nodes with negotiation capabilities. These nodes are similar to, for example, intelligent components discussed in iShopFloor [36].

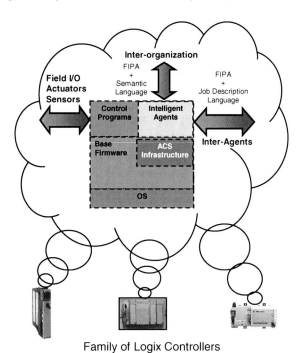

Family of Logix Controllers

**Fig. 2.** Agent extensions to Logix controllers

During the design of the ACS Rockwell took into consideration not only standardization efforts by FIPA [12], but also the fact that its goal is to use this system for control. Thus the ACS is able to interact with FIPA-compliant systems, for example, the Java Agent DEvelopment Framework (JADE) [19]. For industrial control implementation, the FIPA messages are encapsulated inside Common Industrial Protocol (CIP) packets [7]. This encapsulation enables communication on many types of industrial networks, e.g., EtherNet/IP, ControlNet, and DeviceNet. For the content language of the agent communication language (ACL) Rockwell uses either the FIPA Semantic Language (SL) [12] or its proprietary Job Description Language (JDL) [41], with XML, bit efficient, or Lisp representations.

One of the key parts of a MAS is the ability to locate agents based on an agent name or based on a capability that is needed to accomplish some task. This ability allows for more flexible design of control systems. Any new agent in the system is therefore immediately considered for planning and similarly, agent removal will not disrupt the overall functionality of the system if there are agents still present that offer similar capabilities.

The ACS includes full support for these features via a structure of Directory Facilitators, or more generally, *MIiddle-Agents* (MIAGs) [40]. These agents reside in a fault tolerant structure in ACS to overcome possible failures.

## 6   Fault Tolerant Social Knowledge Distribution

Social knowledge is the knowledge that is used to deal with other agents in the MAS. Social knowledge consists of the information about the names of agents, their location (address), their capabilities (services), the language they use, their actual state, their conversations, behavioral patterns, and so on [26].

Social knowledge can be distributed in multiple ways:

- *Centralized* – stored, managed, and offered from one physical or functional point, for instance, the blackboard architecture and federated architectures with one MIAG.
- *Distributed* – statically or dynamically plugged directly into all agents in the system, for instance, acquaintance models [32] or an architecture without any MIAG [15].
- *Hybrid* – social knowledge organized into groups, hierarchies, etc., for instance, a teamwork-based technique [23].

Rockwell designed and implemented a structure of MIAGs called dynamic hierarchical teams (DHT) [40][42]. This structure has a user defined level of fault tolerance and is moreover fixed scalable, i.e., the structure can be extended by fixed known cost. Several other approaches have been used already to deal with fault tolerance and scalability. One proposed approach is a teamwork-based technique [23] that uses a group of MIAGs in which each MIAG is connected to all other MIAGs, thus forming a complete graph. This technique offers fault tolerance but since it uses a complete graph this structure is not fixed scalable.

The main features of the DHT architecture are briefly described in what follows; detailed information can be found in [40].

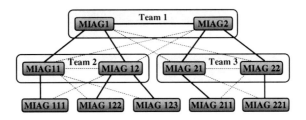

**Fig. 3.** Example of a 3-level DHT architecture

Assume that a MAS consists of MIAGs and end-agents. MIAGs form a structure that can be described by graph theory. Graph vertices represent MIAGs and graph edges represent a possibility for direct communication between two MIAGs, i.e., some set of communication channels. Each MIAG that is not a leaf in the tree should be supported by another MIAG. Groups of these MIAGs are called *teams* (see Fig. 3). Whenever one of the MIAGs from the team fails, other MIAGs from the team can subrogate this agent. During normal operation of the DHT structure all MIAGs use only primary communication channels (solid lines). If a failure of a primary channel occurs then a secondary channel (dashed lines) is used instead.

To describe the DHT structure more precisely, we use graph theory [8] to present the following formal definitions.

*Definition 1:* A graph $G$ will be called a DHT graph if there exist non-empty sets $V_1, ..., V_n \subset V(G)$ such that they are pairwise disjoint and $V_1 \cup ... \cup V_n \neq V(G)$. In that case, the complete subgraph $G_i$ of the graph $G$ induced by the set of vertices $V_i$ will be called a team of $G$ if all of the following is satisfied:

1) $\forall v(v \in V(G)\backslash V_1 \rightarrow \exists j \forall w(w \in V_j \rightarrow \{v, w\} \in E(G)))$ [1]
2) $\forall v(v \in V(G) \wedge v \notin V_1 \cup ... \cup V_n) \rightarrow \exists! j \forall w(w \notin V_j \rightarrow \{v, w\} \notin E(G)))$ [2]
3) $\forall j((j > 1) \wedge (j \leq n) \rightarrow \exists! k((k < j) \wedge \forall v \forall w(v \in V_j \wedge w \in V_k \rightarrow \{v, w\} \in E(G)) \wedge \forall u \forall m(u \in V_m \wedge (m < j) \wedge (m \neq k) \rightarrow \{v, u\} \notin E(G))))$ [3]

*Definition 2:* The graph $G$ is called DHT-$\lambda$ if $G$ is DHT and $|V_i| = \lambda$ for every $i = 1,...,n$, where $\lambda \in \mathbb{N}$.

---

[1] For all vertices $v$ of $G$ except $V_1$ (since there is no team with lower index than $V_1$) there has to be a team such that $v$ is connected to all members of this team.
[2] For all vertices $v$ that are not members of any team there are only connections to one team and there cannot be any other connection from $v$.
[3] All members of each team except $G_1$ are connected to all members of exactly one other team with lower index.

The fault tolerance of an undirected graph is measured by its vertex and edge connectivity. It can be proved [43] that if the graph $G$ is DHT-$\lambda$ then the vertex connectivity $\kappa(G) = \lambda$ and the edge-connectivity $\lambda(G) = \lambda$. The DHT structure in which teams consist of $\lambda$ middle-agents is therefore fault tolerant to a simultaneous failure of at least $\lambda$ - 1 MIAGs and also to a simultaneous failure of at least $\lambda$ - 1 communication channels. It can be also proved [43] that the graph $G$ of type DHT-$\lambda$ is maximally fault tolerant, i.e., there is no bottleneck in the structure of connections among nodes (MIAGs) in the DHT architecture.

Several failure detection mechanisms can be used, not only between MIAGs, but also among other agents, to ensure that an agent detects the failure of a target agent, which oftentimes means that the agent is informed by a third agent in the case of the target's unavailability.

- *Response Timeout Mechanism* – This timeout is set by the originating agent and is used to ensure that the response arrives from the target agent back to the originating agent in a given timeframe.
- *Heartbeat Mechanism* – Also called a keep alive mechanism [3], this mechanism is a periodic checking that the agent did not fail.
- *Meta Agent Observation* – An "overseer" agent that observes the communication among agents and can use a reasoning process to discover possible failures of agents [33]. For example, Fault Tolerance Managers (that even have their backup) and other Adaptive Reconfigurable Mobile Objects for Reliability (ARMOR) in Chameleon architecture [20] work as meta agents and use four level error detection mechanism. Moreover, other types of MAS monitoring techniques such as the Socially Attentive Monitoring (SAM) [21] or the sentinel [15] approach can be used for similar purposes.

The heartbeat mechanism can be implemented in two ways. Assume that an agent that is observed for possible failure is called the *subject* and an agent that is taking care of the subject is called the *observer*. The heartbeat mechanism can be either substantive or enforced.

- *Substantive* heartbeat – The observer initiates the heartbeat mechanism with the subject first. Then the subject periodically sends a status to the observer. Upon each incoming status the observer sets up a timer or resets an already running timer to an initial value. If the timer in the observer expires the subject is declared as having failed (no response or delayed response failure).
- *Enforced* heartbeat - The observer periodically sends a request for a report to the subject. Whenever the subject receives the request, the subject reports back status to the observer. If the observer did not receive the report within a specified timeframe, the subject is declared as having failed (no response or delayed response failure).

For example, advanced techniques such as fault tolerant agent communication language (FT-ACL) [9] use all mentioned failure detection mechanisms to ensure that the interaction does not reach a dead lock.

# 7  Development Environment

As the effort to program agents becomes significant, a natural approach is to develop the notion of an "agent library", essentially a collection of class definitions that describe the behavior both of the agents and of any nonagent components.

Each library is targeted to a specific application domain, e.g., material handling systems, and so can be used to create control systems for multiple similar facilities in the same domain. The control code is generated by the same process from the same templates. In this way, the effort to build the library is amortized over all the facilities (control applications) built with the library, and each facility garners the benefits of having CS software that is structured, well defined, predictable, and tested. Moreover, any additions and/or changes to the library subsequently appear in all affected instances. In other words, instead of changing all affected instances individually, the user applies the change only in one place in the library, thereby not only speeding up the process but also avoiding any inconsistencies and mistakes in programming. The description of control behavior consists of a set of class descriptions for the various component classes. A component may or may not be an agent. If so, in addition to control behavior, it contains a description of its intelligent behavior.

For example, for the development of ACS agents there exists a development environment (DE) that introduces the following dimensions into the development process:

- Allows the user to specify the physical and behavioral aspects of the application in a manner completely independent of the CS.
- Enables the user to specify a multiprocessor control system in a manner completely independent of the application that is to run on it.
- Assists the user in combining an application with a CS.
- Generates the control code and behavior descriptions for each agent in the system.
- Combines the code for all agents assigned to each processor in the system.
- Augments each controller automatically to handle the communications to other controllers as a result of the program distribution.
- Communicates with all the controllers involved in an application, for their programming and configuration, and for subsequent monitoring.

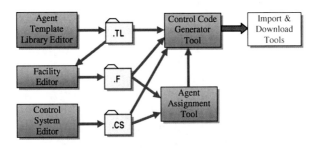

**Fig. 4.** Development process

Fig. 4 shows the general development flow through the system. One foundation for the DE is a library of components called the template library (TL). The library is editable by the user, and each template can contain both low-level control behavior (written in ladder diagram) and higher-level intelligent behavior. The model for the control behavior supports an "object" view of the components in that, for example, inheritance is supported. The TL author can express, for example, that "A Radar is a type of CombatSystem". Each instance of Radar inherits all ladder data definitions and the logic from the CombatSystem template.

The user creates a facility (F) from components of the template library and customizes their parameters. Next, the user establishes a CS that describes all the controllers, I/O cards, and networks, plus their interconnections. After the TL, F, and CS parts are completed, the user generates and compiles the code. After the agent assignment, the user downloads the software into the controllers.

Since TL, F, and CS editors are independent, it is possible to change only the necessary parts when the system needs to incorporate changes. For example, a new controller can be added by the CS editor and subsequently have some agents assigned to it. The system is regenerated in a manner consistent with all modifications.

# 8  Fault Tolerant Agent Programming

There are many aspects of agent programming that significantly affect fault tolerance.

- *Automatic code generation* – provided usually by DE.
- *Layered architecture* – an agent can consist of several layers (for example high-level and low-level) and each level is loosely dependent on other levels.
- *Multiple solutions* – an agent should process multiple solutions (plans) to overcome situations where the best solution is not applicable or the best plan fails to execute.

Automatic code generation can significantly reduce the number of mistakes that the developer can produce. The user will program agents via the DE, allowing a simplified programming process since the resulting code is automatically generated.

Each agent consists of two major parts.

- *High-level* – written in a high-level programming language. This is the core of the agent, the part that is able to communicate with other agents, plan, reason, etc.
- *Low-level* – written in relay ladder logic, IEC 61131-3 [17]. This part is responsible for real-time control to guarantee response times and safety of the control system, and also for generating events to the high-level part of the agent to notify it of some state or condition requiring its attention.

The low-level ensures safe functionality of a control system even when the high-level part is not working properly or when there is a need for fast reaction for some critical situation, perhaps before the high-level part is able to make a more accurate decision.

Both these parts are automatically generated based on templates of agents stored in the library. After creation of any type of agent template it is possible to create as many instances as needed to match the application.

Rockwell has so far been using a declarative style of programming agent behavior. Nevertheless, this approach has been reevaluated and has become clear that the declarative style has low flexibility since any additional feature has to be included into all parts of the system. Thus, a procedural style of programming has been designed and a procedural engine has been created. This engine enables us to use the full power of a programming language (Java or C++ in this case) together with a set of functions and attributes to interact with other agents, trigger planning processes, and so on.

Agent behavior can be designed to provide multiple alternative solutions to achieve the highest possible level of fault tolerance (see [22] for definitions of basic levels of fault tolerance):

- *Full fault tolerance.* An alternative solution(s) that is(are) possible to use when the main solution fails. The system continues to operate without a significant loss of functionality or performance.
- *Graceful degradation.* Next alternative solution(s) with some loss of functionality or performance that is applied when full fault tolerant solution is not applicable.
- *Fail-safe.* The last available solution(s), where vital functions are preserved while others may fail. This type of solution should always be available to ensure fail-safe overall functionality of the system.

Agents in MAS depend on each other, e.g., an agent can use services of another agent and thus for each task agents form an acyclic graph of dependencies. For example [6] defines that "a server u depends on a server r if the correctness of u's behavior depends on the correctness of r's behavior". Therefore presence of multiple agents with the same or overlapping services is important for fault tolerance since a faulty agent can be substituted by another agent, i.e., an agent is dependent on services (provided by multiple agents) and not on a single agent that can fail.

## 9  Validating Agents

Due to emergent behavior of agent systems, there is a strong need for agent testing to ensure that the whole system works properly and that the emergent system behavior is correct. Performing agent testing on simulators, not on the real hardware, is preferred not only to protect the real hardware but also to speed up the whole development process. The simulation has to be attached to the system via some synchronization module to ensure correct data transfer and time synchronization.

To create a fully functional MAS, there is a need for a visualization and debugging tool to observe the internal communication among the agents and the behavior of the system and to discover potential problems. Usually multiple tools for a given implementation of MAS exist. For example, JADE and AGlobe offer a sniffer agent that is able to capture messages sent among agents and display them in a form of a Unified Modeling Language (UML) sequential diagram.

This tool can also present the information from different points of view and the user can also configure running agents by sending various service messages to them. One such tool is the JavaSniffer, a standalone Java application, developed by Rockwell Automation, Inc. It can remotely connect to a running JADE systems, Rockwell's own ACS system, or any other FIPA compliant system. It is able to visualize messages as a low-level UML sequential diagram and provides a high-level view via dynamically created traceable workflow diagrams. XML, Lisp, and BitEfficient message encodings are supported (compatible with FIPA ACL specifications) and SL, XML, and JDL (job description language) content languages are supported. It offers visualization of statistical information, message and agent filtering, automatic log file creation, etc.

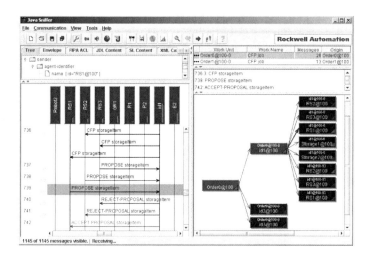

**Fig. 5.** Main window of visualization and debugging tool

The visualization screen is divided into four main sections (see Fig. 5). Each section provides the observer with information at a different level of detail.

- *Message detail view.* The window located at the top left corner provides information about the content of the selected message. The format of the information displayed is dependent on the communication language used.
- *List of messages.* The window located at the bottom left corner displays a list of messages sent among agents as a UML sequential diagram. A scrollable list of agents is located at the top of this window and each row represents one message sent from one agent to another agent, displayed as an arrow pointing from the sender of the message to the receiver.

- *List of work units.* The window located at the top right corner shows the list of work units requested by the agents. Each description of a work unit consists of work unit identification, work name, the number of subsequent messages belonging to this work unit, and the original requester.
- *Workflow view.* The window located at the bottom right corner displays a dynamically created tree of a workflow that belongs to the work unit selected in the list of work units. Any request for and reply to planning, commitment, or execution of the work is visualized as an arrow pointing from the parent (creator of the work) to a child (solver) agent. The arrow represents all messages belonging to a particular part of the work.

All parts of the visualization screen are interconnected. For example, it is possible to identify some problem in the workflow window, select the appropriate part of the conversation among the agents and receive the list of messages involved in the conversation. Any message in the list can be selected and automatically displayed in the message detail view. Also, upon selecting a message in the list of messages, the appropriate workflow and conversation in this workflow are automatically selected. Therefore, the user can seamlessly switch between low-level and high-level visualization views.

## 10  Industrial Application Examples

This section presents two examples of industrial applications [33] where fault tolerance is one of the major requirements. The system has to stay functioning when some part of it is not available or faulty, perhaps with reduced functionality or efficiency.

### 10.1  Chilled Water System

The CWS pilot system is based on the Reduced Scale Advanced Demonstrator (RSAD) model, a reconfigurable fluid system test platform. The RSAD has an integrated control architecture that includes Rockwell Automation technology for control and visualization. The RSAD model is currently configured as a CWS.

The physical layout of the RSAD chilled water system is a scaled down version from a real U.S. Navy ship. There is one chiller per zone, i.e., currently two plants, shown in Fig. 6 as the right-most two boxes. The other 16 boxes represent heat sources onboard ship (for example, combat systems, communication systems, and radar and sonar equipment) that must be kept cool enough, i.e., below the equipment's shutdown temperature, in order to operate. Each water cooling plant is an agent, as well as each heat source, each valve, and some parts of the piping system.

Within the RSAD, immersion heaters provide the energy to increase the temperature of the heat sources. A temperature sensor at each source provides temperature input to the control system. The main circulation piping is looped to provide alternative paths from any chilled water plant to any heat source.

This application is suitable for a MAS since survivability is one of the main requirements for shipboard systems. Reusability is another aspect that is addressed in a MAS by reconfiguration of present agent types. The agents are distributed according

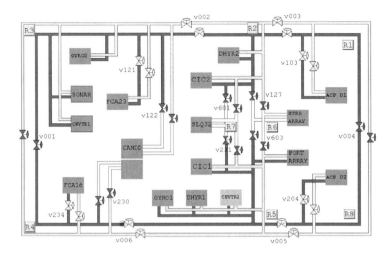

**Fig. 6.** Chilled water system

to the physical location of the hardware equipment, i.e., an agent is placed as closely as possible to the corresponding equipment. When some part of the system is damaged, usually only agents that are located in this part are lost. Single-point-of-failure nodes, from the hardware distribution perspective, are avoided. Other rules need to be satisfied to be free from single-point-of-failure nodes, for example, use no remote I/O, deploy the smallest possible number of agents per component and controller, and use no mapped inter-controller data. With these rules in mind, the 68 agents for this application are deployed within 23 industrial controllers.

Each agent is associated with capabilities and each capability is associated with a specific set of operations. The negotiation among agents uses local planning and negotiated planning (i.e. cooperation). The agents use their local "world observations" to determine their actions which are then translated into execution steps. In the negotiated planning, the agents discover each other's capabilities. When some part of the system is damaged and corresponding agents are not present, they are not further used in the planning process, i.e., the MAS dynamically reacts to the changes in system configuration. Also, when an agent or some part of the MAS is disconnected from the rest of the system, it is still able to make local decisions based on local knowledge and react accordingly. Moreover, active distributed diagnostics is present in agents, which are able to detect and localize problems with water distribution (e.g., water leakage and pipe blockage). This knowledge is used, for example, to isolate water leakage and use another route through the piping system.

Agent behavior has been designed to provide multiple alternative solutions to increase fault tolerance (see Section 6). For example, the main solution for getting water to a service in the CWS can be to ask each adjacent valve whether there is water present next to it and whether the valve is able to open. An alternative solution is to trigger the water path planning process. A possible further solution is to turn all non-vital systems off and retry the path-planning process, causing some graceful degradation since there will be a significant loss of performance. A fail-safe solution can be to turn all available chillers on, turn all non-vital systems off, and open all valves that are not isolating water leaks.

A broken section in the piping system is found step by step by opening recently closed valves. As each valve is opened in turn, the leak detection algorithm determines whether the leak has been found, and if so, instructs the last opened valve to ask its relevant neighbors to isolate the other sides of the broken section.

In case the system faces multiple pipe section destruction, the sections are isolated from the rest of the system step by step. After finding the first broken section the cooling is reestablished and the next broken section is searched for when the second leakage is detected.

## 10.2  Material Handling System

The Manufacturing Agent Simulation Tool (MAST) [44] provides simulation with embedded MAS aimed at the manufacturing domain, more precisely, at the transportation of (semi-)products or discrete materials on the factory's shop floor using a network of conveyor belts and/or automated guided vehicles (AGVs).

To show the robustness and flexibility of the agent solution, attention is paid to failure detection and recovery. A failure of any component can be emulated (e.g., a failure of a conveyor belt) causing the agents to start negotiations on alternative transportation paths while avoiding the broken component. MAST consists of the following parts:

- *The agent control part* that contains a library of Java/JADE classes representing the material handling components
- *The emulation part* that provides the agents with the emulation of the physical manufacturing environment.
- *The control interface* is a bidirectional connection from each emulation object to the appropriate agent (sensor signals) and vice versa (actuator signals).
- *The GUI* (see Fig. 7) for the graphical drag-and-drop design of the material handling system as well as for the visualization of the simulation.

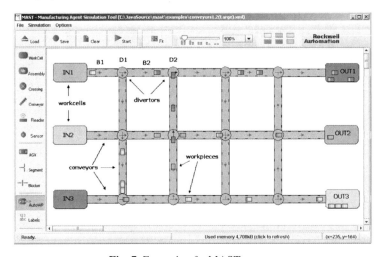

**Fig. 7.** Example of a MAST system

Via the GUI, the user can send workpieces between the work cells, introduce failures of different components and even change the structure of the system at runtime – any component can be removed (or disconnected from its neighbors) or a new component can be added without the need to stop the simulation.

## 11 Conclusion

This chapter summarizes the major areas of MAS where fault tolerance plays important role. Traditional fault tolerance techniques can be directly applied to all these areas, but there are also MAS specific techniques that can improve fault tolerance, such as social knowledge management, development environment utilization to minimize agent programming mistakes, and how to program agents to increase robustness of an agent. Important role play also validating agents that are used to observe behavior of the whole system and to discover potential problems. Agent simulation is needed to discover problems not only in basic agent behavior, but also in emergent behavior of the whole MAS.

## References

1. AgentLink III, http://www.agentlink.org/
2. AGlobe, Agent Technology Center, Czech Technical university, http://agents.felk.cvut.cz/aglobe/
3. Aguilera, M.K., Chen, W., Toueg, S.: Heartbeat: A Timeout-free Failure Detector for Quiescent Reliable Communication. In: Mavronicolas, M. (ed.) WDAG 1997. LNCS, vol. 1320, pp. 126–140. Springer, Heidelberg (1997)
4. Byrne, C., Edwards, P.: Refinement in Agent Groups. In: Weiss, G., Sen, S. (eds.) IJCAI-WS 1995. LNCS, vol. 1042, pp. 22–39. Springer, Heidelberg (1996)
5. Chen, L., Avizienis, A.: N-version Programming: A Fault-Tolerance Approach to Reliability of Software Operation. In: Digest of Papers of the 8th Annual International Conference on Fault-Tolerant Computing, Toulouse, France (1978)
6. Christian, F.: Understanding Fault-Tolerant Distributed Systems. Communications of the ACM 34, 56–78 (1993)
7. CIP: Common Industrial Protocol, http://www.ab.com/networks/cip_pop.html
8. Diestel, R.: Graph Theory. Graduate Texts in Mathematics, vol. 173. Springer, New York (2000)
9. Dragoni, N., Gaspari, M.: Crash failure detection in asynchronous agent communication languages. Autonomous Agents and Multi-Agent Systems 13(3), 355–390 (2006)
10. Durfee, E.H.: Distributed problem solving and planning. In: A Modern Approach to Distributed Artificial Intelligence, ch. 3, The MIT Press, San Francisco (1999)
11. Fedoruk, A., Deters, R.: Improving fault-tolerance by replicating agents. In: Proceedings of the first international joint conference on Autonomous agents and multiagent systems: part 2, Session 8B, scalability and robustness, pp. 737–744. ACM, New York (2002)
12. FIPA: The Foundation for Intelligent Physical Agents Geneva, Switzerland (1997), http://www.fipa.org
13. Giorgini, P., Kolp, M., Mylopoulos, J.: Multi-Agent Architectures as Organizational Structures. Autonomous Agents and Multi-Agent Systems 13(1), 3–25 (2006)

14. Havel, I.M.: Artificial Intelligence and Connectionism: Some Philosophical Implications. In: Trappl, R., Mařík, V., Štěpánková, O. (eds.) Advanced Topics in Artificial Intelligence. LNCS, vol. 617, pp. 25–41. Springer, Heidelberg (1992)
15. Hägg, S.: A Sentinel Approach to Fault Handling in Multi-Agent Systems. In: Foo, N.Y., Göbel, R. (eds.) PRICAI 1996. LNCS, vol. 1114, pp. 181–195. Springer, Heidelberg (1996)
16. Hayden, S., Carrick, C., Yang, Q.: Architectural Design Patterns for Multiagent Coordination. In: Proceedings of the 3rd International Conference on Autonomous Agents, Agents 1999, Seattle, USA, pp. 10–21 (1999)
17. IEC (International Electrotechnical Commission), TC65/WG6, 61131-3, 2nd ed., Programmable Controllers - Programming Languages, April 16 (2001)
18. JACK, Agent Oriented Software,
   http://www.agent-software.com.au/products/jack/
19. JADE: Java Agent DEvelopment Framework, Telecom Italia Lab, Torino, Italy,
   http://sharon.cselt.it/projects/jade/
20. Kalbarczyk, Z.T., Iyer, R.K., Bagchi, S., Whisnant, K.: Chameleon: A Software Infrastructure for Adaptive Fault Tolerance. IEEE Transactions on Parallel and Distributed Systems 10, 560–579 (1999)
21. Kaminka, G.A., Tambe, M.: What's Wrong With Us? Improving Robustness through Social Diagnosis. In: Proceedings of the Fifteenth National Conference on Artificial Intelligence (AAAI-1998), Madison, Wisconsin, pp. 26–30 (1998)
22. Kit, E.: Software Testing in the Real World: Improving the Process. Addison-Wesley, Reading (1995); ISBN 0-201-87756-2
23. Kumar, S., Cohen, P.R., Levesque, H.J.: The Adaptive Agent Architecture: Achieving Fault-Tolerance Using Persistent Broker Teams. In: Proceedings of the 4th International Conference on Multi-agent Systems (ICMAS-2000), Boston, MA, pp. 159–166 (2000)
24. Kumar, S., Cohen, P.R.: Towards a Fault-Tolerant Multi-Agent System Architecture. In: Proceedings of the 4th International Conference on Autonomous Agents, Barcelona, Spain, pp. 459–466 (2000)
25. Marin, O.: The DARX framework: Adapting Fault Tolerance for Agent Systems. Ph. D. Thesis, Université du Havre (2003)
26. Mařík, V., Pěchouček, M., Štěpánková, O.: Social Knowledge in Multi-Agent Systems. In: Luck, M., Mařík, V., Štěpánková, O., Trappl, R. (eds.) ACAI 2001 and EASSS 2001. LNCS (LNAI), vol. 2086, pp. 211–245. Springer, Heidelberg (2001)
27. Maturana, F.P., Staron, R.J., Tichý, P., Šlechta, P., Vrba, P.: A Strategy to Implement and Validate Industrial Applications of Holonic Systems. In: Mařík, V., William Brennan, R., Pěchouček, M. (eds.) HoloMAS 2005. LNCS (LNAI), vol. 3593, pp. 111–120. Springer, Heidelberg (2005)
28. Maturana, F.P., Tichý, P., Šlechta, P., Staron, R.J., Discenzo, F.M., Hall, K.H., Mařík, V.: Cost-Based Dynamic Reconfiguration System for Intelligent Agent Negotiation. In: Proceedings of IEEE/WIC International Conference on Intelligent Agent Technology (IAT), pp. 629–632. IEEE Computer Society Press, Los Alamitos (2003); ISBN 0-7695-1931-8
29. Minski, Y., Renesse, R., Schneider, F., Stoller, S.: Cryptographic Support for Fault-Tolerant Distributed Computing. In: Proceedings of the 7th ACM SIGOPS European Workshop, Connemara, Ireland, pp. 109–114 (1996)
30. Nimis, J., Lockemann, P.C.: Robust Multi-Agent Systems: The Transactional Conversation Approach. In: The 1st International Workshop on Safety and Security in Multiagent Systems (SASEMAS 2004), New York City, NY (2004)

31. Obitko, M., Mařík, V.: Integration of Multi-Agent Systems: Architectural Considerations. In: The 11th IEEE International Conference on Emerging Technologies and Factory Automation (ETFA 2006), Prague, pp. 1145–1148 (2006)
32. Pěchouček, M., Mařík, V., Štěpánková, O.: Role of Acquaintance Models in Agent-Based Production Planning System. In: Klusch, M., Kerschberg, L. (eds.) CIA 2000. LNCS (LNAI), vol. 1860, pp. 179–190. Springer, Heidelberg (2000)
33. Pechoucek, M., Marik, V.: Industrial Deployment of Multi-Agent Technologies: Review and Selected Case Studies. International Journal on Autonomous Agents and Multi-Agent Systems, 1387–2532 (2008)
34. Pěchouček, M., Macůrek, F., Tichý, P., Štěpánková, O., Mařík, V.: Meta-agent: A Workflow Mediator in Multi-Agent Systems. In: Watson, I., Gordon, J., McIntosh, A. (eds.) Intelligent Workflow and Process Management: The New Frontier for AI in Business IJCAI-1999, pp. 110–116. Morgan Kaufmann, San Francisco (1999)
35. Pleisch, S., Schiper, A.: Approaches to fault-tolerant and transactional mobile agent execution - an algorithmic view. ACM Computing Surveys (CSUR) 36(3), 219–262 (2004)
36. Shen, W., Lang, S.Y.T., Lang, L.: iShopFloor: An Internet-Enabled Agent-Based Intelligent Shop Floor. IEEE Transactions on System, Man, and Cybernetics, Part C 35(3), 371–381 (2005)
37. Sheory, O.M.: A Scalable Agent Location Mechanism. In: Wooldridge, M., Lesperance, Y. (eds.) ATAL 1999. LNCS, vol. 1757, pp. 162–173. Springer, Heidelberg (2000)
38. Sheory, O.M., Sycara, K., Jha, S.: Multi-Agent Coordination through Coalition Formation. In: Rao, A., Singh, M.P., Wooldridge, M.J. (eds.) ATAL 1997. LNCS, vol. 1365, pp. 143–154. Springer, Heidelberg (1998)
39. Singh, N.: A Common Lisp API and Facilitator for ABSI. Technical Report Logic-93-4, Computer Science Department, Stanford University (1994)
40. Tichý, P.: Middle-agents Organized in Fault Tolerant and Fixed Scalable Structure. Computing and Informatics 22, 597–622 (2003)
41. Tichý, P., Šlechta, P., Maturana, F.P., Balasubramanian, S.: Industrial MAS for Planning and Control. In: Mařík, V., Štěpánková, O., Krautwurmová, H., Luck, M. (eds.) ACAI 2001, EASSS 2001, AEMAS 2001, and HoloMAS 2001. LNCS (LNAI), vol. 2322, pp. 280–295. Springer, Heidelberg (2002)
42. Tichý, P.: Fault Tolerant and Fixed Scalable Structure of Middle-Agents. In: Dix, J., Leite, J. (eds.) CLIMA 2004. LNCS (LNAI), vol. 3259, pp. 53–70. Springer, Heidelberg (2004); ISBN: 3-540-24010-1
43. Tichý, P.: Social Knowledge in Multi-agent Systems. Dissertation thesis, Czech Technical University, Prague (2004)
44. Vrba, P., Mařík, V.: Simulation in Agent-based Manufacturing Control Systems. In: Proceedings of the IEEE International Conference on Systems, Men and Cybernetics, Hawaii, USA, pp. 1718–1723 (2005)
45. Wensley, J.H.: SIFT Software Implemented Fault Tolerance. In: Proceedings of Fall Joint Computer Conf., AFIPS, vol. 41, pp. 243–253 (1972)
46. Wooldridge, M.J., Jennings, N.R.: Intelligent Agents: Theory and Practice. The Knowledge Eng. Rev. 10(2), 115–152 (1995)

Dr. Pavel Tichý received his M.S. degree in electrical engineering from the Czech Technical University (CTU) in Prague in 1995, and his Ph.D. in artificial intelligence from CTU in 2004. He has been a contractor and later an employee of the Rockwell Automation research center in Prague since 1995. He specializes in the fields of artificial intelligence and multi-agent systems.

Dr. Raymond J. Staron received a B.S. degree in electrical engineering from Case Western Reserve University (CWRU) in Cleveland, Ohio, in 1974, an M.S. degree from the Massachusetts Institute of Technology, Cambridge, Massachusetts, in 1976, and a Ph.D. in computer science from CWRU in 1993. He has been working for the Allen-Bradley Company and Rockwell Automation for 31 years.

# Chapter 9

# Timing Agent Interactions for Efficient Agent-Based Simulation of Socio-Technical Systems

Seung Man Lee[1] and Amy R. Pritchett[2]

[1] Senior Research Engineer, Cognitive Engineering Center, Georgia Institute of Technology,
Atlanta, Georgia 30332-0150, USA
seungman.lee@gatech.edu
[2] David S. Lewis Associate Professor, Cognitive Engineering Center,
Daniel Guggenheim School of Aerospace Engineering,
Georgia Institute of Technology,
Atlanta, Georgia 30332-0150, USA
amy.pritchett@ae.gatech.edu

**Abstract.** In recent decades, agent-based modeling and simulation (ABMS) has been increasingly used as a valuable approach for design and analysis of dynamic and emergent phenomena of large-scale, complex multi-agent systems, including socio-technical systems. The dynamic behavior of such systems includes both the individual behavior of heterogeneous agents within the system and the emergent behavior arising from interactions between agents within their work environment; both must be accurately modeled and efficiently executed in simulations. An important issue in ABMS of socio-technical systems is ensuring that agents are updated together at any time where they must interact or exchange data, even when the agents' internal models use fundamentally different methods of advancing their internal time and widely varying update rates. This requires accurate predictions of interaction times between agents within the environment. Predicting the time of interactions, however, is not a trivial problem. Thus, timing mechanisms that advance simulation time and select the proper agent to be executed are crucial to correct simulation results. This chapter describes a timing and prediction mechanism for accurate modeling of interactions among agents which also increases the computational efficiency of agent-based simulations. An experiment comparing different timing methods highlighted the gains in computational efficiency achieved with the new timing mechanisms and also emphasized the importance of identifying correct interaction times. An intelligent timing agent framework for predicting the timing of interactions between heterogeneous agents using a neural network and a method for assessing the accuracy of interaction prediction methods based on signal detection theory are described. An application of agent-based modeling and simulation to air transportation systems serves as a test case and the simulation results of different interaction prediction models are presented. The insights of using the framework and method to the design and analysis of complex socio-technical systems are discussed.

D. Srinivasan & L.C. Jain (Eds.): Innovations in MASs and Applications – 1, SCI 310, pp. 247–276.
springerlink.com                                © Springer-Verlag Berlin Heidelberg 2010

# 1   Introduction

Computational agent-based modeling and simulation (ABMS) have emerged as critical tools for the design and safety analysis of large-scale complex multi-agent systems, including socio-technical systems, in which a large number of components, such as humans, machines, and technical systems, interact with each other to accomplish their tasks and goals [1]. In agent-based simulation, each of the entities within the system can be defined and modeled as an agent requiring a different simulation model. For instance, aircraft in the National Airspace System (NAS) may require a high-fidelity continuous-time simulation model to predict the dynamic behavior of the aircraft, certain stochastic events such as aircraft entry into the airspace, machine failures, and malfunctions require discrete-event simulation models, and human operators such as pilots and air traffic controllers require computational human performance models. These models must interact with one another in significant ways to create the performance of the entire system. ABMS can potentially be a cost-effective method for evaluating new technologies and operating procedures during early stage design of large-scale socio-technical systems, for assessing sensitivity of overall system performance to the performance of individual agents, and for providing a safe strategic analysis tool for examining the impact of potentially safety-critical changes [2]. ABMS also allows for experimentation to evaluate system designs and to identify the agent behaviors corresponding to interesting system-level emergent phenomena.

However, existing conventional modeling and simulation methods are not sufficient to fully understand and reflect the complex dynamic characteristics of large-scale, complex socio-technical systems [3],[4, 5]. Although most real-world complex systems are often characterized by a combination of different types of models, such as continuous-time, discrete-event, and human performance models, and their interactions, these systems have been mainly simulated by either entirely discrete-event simulation models capable of capturing and predicting stochastic effects within the system of study, or entirely continuous-time simulation models capable of predicting the dynamic behavior of physical systems such as aircraft trajectories and mechanical system performance. The different modeling and simulation approaches used for continuous-time, discrete-event, and human performance models must be integrated to represent their heterogeneous dynamic behavior. This dynamic behavior includes both the individual behavior of entities within the system and their interactions. While the individual behavior of agents may be simple, emergent behavior at the system level may not be predictable.

Different simulation models in agent-based simulation use different timing mechanisms to advance the simulation and also tend to require substantially different update rates. For example, a continuous-time model of aircraft flight dynamics might require very small time steps and frequent update rates, whereas a discrete-event model for generating aircraft objects into the airspace might require relatively large update intervals. This disparity between the various timing mechanisms of different simulation models must be efficiently resolved in agent-based simulation. Despite these differences, agent-based simulation must also ensure that agents in the simulation are updated together at any time where they must interact or exchange data, even when dramatically different time steps are used. This requires accurate predictions of the timing of agent interactions. Predicting these interactions, however, is not a trivial problem. Using established methods, better predictions require more extensive computations and knowledge about agents' internal dynamics and future actions. This also imposes an obvious development cost not only in modeling the agents but also in reconfiguring the simulation to include new agents, as well as reducing computational efficiency. Thus, it is important to predict when an interaction between agents may next occur without requiring substantial computations and without a large development effort each time the simulation is reconfigured to include new agents or new types of agent interactions.

This chapter discusses mechanisms for accurately timing agent interactions in ABMS to support the design, analysis and evaluation of large-scale complex socio-technical systems, such as air transportation systems. Timing mechanisms for agent-based simulation are described, including an intelligent timing mechanism for accurately predicting the timing of interactions among heterogeneous agents using a neural network and a method for assessing the accuracy of interaction prediction methods based on signal detection theory. An application of agent-based simulation to air traffic control is illustrated as a test case and the simulation results of different interaction prediction models are also presented. The insights of using this method and framework to the design and analysis of complex socio-technical systems are discussed in the conclusion.

## 2  ABMS of Socio-Technical Systems

There has been considerable recent interest in multi-agent systems (or often called agent-based systems), which comprise multiple humans and autonomous software and/or hardware components (agents) cooperating within an environment to perform some tasks. Multi-agent systems are often extremely complex and it can be difficult to formally verify their properties [6]. In this context, simulation has traditionally played an important role in the design and analysis of agent architectures and agent-based systems, allowing the agent designer and researcher to investigate the dynamic behavior of a system or to probe the relationship between agent architectures, environments and behavior [7, 8]. In this chapter we focus on an agent-based modeling and simulation framework as a vehicle for studying complex multi-agent systems, specifically socio-technical systems.

A socio-technical system can be defined as a system in which many humans and technology interact with each other within their work environment to accomplish their goals and tasks, and is a construct used in studies of human-computer interaction and cognitive systems engineering [9],[10],[11]. Examples of such systems include economic and financial systems, health care systems, manufacturing systems, military systems, and ground/air transportation systems. The dynamic behavior of such a system, to a great extent, can be characterized by both the individual behavior of entities within the system and the aggregate and emergent behavior of the entities interacting with each other. The task of performing design and analysis of the large-scale complex socio-technical systems, therefore, requires a more robust and flexible modeling and simulation approach than is currently available due to the emergent dynamics and tightly coupled nature of these systems [12].

## 2.1    Agent-Based Modeling and Simulation

Agent-based modeling and simulation (ABMS) has been widely used for modeling and simulation of complex socio-technical systems. Compared to other approaches, such as traditional discrete event simulation and object-oriented simulation, ABMS has a number of useful properties. For instance, it supports structure preserving modeling of the simulated reality, simulation of pro-active human behavior, parallel computations, and very dynamic simulation scenarios [13]. In ABMS, the specific behaviors of simulated entities are modeled and implemented in terms of agents and the structure is viewed as emergent from the interactions between the individual agents while traditional simulations attempt to model the averaged characteristics of the whole systems with mathematical models.

The overall dynamic behavior of such a complex system typically emerges from the interactions among components. The collective and emergent behavior of individual system components including hardware, software, and human operators, therefore, can be modeled and simulated as an interaction among agent models [3],[14]. ABMS has strong roots in the fields of multi-agent systems (MAS) and robotics from the field of AI. However, ABMS is not only tied to designing and understanding "artificial" agents but also modeling human individual behavior and social behavior [15]. Historically, agent-based modeling concentrated on creating intelligent agents towards achievement of autonomy, an artificial intelligence perspective on emulating humans and designing autonomous technologies [16]. More recently, researchers have applied "multi-agent" simulation of many interacting agents in complex ways. Such multi-agent simulations have two concerns: modeling individual entities as autonomous and interactive agents, and simulating the system behavior that emerges from the agent's collective interactive actions. The agent-based simulation can integrate agent models of human cognition and performance, physical models of technology behavior, and models of their operating environment. Simulation of these individual models interacting together, capable of supporting a wide variety of agents and environments, enables one to predict the impact of completely new transformations in the organizational structure, operating procedures and technologies of larger and more complex multi-agent systems. As such, agent-based modeling and simulation has been increasingly applied for examining complex socio-technical systems in a wide range of areas including social sciences [17],[18], telecommunications, manufacturing [19], supply chain systems [20], business processes [21], financial markets [22], environment [23], transportation [24],[25], and military simulations.

However, to be a good design and analysis tool simulation requires suitable models of elements, such as models of human behavior and performance and dynamic models of physical machines, to predict the evolution of complex interactions of the elements. It requires a deep knowledge of engineering principles and cognitive psychology to develop models of mechanical components and human performance. This requires the development of an agent-based simulation able to simulate collective and emergent behavior of heterogeneous agents including models of hardware, software, and human performance [26].

As noted earlier, in simulating socio-technical systems by agent-based simulations, the overall behavior of the systems can be considered as emergent phenomena. The global phenomena caused by locally interacting agents are called *emergent* properties of the system. The emergent property is formally defined here as a system property in which system behaviors at a higher level of abstraction are caused by behaviors at a lower level of abstraction which could not be predicted, or made sense of, at that lower level. The emergent behavior can dominate overall system behavior and performance. Even when the behavior of agents at the individual level may be simple, the emergent behavior at the system level may not be at all obvious. Thus, emergent properties are often unpredicted. As traditional simulation methods typically lack the capabilities to capture both the individual behavior of each entity and the system-wide emergent behavior, it is important that new simulation methods be established. The enormous complexity and complicated interactions among entities within large-scale socio-technical systems require ABMS approaches in order to understand their dynamic and emergent behavior [27]. In this case our levels of abstractions are the agents (typically humans) and the emergent system-wide behavior. Agent-based simulation provides interesting insights at both levels of abstraction. In addition to the system-wide behavior, the agents respond to their environment and each other in agent-based simulations. While we can model what the agents' responses would be to a variety of conditions, only simulation can predict what specific conditions they will need to respond to.

Thus, agent-based simulations are uniquely suitable for evaluating the 'macro' level emergent behavior of multi-agent systems as well as the 'micro' level behavior of individual agents as they are not necessarily built upon or parameterized by the structures used within the system. The ABMS approach shows how processes evolve over time and how interactions and changes occurring at the micro-level affect overall system behavior over time. Both micro-level (agent) and macro-level (system-wide) behaviors can be simulated simultaneously, highlighting system-wide issues arising from a change in individual agents' behavior, including new technologies, and changes in environmental structures – as well as identifying unreasonable demands that system dynamics may place on individual agents. Different systems can be designed by manipulating not only individual agent configurations but also interrelationships between agents in their work environment.

Specifically, in this chapter, the air transportation system is viewed as a large-scale, complex socio-technical system composed of multiple agents such as controllers, pilots, airline dispatchers, aircraft, airports, and technical devices. Thus, the individual behavior of these different entities and overall behavior of the system can be modeled by a combination of agent models, environment models, and their interactions. More informally, our agent-based simulations are not based on any high-level system models; instead, we put agent models in a rich environment, simulate them in a realistic scenario, and see what system behavior comes out.

### 2.1.1   Notion of Agents in ABMS

Recent developments in software engineering, artificial intelligence, human-machine systems, and simulation science have placed an increasing emphasis on concepts of agents [28],[29],[30],[31]. There is no universal agreement on the precise definition of the term *agent*. The term *agent* has been used to mean anything between a mere subroutine or object and an adaptive, autonomous, intelligent entity; an independent component's behavior can range from primitive reactive behavior to complex adaptive artificial intelligence (AI) [6],[32],[27]. A computer science view of agency emphasizes the essential characteristic of *autonomous* behavior which requires agents to be proactive rather than purely reactive [31]. However, the term agent in agent-based modeling and simulation needs to be distinguished from the term *intelligent agent* used in the domain of software engineering and artificial intelligence. The agent in agent-based simulation has a weak notion of agency. Thus, this paper defines an agent in ABMS as an entity with (1) autonomy, i.e., the capability to carry out some set of activities, (2) interactivity with other agents, i.e., the need and ability to interact with other agents to accomplish its own tasks and goals, and (3) interactivity with the simulated environment.

In agent-based modeling and simulation, each entity satisfying the above properties can be defined and modeled as an agent. Agents can be physical entities or technologies such as aircraft, ground vehicles, technologies, and weather, or agents can be task-oriented entities such as strategic planning, scheduling, monitoring, communications, and decision making activities. Such agents interact with each other within the system to accomplish their tasks. Thus, each agent is itself typically a complex system with capabilities of sensing, interpreting, planning, inference, decision making, and taking an action etc., requiring considerable computational resources, and multiple agents may be required to evaluate the overall emergent behavior of the system as a whole or even the behavior of a individual agent [33],[34].

Agents used in agent-based simulation can also include rich cognitive human performance models and sophisticated communication and interaction mechanisms [34],[35]. By representing the human as an agent in an agent-based simulation, it is possible to examine the interaction of human behavioral tendencies with system state over a wide range of circumstances. For example, studies have examined using human performance models as agents in large-scale agent-based simulations to evaluate air transportation safety and capacity issues [36],[37],[38],[38]. Agent models can identify the impact of the accuracy, speed, and variability of human performance, which are critical to the emergent behavior of the larger system. This approach adds fidelity both to the human performance models and to the larger simulation; correspondingly, agent-based simulation brings to these human performance models a dynamic representation of their environment, including detailed models of the physical and technical systems, and the opportunity to dynamically interact with other humans. Even simple agent models of human cognition and performance provide valuable insight when used in the context of ABMS: they allow evaluation as to system performance resulting from well-known aspects of human behavior. With simple normative agent models, for example, agent-based simulation can observe whether the system will function as desired when all components act exactly as procedures, regulations and organizational structures mandate – and highlights areas where individuals' flexibility and creativity are still required to operate the system.

The level of detail to which each human agent needs to be modeled depends upon the purpose of the simulation model. Too detailed a simulation must be expected to cost more and may only complicate the evaluation procedure. Too shallow a simulation provides insufficient or misleading information. Given the complexity of human performance models, creating the ability to interact with other simulation models (e.g., communicate with other agents and synchronize their time advance with the other agents) can require significant adaptations. Ultimately, it is hoped that these simulations will have sufficient fidelity in their agents' ability to reason and react to unexpected situations to examine a wide range of potential hazardous situations. It should also be noted, however, that even examining simple representations of agents' behavior in normal circumstances can identify potential weakness or inconsistencies in performing standard operating tasks and procedures given a work environment.

## 2.2  Timing Mechanisms and Interactions between Agents

Heterogeneous agents in socio-technical systems, including agent models that can produce human-like behavior and other physical and technological agents, need to interact and communicate with each other in significant ways to accomplish their tasks and goals. The agent interactions can be with each other or with the environment, and thus the prediction of interaction times also fosters better modeling and simulation of interaction with the environment in support of ecological modeling approaches. In the real world, many interactions between entities can be easily perceived and detected by themselves or by third entities; for example, in air traffic control, controllers can easily monitor a spatial radar display to perceive which pairs of aircraft are proximate. In ABMS, on the other hand, this process may require pairwise comparisons between all pairs of aircraft whose states are expressed at the same point in simulated time. When the agents update asynchronously, this difficulty is compounded by their immediate representations of state corresponding to different points in time.

Thus, a simulation timing mechanism that advances simulation time and selects the subroutine or object to be executed is a vital component of every simulation [39], [40]. In agent-based modeling and simulation timing mechanisms need to address several issues beyond those found in traditional discrete-event or continuous-time simulations. First, timing mechanisms need to properly handle heterogeneous agent models, which may have considerably different methods for representing dynamics and the points in time where they should be updated but may not use the same underlying representation of time. Second, agents must be simultaneously updated when they interact to capture important interactions between agents and to improve the computational efficiency of agent-based simulation.

Timing mechanisms can be typically defined as synchronous or asynchronous. While synchronous timing methods require all agents in the simulation to update at the same time, asynchronous timing methods allow each agent to update independently. For large-scale or repeated runs of the simulation, synchronous timing methods are usually computationally inefficient since the timing method requires all agents to update at every time step whether each needs to or not. In addition, synchronous timing mechanisms may or may not provide more accurate simulation results as very small time steps may reduce some types of error (e.g., numerical loss in numerical integration of ordinary differential equations in continuous-time models) but increase others (e.g., round-off errors corresponding to repeated operations on small increments to continuously-valued states). For example, cognitive agent architectures such as SIM_AGENT [41], SOAR [42], and ACT-R [43]  maintain a single, centralized representation of the agent's internal state, e.g., a working memory or goal stack, making synchronous timing mechanism relatively straightforward. On the other hand, asynchronous timing mechanisms can provide correct modeling and simulation results if each agent is updated only when needed for its internal dynamics and interactions with other agents, and can be much more computationally efficient than synchronous timing methods [44], [45].

At first glance, the best efficiency may be assumed to correspond to the optimistic prediction of interaction times between agents. However, several main issues limit the size of predicted interaction times. First, it is typically difficult to detect interactions between agents with larger predicted interaction times since the simulation does not know what stochastic events may occur during these intervals. Second, some types of agent models are incapable of rolling back to their state should an optimistic prediction 'miss' an interaction. Typically, rollback mechanisms have most commonly been applied to systems with purely discrete dynamics or very simple continuous-time models, and at the cost of additional computational overhead due to the rollback mechanisms [46], [47]. Some agent models can not rollback, such as complex continuous-time models, expert systems with complex dependencies, etc [48]. Likewise, sometimes post-hoc assessment of whether an interaction was needed may not be feasible; for example, in an air traffic control simulation it can be as hard to estimate whether two aircraft did hit as to estimate whether they will hit, short of recording and compare their flight tracks over very small intervals.

If the simulation cannot rollback to the time of a recent interaction, then estimates of interaction times need to be conservative so that the interactions between agents are not missed. However, if a predicted interaction time is conservatively set to be very short, the simulation will synchronize agents too frequently. While reducing the risk of missed interactions, it will also reduce computational efficiency, and the unnecessary updates of agents may adversely impact the accuracy of their functioning by asking them to change their states and interact with their environment at an unrealistically high frequency. While overly conservative predicted interaction times can be computationally inefficient due to the unnecessary agent updates they incur, they often do not require complex computations. On the other hand, more accurate predicted interaction times typically require more computationally extensive calculations; at an extreme, the most accurate predictor would need to internally simulate all agents. As such, the value of better predictions can reach a point of diminishing returns where the additional computations used to predict interaction times more accurately may offset any savings in computations through reduced numbers of updates of the agent models. When the predictions are based on a model of system dynamics, they are domain or scenario specific. If a simulation requires a different type of prediction for each type of interaction, then there is an obvious development cost for any changes to the simulation configuration, and the predictors are limited to specific scenarios and agents. Therefore, in an agent-based simulation where a large number of heterogeneous agents may be involved, developing a predictor that is computationally efficient, easy to implement, and accurate is not a trivial problem.

In previous studies [12], several timing mechanisms for agent-based simulation have been developed and compared their computational efficiency. Specifically, *asynchronous with resynchronization*, allows agents to update asynchronously following their own update times, and also estimates when interactions may occur, requiring the relevant agents to jointly update at these times. With this timing mechanism, accurately predicting agent interaction times provides more accurate simulation results and also increases computational efficiency by reducing unnecessary updates of agents during the simulation.

However, deciding when an update may be required for correct interactions is generally non-trivial once the simulation contains stochastic elements. For example, in an agent-based simulation of air traffic control, loss of separation between aircraft cannot be efficiently predicted from dead-reckoning as aircraft may, at any time, maneuver in response to controller clearances or pre-established flight plans; such deterministic methods of estimating interactions are only valid when they are re-evaluated between all potentially-impacted agents in response to every event, with significant computational cost. Such constant deterministic calculations may instead be viewed as the origin of training data for more intelligent predictors; once trained, the intelligent predictors (providing a statistical estimate of interaction time) may be used in lieu of exact deterministic predictors for computational efficiency.

## 3  Timing Mechanisms and Interaction Prediction Methods

As just noted, timing mechanisms for an agent-based simulation incorporating different types of agent models need to address several issues. First, timing mechanisms for agent-based simulation need to adequately deal with different types of agent models, including discrete-event models and continuous-time models, often with considerably different update rates. The second issue is the computational efficiency of accurate and timely interactions. Time constitutes an important component in the behavior of the agents and their interactions. Thus, timing mechanisms for agent-based simulation must ensure that agents in the simulation are synchronized at any time when agents must interact with each other or exchange data between agents. However, methods of deciding when an update may be required for correct interactions or measurements are generally non-trivial once the simulation contains stochastic elements. This section focuses on various aspects of timing mechanisms for the agent-based simulation of large, complex systems and describes their implementation using an object-oriented approach.

### 3.1  Timing Mechanisms for Agent-Based Simulation

Based on the characteristics of different timing methods [4], several timing mechanisms can be developed to advance time in agent-based simulation: synchronous fixed time step, synchronous variable time step, and asynchronous with resynchronization mechanisms.

### 3.1.1  Synchronous Fixed Time Step
This timing method requires all agents to update at every predetermined fixed time step. The advantage of this timing method is that synchronization will always be maintained since all different types of agent models are updated at the same time. By setting a sufficiently small time step, this method provides accurate results that can be guaranteed not to miss any important measurements or interactions, without requiring predictions of interactions between agents. The obvious drawback is that a small time step results in a large number of updates of every agent, whether needed or not. Thus, this method degrades the overall simulation speed. This method is commonly used in current vehicle simulation techniques, where the time step may be fixed by conservative analysis of the fastest dynamics in the system.

In this timing method, it is assumed that all agents within the system are executed at the same time, but not concurrently in reality. Even though some agents need to be updated concurrently, it is difficult to practically implement the concurrency of updating agents on a single machine (or even on parallel machines). Therefore, a potential problem is the ordering of agents to be executed at the predetermined fixed time steps. A set of rules must be embedded into this method to decide which agent needs to be processed first in order to satisfy cause and effect constraints.

### 3.1.2  Synchronous Variable Time Step
The synchronous variable time step timing method requires all agents in the simulation to update at the same time, usually selecting the most restrictive time step demanded by any of the agents at that time, rather than always using a small predetermined time step. Thus, the time step varies from one time step to the next to meet the needs of the simulation. This method still forces some agents to update unnecessarily even though this timing method offers more efficiency and better flexibility than the synchronous fixed time step method. This method also still has a ordering problem to decide the ordering of updating agents at a given simulation time step to produce temporal aspects of the system correctly.

### 3.1.3  Asynchronous with Resynchronization Mechanisms
In theory, asynchronous timing mechanisms should be much more computationally efficient than synchronous timing methods since asynchronous timing methods do not require all agents to update at every time step. However, a completely asynchronous method is not appropriate for agent-based simulation because it may not fully capture interactions between agents. In an agent-based simulation, it is common for one agent to collect state information from other agents. For example, in the simulation of the NAS, an air traffic controller agent might need the current values of the location, speed, and heading state variables of aircraft to determine a desired speed for the aircraft to avoid loss of separation. Therefore, it is necessary to resynchronize all aircraft at the time when the air traffic controller evaluates the current traffic situation.

In order to capture important interactions between agents, periodic synchronizations are required in asynchronous timing mechanisms. Thus, this research proposes a new timing method, *asynchronous with resynchronization*. This timing method allows agents to update asynchronously following their own update times, but also makes conservative estimates of when interactions may occur in the future, and requires the relevant agents to jointly update at these resynchronization intervals. The new timing mechanism involves two varieties: *asynchronous with complete resynchronization* and *asynchronous with partial resynchronization* timing mechanisms.

*Asynchronous with Complete Resynchronization*
The asynchronous with complete resynchronization mechanism allows all agents in the simulation to update independently until any agent requires resynchronization. With this timing method, the state updater object synchronizes all agents at each resynchronization interval. This method is shown schematically in Fig. 1 for a simulation with three agents. For example, $agent_1$ and $agent_2$ update at their own rates until $agent_3$ requires resynchronization with at least one other agent. In large-scale agent-based simulation, the asynchronous with complete resynchronization method can

reduce a lot of unnecessary updates of agents since all agents are not required to update at every single time step. However, this method still leads to unnecessary updates since all agents are required to update at the resynchronization times. This
method also still has an ordering problem with which agent needs to be updated first
in the resynchronization (or last, since only the last agent updated can have current
information about all the other agents).

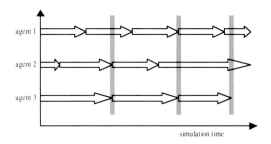

**Fig. 1.** Schematic of Agent Updates Using Asynchronous with Complete Resynchronization

*Asynchronous with Partial Resynchronization*
At the times of resynchronization, only a subset of the agents may be required to
update due to the interactions or measurements involving them. Therefore, a better
approach is to update only those agents interacting with each other at resynchronization times. The asynchronous with partial resynchronization timing method allows
agents in the simulation to update at their own update times independently until an
agent specifically requires some of the other agents to be resynchronized. With this
timing method, the state updater agent synchronizes only those agents that another
agent requires to also update at the resynchronization time. This method is shown
schematically in Fig. 2: agent 2 requires only agent 1 to update at time 5; agent 3 requires only agent 2 to be updated at time 20; and agent 3 requires both agent 1 and
agent 3 to update at time 40. This timing method can improve computational efficiency by updating just some of the agents interacting with each other at resynchronization intervals. Even though only some of agents interact with each other at a resynchronization time, ordering of updating those agents needs to be established correctly
based on the dependencies among those agents.

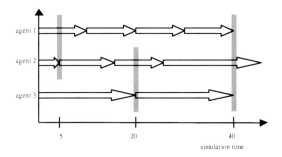

**Fig. 2.** Schematic of Agent Updates Using Asynchronous with Partial Resynchronization

## 3.2  Performance Measurements of Timing Mechanisms

Typically, measuring the performance of a simulation is not a trivial problem. Specifically, once the simulation includes stochastic elements, it can be difficult to compare with certainty the relative speed of different timing mechanisms. The total run-time of simulation can be considered as a standard performance measure of computational efficiency in many computer science and software engineering domains. However, the total run-time may not be a good single measurement of computational efficiency of agent-based simulation because it doesn't capture the accuracy of interactions in an agent-based simulation and may also depend on other factors such as the number of updates of agents, the fidelity of agents, and the interactions of agents. For example, if the agent models are very simple and there are no agent interactions then there might be no significant gain in computational efficiency with asynchronous simulations. On the other hand, if agents requires large amount of computation time and there are many interactions between agents then the total run-time of asynchronous simulation will be much smaller than the total run-time of synchronous simulation by reducing the large number of unnecessary updates of agents. The total run-time also depends on hardware configurations of the particular computer used, such as CPU speed, bus speed, and memories, and on software design such as how to implement the algorithm, data structures, and number of calculations.

As described earlier in this chapter, overall efficiency is achieved when each agent updates only when needed for accurate modeling of its interior dynamics, correct interaction between agents, and timely measurement. Each agent requires some time to execute its internal dynamics. The cost of computation is proportional to the number of agent executions performed. Therefore, any unnecessary updates of agent may be considered wasted use of processor time, manifested as computational inefficiency. Thus, the number of updates (executions) of agents, especially the most computationally intensive agents, can also serve as a measure of computational efficiency.

Predicting interaction times must balance a tradeoff between conservative predictions to avoid missing interactions and optimistic predictions to reduce unnecessary updates of agents. In complex systems, these tradeoffs are often difficult to quantify and evaluate. Our method recasts the prediction of interaction times as a signal detection problem, which allows methods from signal detection theory (SDT) to be applied to the evaluation of interaction prediction methods [49]. The basic idea is to apply Receiver Operating Characteristic (ROC) analysis [50],[51], a classic methodology of signal detection theory, now commonly called System Operating Characteristic (SOC) analysis when applied to systems other than receivers. SOC analysis is common in medical diagnosis and radiology [52] and has recently begun to be used more generally in machine learning, data mining, AI, and aviation research communities [53],[54].

The performance of a prediction algorithm can be evaluated by a decision matrix as shown in Table 1. In SDT, a false alarm (FA) occurs when an affirmative prediction is made in conditions where, in fact, no event is occurring; the probability of false alarm $P(FA)$ is defined as the conditional probability that an interaction is triggered given that no interaction is needed. Similarly, the probability of correct detection $P(CD)$ is defined as the conditional probability that a positive prediction is made given that an event is occurred. The probability of missed detection $P(MD)$ can be defined in this case as the probability that the simulation fails to predict an interaction between agents before it occurs. Finally, the probability of a correct rejection can be defined as the probability that the simulation does not predict an interaction within a specified interval.

**Table 1.** Decision Matrix

|  | Predictive Positive | Predictive Negative |
|---|---|---|
| Actual Positive | Correct Detection | Missed Detection |
| Actual Negative | False Alarm | Correct Rejection |

Note that missed detections and correct rejections are redundant with correct detections and false alarms. The probability of missed detection $P(MD) = 1 - P(CD)$ and the probability of correct rejection $P(CR) = 1 - P(FA)$. Therefore, accuracy in predicting interaction times can be perfectly described through knowledge of their probabilities of correct detection and false alarm.

Let $t_{interaction}$ be the time that an interaction actually occurs and $\hat{t}_{prediction}$ be the predicted time of interaction. The interaction times are continuous and can never be identically equal, so $\Pr(\hat{t}_{prediction} = t_{interaction}) = 0$; instead, if a prediction is within an allowable time range $\Delta t_b$ before an actual interaction time and $\Delta t_a$ after an actual interaction time, the prediction is considered to have accurately triggered an interaction. The allowable time ranges ($\Delta t_b$ and $\Delta t_a$) can be chosen based on accuracy requirements of the simulation.

In summary, the probabilities can be expressed by following equations and can be estimated experimentally through simulation.

$$P(CD) = \Pr(t_{interaction} - \Delta t_b \leq \hat{t}_{prediction} \leq t_{interaction} + \Delta t_a)$$

$$= \frac{CD}{(CD + MD)} = 1 - P(MD) \tag{1}$$

$$= \frac{number\ of\ interactions\ predicted\ correctly}{total\ number\ of\ actual\ interaction\ events}, \text{ and}$$

$$P(FA) = \frac{FA}{(FA + CR)} = 1 - P(CR) \tag{2}$$

$$= \frac{total\ number\ of\ updates\ triggered\ incorrectly}{total\ number\ of\ unnecessary\ updates + total\ number\ of\ correct\ rejections}$$

In predicting the time of an interaction, any algorithm may choose to scale its predicted time to be more or less conservative; this is equivalent to the choice of scale factor, *theta* ($\theta$) (traditionally defined by SDT as a threshold). SOC curves are used to illustrate the fundamental tradeoff between false alarms and correct detections, as shown in Fig. 3. The SOC curve describes the locus of probability of correct detection (y-axis) against the probability of false alarm (x-axis) for all values of $\theta$.

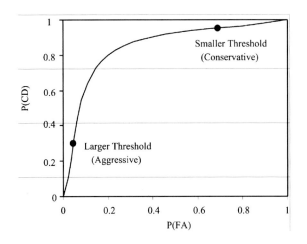

**Fig. 3.** Schematic SOC curve with prediction threshold values

A perfect method for predicting the time of interactions would have an SOC curve that passes through the extreme upper left corner, corresponding to an algorithm capable of $P(CD)$ of 1.0 and $P(FA)$ of 0.0. This would create a simulation that always synchronizes agents only when they need to interact, allowing for maximum computational efficiency. However, due to uncertainties in the dynamic behavior of agents, SOC curves generally do not pass through this point. Instead, as shown in Fig. 3, if predictions are weighted to be conservative (the threshold is relatively small), individual agents will be updated more often during the simulation. This approach can reduce the number of missed detections at the expense of an increased rate of false alarms, at the expense of computational efficiency. If predictions are allowed to become aggressive (the threshold is relatively large), fewer updates will occur (fewer false alarms), at the expense of increased missed detections. Here lies the fundamental tradeoff between missed detection and false alarm in setting the interaction times: reducing $P(MD) = 1 - P(CD)$ will increase $P(FA)$ while $P(FA)$ reducing will increase $P(MD)$. Thus, we propose that this application of SOC analysis is a useful method to evaluate the consequences in terms of $P(CD)$ and $P(FA)$ of making predictions more or less conservative within the simulation.

## 3.3   Prediction of Agent Interactions

The prediction of interaction times of agents is also very critical in implementing efficient timing mechanisms for agent-based simulations. This describes the issues of predicting interaction times between agents and limitations of developing a measurement agent to more accurately predict the interaction times without significant computation. Specifically, a generic interaction timing component can be developed to predict the interaction times between all combinations of pairs of designated sets of agents, as shown schematically in Fig. 4. This interaction timing component (ITC) maintains a list of modular interaction detection subcomponents, each of which is associated with one pair of agents to monitor; each predicts the time of their next interaction and calls for two relevant agents to be synchronized at that time. The ITC relates these agent update times to the broader simulation architecture.

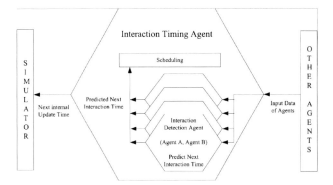

**Fig. 4.** Schematic of interaction timing component

The ITC uses an object-oriented modular structure that makes it easy to develop new interaction detection subcomponents for predicting different types of interactions. The ITC structure provides the interface to the broader simulation environment and establishes the intelligent detection subcomponent needed for all pairs of relevant agents. This modular interaction detection subcomponent structure has the direct benefit of allowing for easy implementation of new prediction models of interactions in the simulation, as a developer only needs to create a prediction function specific to an interaction between two agents without modifying the agent models themselves and without needing to duplicate the infrastructure provided by the ITC. Similar structures can also be easily created for interactions involving more than two agents.

## 3.4   Intelligent Interaction Timing Prediction Using a Neural Network

The accuracy and computational efficiency of the simulation depend on the prediction of agents' interaction times. This section details a method for automatically determining the interaction times that needs neither extensive computation during run-time nor scenario-specific development of prediction algorithms. Specifically, this section examines building intelligence into the interaction timing component itself to first learn from the dynamics of the simulation such that it can then predict the time of interactions between agents. This resulting 'intelligent ITC' also has the benefit of reducing the time and cost required to develop new large-scale simulations, as de-tailed, scenario-specific predictors of interaction times are no longer be needed.

Our method of predicting interaction times trains a back-propagation neural net-work (BPNN) to predict interaction times between agents within the ITC, as shown in Fig. 5. A BPNN is trained to predict when interactions will occur from a set of train-ing data created through the preliminary simulations with a conservative prediction method. Once the neural network is trained, it can be embedded into the simulation to automatically predict interaction times during run-time without expensive computa-tions. This intelligent ITC control structure has the direct benefit of allowing for com-paratively easy implementation of new interactions in the simulation: when a different interaction between agents needs to be modeled, the neural network can be re-trained.

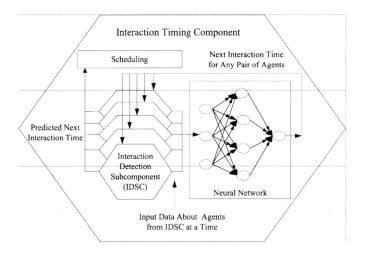

**Fig. 5.** Structure of intelligent interaction timing component using a neural network

### 3.4.1  Procedures for Applying BPNN

The concept of backpropagation neural network was first formalized by Paul Werbos for forecasting analysis [55] and later by Rumelhar and McClelland [56]. The backpropagation algorithm is the most popular supervised learning method for multilayer feedforward neural networks due to its simplicity and reasonable speed. The most appealing feature of backpropagation is its adaptive nature, which allows complex processes to be modeled through learning from training data by updating the weights of connecting neurons in each layer. Backpropagation and its variants have been applied to a variety of problems, including pattern recognition [52], signal processing, image compression, speech recognition, and nonlinear system modelling and control [57].

Traditional backpropagation algorithms have some disadvantages, such as slow con-vergence and trapping in local minima. Therefore, in the backpropagation with momen-tum algorithm used here, a momentum parameter is used in scaling the adjustments from a previous iteration and adding to the adjustments in the current iteration. In most

cases, neural networks with a single hidden layer are capable of approximating all continuous functions [58]. Thus, our method of predicting interaction times uses the neural
network with a single hidden layer and a single output neuron, and the network is
trained with the backpropagation with momentum learning algorithm.

This algorithm uses two phases: one to propagate the input pattern, and the other to
adapt the output. After an input pattern has been applied as a stimulus to the first layer
of network, it is propagated through each upper layer until an output is generated.
This output value is then compared to the desired output, and an error is computed for
each output unit. Specifically, the measure of the error on a training pattern is the sum
of the squares of the errors for all output units as defined by Equation (3). In this case,
$d_j^p$ is the desired value of the $j$th output neuron for input pattern $p$, and $o_j^p$ is the
actual output value of the $j$th output neuron for pattern $p$:

$$E^p = \frac{1}{2} \sum_{j \in output\ units} (d_j^p - o_j^p)^2 \tag{3}$$

The actual output from each neuron $j$ for pattern $p$ is the transfer function $f_j(\bullet)$
acting on the weighted sum on the connection from neuron $i$ to neuron $j$, which can be
written as:

$$o_j^p = f_j \left( \sum_i w_{ij} \cdot o_i^p \right)$$

$$= \frac{1}{1 + e^{-\left( \sum_i w_{ij} \cdot o_i^p \right)}}, \tag{4}$$

where $w_{ij}$ denotes a weight between neuron i and neuron j.

The output value of our network is non-linearly scaled between 0 and 1 based on the
sigmoid function (4). The output value at the output layer is then scaled to the correct
magnitude of a predicted interaction time. The backpropagation algorithm generalizes
the delta rule. The connection weights between nodes are adjusted backwards based on
a generalized delta learning algorithm to reduce the total error. The errors at the output
layer are propagated backward from the output layer to each node in the hidden layer
during training phase. To determine the direction in which to change the weights, calculate the negative of the gradient of $E^p$ with respect to the weights connected to output
units. In the BPNN with momentum, a percentage of the last weight adjustments at time
$(t\text{-}1)$, called the momentum term, is added to current weight adjustments at time $t$, as
defined in Equation (5). The adjustments to the weights when training the BPNN with
momentum consider a learning rate ($\eta$), which has a significant effect on the network
performance, and a momentum term ($\alpha$), which increases the speed of convergence. A
low learning rate can typically ensure more stable convergence while a high learning
rate can speed up convergence in some cases.

$$\Delta^p w_{ij}(t) = \eta \left( -\frac{\partial E^p}{\partial w_{ij}} \right) + \alpha \cdot \Delta^p w_{ij}(t-1) \tag{5}$$

The following three main phases are followed in using the intelligent ITC with the BPNN with momentum.

*1) Training Data Selection Phase:* During preliminary simulation runs with a conservative prediction algorithm, a set of training data (pairs of input variables and desired output value) for BPNN is collected. In this case, the training data describes the state of the agents whenever an interaction occurs. The desired output value is an interaction time between a pair of agents. The performance of BPNN depends strongly on the quality of training data. The best training procedure is to compile a wide range of training data (for more complex problems, more training data is required) that exhibits all the different characteristics of the problem of interest [59].

As shown in Fig. 6, this can be thought of as the space of training data. Determination of this space requires an understanding of the domain being simulated and the purpose of the simulation, such that the predictor will be trained sufficiently on all conditions of accuracy.

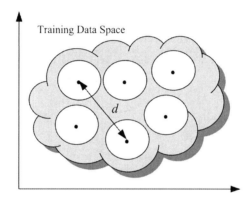

**Fig. 6.** Training data space and distance between input points of training data

Too many redundant elements in the collected training data set can degrade the performance of the network in terms of accuracy and training time. Therefore, we developed an algorithm to efficiently select good training data during the preliminary simulation runs. The idea is to measure a distance (called deviation) between existing training data and a new instance of training data. The deviation of each element of training data is determined by calculating the difference of new input variables (feature values) from the normalized input variables of the existing training data, summing up the squares of the difference measurements, and then taking the square root of the sum as follows:

$$\text{Deviation } (d) = \sqrt{\sum_{i=1}^{n} (x_i - y_i)^2} \quad , \tag{6}$$

where $x_i$ = the $i$th input variable of the existing training data
$y_i$ = the $i$th input variable of the current training data
for i = 1,2, .. ,$n$.     $n$ = the number of input variables

Whenever a new training data is collected during the preliminary simulation, its deviation is compared with that of all accepted existing training data. If the deviation is larger than a predefined minimum ($\varepsilon$) for all existing training data then the new training data is accepted as representing an aspect of the training space hitherto not captured in. The selection of the minimum deviation ($\varepsilon$) will affect the performance of the neural network. For example, if the minimum deviation is too small, many nearly-identical elements of training data will be accepted, resulting in a slow convergence rate in training the neural network and a tendency to converge on a local minimum. On the other hand, if the minimum deviation is too large, the input data space will be too sparse, which means the performance of neural network may be degraded since the neural network is not well trained with a complete set of input patterns that covers the entire input space. Thus, while the algorithm can be applied automatically, its parameter $\varepsilon$ must be chosen based on understanding of its implications for training BPNN.

*2) Training Network Phase:* Next, BPNN must be trained using the selected set of training data. The backpropagation with momentum learning algorithm is used to efficiently modify the different connection weights to minimize the errors at the output [60]. Most agent-based simulations need to predict the interaction times conservatively to avoid missing any important interactions between agents. Therefore, the generalized delta learning rule in the BPNN can be modified to penalize more strongly over-estimates of interaction time as follows:

$$\Delta^P w_{ij}(t) = \eta \cdot \delta_j^P \cdot o_i^P + \alpha \cdot \Delta^P w_{ij}(t-1), \qquad (7)$$

$$\text{where } 0 \le \eta \le 1, \quad 0 \le \alpha \le 1$$

$$\delta_j^P = \lambda \cdot o_j^P \cdot (1 - o_j^P) \cdot (d_j^P - o_j^P) \qquad if \ (d_j^P - o_j^P) < 0$$

$$\delta_j^P = o_j^P \cdot (1 - o_j^P) \cdot (d_j^P - o_j^P) \qquad if \ (d_j^P - o_j^P) > 0$$

This modification imposes a penalty ($\lambda$) when the predicted output value of input pattern is overestimated to control how conservatively to train the neural network. The stopping criteria for learning need to be also defined: If the average error per input pattern drops below a specified error tolerance, then the learning process can be stopped. In the training phase, the training data is split into two parts: a training set and a test set. Once the neural network is trained with training data set, the total error of the neural network can be computed with the test data set. If this error meets the specified error tolerance, then the neural network may be considered validated; if it does not, then the network needs to be retrained with a different set of training data or network parameters to meet the specified error tolerance. Note that the network will never exactly learn the ideal function between input and output values, but rather it will asymptotically approach the ideal function.

*3) Simulation Phase:* During the course of the simulation, interaction detection sub-components provide the state of pairs of agents to the BPNN as input values of the neural network. The neural network then gives back a predicted next interaction time of the pair. The intelligent ITC selects a minimum value among the next interaction times of all interaction detection agents and interacts with the timing mechanism of the simulation to synchronize those agents at their predicted interaction time.

### 3.4.2  Main Issues of Using the BPNN

There are several issues to be considered when designing and training a neural network. The goal is to achieve a good performance of the network, which means the network can predict desired target values for any inputs within a specified operating space, even if not specifically identically captured in the training set.

First, sufficient information about the agents must be provided to the neural network to discriminate outputs with the desired degree of accuracy. Likewise, eliminating redundant or ineffective inputs can improve accuracy. The wise choice of inputs to the neural network requires expertise in the domain being simulated; with it the complexity and training time of the network can be reduced and predictive performance of the network can be improved [61], as manifested in shifting its SOC curve closer to the ideal point corresponding to $P(CD) = 1$ and $P(FA) = 0$.

Second, it is a necessary condition for good performance that the training sample set should be sufficiently large. If an untrained input pattern is within the training data space and close to elements of the training data set, the trained network might be able to interpolate reliably. However, if an untrained input pattern is outside of the training data space or lies far away from elements of the training data set, the performance of the network can be unreliable. Unfortunately, there is no firm method for guaranteeing a priori that the training data set is sufficient to train the neural network. Selection of a set of significant features and suitable training data is often done experimentally and requires knowledge about the domain or system. The use of a separate set of data spanning the training space to validate the neural network provides a necessary, but only post hoc, assessment of the training data set.

Third, the training time can be significant and a number of techniques may be warranted to improve the training process. These include early stopping that is used for improving the generalization of a neural network during training and preventing it from overtraining, cross-validation for estimating generalization error based on re-sampling and varying the size of the training data [62]. The effort required for training will be merited when the simulation will be used repeatedly and the trained neural net will provide beneficial levels of computational efficiency and prediction accuracy.

Other issues relate to the network architecture, including the number of hidden layers and the number of units for each hidden layer. For applications where a nonlinear model is needed, typically one hidden layer with arbitrary large number of units is sufficient for a good approximation. However, there is no general 'rule of thumb' that gives the most appropriate number of hidden units for training the network.

## 4  Test Scenario: Simulation of Air Traffic Control System

To demonstrate the computational performance of different timing mechanisms, an experiment was conducted using agent-based simulation architecture [63], [64] and the intelligent ITC using the BPNN described in the previous section. An agent-based simulation model was developed for a stream of arriving aircraft flying the Macey Two Standard Terminal Arrival Route (STAR) into Atlanta Hartsfield-Jackson International Airport. Three different arrival paths merge into one traffic stream to the airport, defined by a sequence of waypoints for aircraft. Several different types of agent models were included in the simulation, including an air traffic controller and aircraft. Although a specific airport was modeled, this simulation model could easily be modified to nearly any airport traffic control center by changing the traffic flow pattern (and retraining the ITC) without any changes to the agent models themselves.

One of the main interactions of concern is the loss of minimum separation between aircraft. A *conflict* can be defined as two or more aircraft violating their minimum allowed horizontal distance and altitude separation between aircraft. In our agent-based simulation, the interaction timing agent predicts the conflict time at which a minimum separation distance between aircraft is violated using either a BPNN or a simple conservative prediction algorithm.

To predict aircraft conflicts, it is necessary to project the future positions of aircraft over time. However, due to variability and uncertainty in the aircraft trajectory, linear extrapolations of the aircraft's location at future times are frequently erroneous. For example, 'dead-reckoning' linearly extrapolates the position of aircraft and thus can be used to predict the conflict times between them. However, dead-reckoning will not account for interactions predictable in a stochastic sense as, in the air traffic environment modeled here, stochastic events such as air traffic controller commands and aircraft maneuvers to follow a defined arrival route can occur at any time. Instead, knowledge of the pattern of the traffic flow, including likely air traffic commands, can be learned to improve the statistical accuracy of the predictor even accounting for likely changes in any one aircraft's trajectory. Analytic prediction algorithms, therefore, would need to be redesigned for every arrival route, while an intelligent ITC would need only for its neural network to be retrained.

## 4.1  Agent Models

For the agent-based simulation, several different types of agent models (heterogeneous agents) were developed, each with different methods for internally representing time and rates at which they do so. These models include:

*1)  Waypoint-Following Aircraft Agent:* Each waypoint following aircraft can be given a desired trajectory defining the planned route of an aircraft by a list of waypoints. A dynamic model based on control algorithms and ordinary differential equations provides the necessary guidance and equations of motion to yield the actual trajectory of the aircraft. It requires very small time steps, which is computational intensive. It can be commanded by the air traffic control (ATC) controller agent to change its speed and heading.

*2)  Random Aircraft Generator:* The random aircraft generator enters waypoint-following aircraft agents stochastically into the three arrival routes with an inter-arrival time specified by a statistical distribution. The stochastic input implies that the amount of traffic in the simulation also varies stochastically; traffic can be very intense for short periods of time yet can drop off to nearly no traffic on other occasions. This property is also true of actual operations.

*3) Air Traffic Controller Agent:* The air traffic controller monitors aircraft movement and ensures that a safe separation distance is maintained between the in-flight aircraft under their control. In this scenario, a simple ATC agent determines aircraft sequences within merging arrival streams and then commands desired speeds or headings to aircraft to maintain proper spacing. Several instances of ATC agents were created, one for each of the airspace sectors the aircraft would fly through. ATC agents monitor their control sectors for aircraft entering or leaving their control sectors. The ATC agent also routinely requires aircraft to synchronize when it needs information about the current state of all aircraft in its sector, mimicking a scan of the radar screen.

*4) Conflict Detection Component:* A conflict detection component was developed and implemented using the ITC structure described earlier. The conflict detection component predicts conflict times between aircraft from the current states of aircraft. Two different prediction methods were used: a simple conservative prediction algorithm and the intelligent prediction method using BPNN.

## 4.2   Conservative Prediction Method of Conflict Detection Times

For the conservative prediction method of conflict detection, a simple algorithm was used to predict the worst-case time to conflict between aircraft ($\hat{t}_{CPM}$) by assuming two aircraft are flying directly at each other, regardless of their actual convergence angle, as follows:

$$\hat{t}_{CPM} = \frac{distance\ between\ two\ aircraft\ \text{-}\ minimum\ separation\ distance}{airspeed\ of\ one\ aircraft + airspeed\ of\ the\ other\ aircraft} \tag{8}$$

Let $\theta$ be a threshold value which can be varied to set the conservativeness of the prediction time. Then, the next interaction time of the ITA using the conservative method can be defined as:

$$\text{Next Predicted Interaction Time} = current\ time + \theta \cdot \hat{t}_{CPM} \tag{9}$$

## 4.3   Intelligent Prediction Method of Conflict Detection Times Using the BPNN

The input values for the BPNN were determined by knowledge of system dynamics. Andrews [65] described the motion of a pair of aircraft as a dynamic system captured by five state variables: range ($r$), relative bearing seen from aircraft 1 ($\beta_1$), relative bearing seen from aircraft 2 ($\beta_2$), airspeed of aircraft 1 ($V_1$), and airspeed of aircraft 2 ($V_2$), as shown in Fig. 7. These five state variables are used as input variables of the neural network to predict the time to the loss of minimum separation between a pair of aircraft. Through a preprocessing of input patterns, the five state values are normalized to a value ranging from 0 to 1.

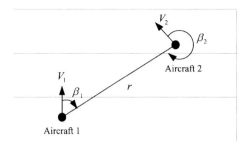

**Fig. 7.** Input variables of neural network for prediction of conflict times between aircraft

The output of the neural network is $\hat{t}_{IPM}$, the predicted time to the conflict between aircraft by the intelligent prediction method using the BPNN, measured relative to current time. If there is no predicted conflict between aircraft, then the output value is limited to an upper bound; if two aircraft are already too close to each other, then the desired output value is set to a lower bound. During the simulation, this predicted time is converted to an absolute time at which the pair of aircraft should be next evaluated for a conflict. The next predicted interaction time can be made more or less conservative by multiplying by the threshold value ($\theta$):

$$\text{Next Predicted Interaction Time} = current \ \ time \ + \theta \cdot \hat{t}_{IPM} , \quad (10)$$

$$\text{where}$$

$$\hat{t}_{IPM} = predicted \ target \ value \cdot (upper\_bound - lower\_bound) + lower\_bound$$

In the training data selection phase, a total of 1173 input training patterns were selected with a minimum deviation of 0.02 through the preliminary simulation with the conservative prediction method. In the learning phase, a penalty value of 5 was used in the modified delta learning rule to train the BPNN conservatively. Generally, the performance of the neural network depends upon many factors such as network topology selection, learning parameters, and completeness of representation of the problem space in the training patterns [43]. The parameters to train the BPNN in this simulation were obtained through preliminary simulations. Once identified, these parameters can be re-used for automatic re-training of the neural network to similar configurations.

### 4.4 Performance Analysis of Prediction Methods

Typically, each agent requires some time to execute its internal dynamics. The cost of computation scales with the number of agent executions performed. Therefore, the number of updates of the most computationally intensive agents can serve to measure computational efficiency. In this test scenario, the waypoint following aircraft agents are both the most computationally expensive and the ones most likely to be asked to unnecessarily synchronize. Thus, the average number of updates per aircraft in flying the arrival route will be used as the measure of computational efficiency: fewer updates correspond to better performance.

The computational efficiency of the intelligent prediction method using BPNN was compared with the conservative prediction method, as shown in Fig. 8. The ideal minimum number of updates represents the number of updates the aircraft agent should be executed for their dynamics and for interactions with other agents. In the synchronous simulation, each agent predicts its own next update time based on its internal dynamics or a predefined fixed update time. This method requires all agents in the simulation to update at the same time, using the worst-case (smallest) next update time of any agents. The asynchronous with complete resynchronization simulation allows for agents to be updated independently following their own update times until an agent requires a complete synchronization given for any reason; for example, the air traffic controller agents or conflict detection component are allowed to command a complete resynchronization at times when they need to get current state information of aircraft agents or the next air traffic command might be warranted. The conservative prediction method was used in the conflict detection component with this timing mechanism. However, there are still many unnecessary updates of aircraft agents because this timing method requires a complete resynchronization of all agents at the conservatively predicted interaction times, as shown in Fig. 8, rather than just resynchronization of those agents involved in a predicted interaction.

On the other hand, with the asynchronous with partial resynchronization timing mechanism, the controller agents require only the aircraft in their sector to be updated when they "scan their radar", and the conflict detection component requires only the two aircraft involved in a potential loss of separation to be updated. The computational efficiency of the simulation is significantly improved overall with both asynchronous with partial resynchronization timing methods; the asynchronous with partial resynchronization using intelligent prediction method is the most computationally efficient. Further computational efficiency gains can be achieved by predicting the interaction times between agents more accurately and efficiently, as shown by the further reductions in aircraft updates using intelligent interaction predictions.

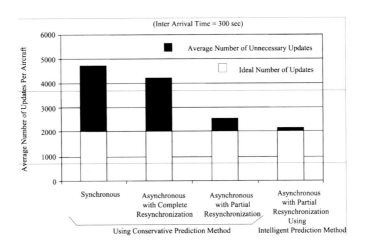

**Fig. 8.** Comparison of performance of different prediction methods

## 4.5  SOC Curves of the Prediction Methods

SOC curves for both prediction methods are shown in Fig. 9. In each case, the conflict interval defined the period of time within which the prediction must fall after the start of the conflict; this corresponds to $\Delta t_a$ as described in the section 3.2; for this application the prediction was not wanted before the start of the conflict so $\Delta t_b$ was set to zero. Several conflict intervals were used in defining the criteria for a correct detection. The SOC curves show a significant benefit is gained with the intelligent prediction method using BPNN compared to the conservative prediction method. Especially for larger conflict intervals, appropriate threshold setting for the BPNN can achieve a high probability of correct detections with a low probability of false alarms. On the other hand, the conservative prediction method is limited in its ability to reduce the number of false alarms (unnecessary updates of agents) by varying its threshold value. Even though the same probability of correct detections can be achieved by adjusting the threshold value of the prediction methods, the incidence of false alarms is much lower with the intelligent prediction method. Therefore, once its BPNN is trained, the intelligent prediction method provides more accurate interaction times with greater computational efficiency and without requiring sophisticated prediction algorithms to be developed.

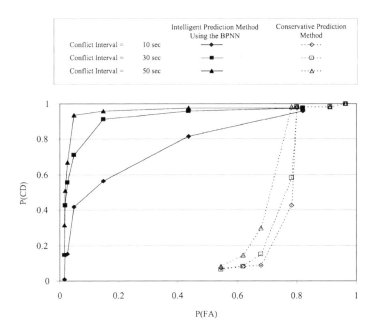

**Fig. 9.** SOC curves with conservative and intelligent prediction using BPNN methods

## 5  Conclusions and Discussion

Agent-based modeling and simulation can be a powerful method for the design and analysis of large-scale multi-agent systems to capture both the individual behavior of agents within the system and the emergent behavior of heterogeneous agents interacting with each other in significant ways. The main advantage of agent-based simulation is that it facilitates the simulation of emergent (macro-level) behavior as well as individual (micro-level) agent behavior in highly dynamic situations, which is hard to capture with traditional simulation methods. The emergent behavior of a system represents the intricate inter-relationships that arise from the interaction of agents. This type of simulation generates system dynamics from the actions of heterogeneous agents and their interactions. From these, it predicts the behavior of the socio-technical system as a whole, and the corresponding demands the environment will place on the agents.

Typically, large-scale agent-based simulation requires considerable computational power. Timing mechanisms capable of adding accuracy and computational efficiency to the agent-based simulation were demonstrated in this chapter. A new timing method, *asynchronous with resynchronization*, was proposed and detailed in this chapter. A key idea of the newly developed timing mechanisms is that it is not necessary for the simulator to update all agents whenever only a single agent is updated. The accuracy and computational efficiency of the asynchronous with resynchronization timing mechanisms can be improved by better estimation of the resynchronization intervals. In summary, setting the resynchronization intervals for the asynchronous with resynchronization timing mechanism must balance a tradeoff between early resynchronization to increase accuracy and extending a resynchronization interval to improve computational efficiency.

This chapter showed that the new timing mechanisms greatly reduce the unnecessary updates of agents, resulting in much more efficient performance of the agent-based simulation in the air traffic test case. The evaluation of different prediction methods of interactions was cast as equivalent to a signal-detection problem, allowing the performance of interaction prediction algorithms to be examined directly by SOC curves. Once a SOC curve is constructed, the accuracy and computational efficiency of the simulation can be adjusted by setting the threshold value that best suits the needs of the simulation within the tradeoffs between computational efficiency and accuracy.

It was shown that such accuracy and efficiency can be achieved when agents are updated only when required for their internal dynamics, for correct interactions with other agents, and for timely measurements. Even with a simple method of conservatively estimating resynchronization intervals, significant computational benefits were found in the test case simulation. Timing mechanisms for agent-based simulation, therefore, must ensure that agents in the simulation are synchronized at any time when they must interact or exchange data between agents. The new timing mechanisms can scale to the large-scale simulations involving large number of agents interacting with each other. In particular, when agents have widely varying update times and need to interact frequently, the asynchronous with partial resynchronization method can greatly improve computational efficiency.

This chapter also described the development of an intelligent timing component capable of taking a simple interaction detection subcomponent containing only a prediction function and applying this agent to all pairs of relevant agents in the simulation. This interaction timing component has the direct benefit of allowing for easy implementation of new predictions in the simulation, as it monitors all combinations of relevant agents and interacts with the larger simulation architecture, including commanding the relevant agents to resynchronize at appropriate times. Improvements in computational efficiency have been demonstrated due to the ability of the interaction timing component to declare when resynchronizations are needed for agent interactions. To create a flexible, reconfigurable, yet accurate and computationally efficient prediction mechanism, this research also demonstrated the application of a backpropagation neural network (BPNN) in predicting the time of interactions between agents. An intelligent timing component using the BPNN was presented to predict the interaction times between agents. This modular intelligent agent framework could be easily adapted to simulating other configurations or even other application domains without extensive development efforts; only the BPNN would need to be re-trained. In the test case, unnecessary updates of agents were found to be significantly reduced with the intelligent prediction method by accurately capturing and predicting the interactions between agents using the BPNN.

However, there are still several issues meriting further study. First, in this study, the training data for the prediction of interactions was obtained through the off-line preliminary simulation. The learning structure of the BPNN can be improved by allowing the neural network to gather training data and adjusting the weights of the neural network adaptively during normal simulation runs. Further research on a more general method to select input features and input patterns, to train a neural network to improve the performance of the network, and to choose the topology of the neural network for better training and performance of the network, would facilitate use of the intelligent ITC in a range of simulations. Second, as described earlier, the computational efficiency and accuracy of the prediction methods might depend on the predictability of simulation environment. For example, if numerous stochastic events (such as human errors and machine failures) can occur, the performance of interaction predictions might degrade. Further research is required to evaluate the impact of the predictability of system dynamics on the performance of such intelligent timing and prediction mechanisms.

# References

1. Macal, C.M., North, M.J.: Tutorial on Agent-Based Modeling and Simulation Part 2: How to Model With Agents. In: Proceedings of the 2006 Winter Simulation Conference, IEEE, Los Alamitos (2006)
2. Josylyn, C., Rocha, L. (eds.): Towards Semiotic Agent-Based Models of Socio-Technical Organizations. In: Sarjoughian, H., et al. (eds.) Proceedings of AI, Simulation and Planning in High Autonomy Systems, pp. 70–79 (2000)
3. Barrett, C.L., et al.: Science & Engineering of Large Scale Socio-Technical Simulation, Los Alamos National Laboratory (2001)
4. Law, A.M., Kelton, W.D.: Simulation modeling and analysis, 3rd edn. McGraw-Hill series in industrial engineering and management science, vol. xxi, p. 760. McGraw-Hill, Boston (2000)
5. Zeigler, B.P., Kim, T.G., Praehofer, H.: Theory of modeling and simulation: integrating discrete event and continuous complex dynamic systems. 2nd edn., vol. xxi, p. 510. Academic Press, San Diego (2000)

6. Jennings, N.R., Wooldridge, M.: Applications of Intelligent Agents. In: Jennings, N.R., Wooldridge, M. (eds.) Agent Technology: Foundations, Applications, Markets, pp. 3–28. Springer, Heidelberg (1998)
7. Cohen, P.R., et al.: Trial by fire: Understanding the design requirements for agents in complex environments. AT Magazine, 32–48 (1989)
8. Atkin, S.M., et al.: AFS and HAC: Domain general agent simulation and control. In: Software Tools for Developing Agents: Papers from the 1998 Workshop, pp. 1–10. AAAI Press, Menlo Park (1998)
9. Norman, D.A., Draper, S.W.: User centered system design: new perspectives on human-computer interaction, vol. xiii, p. 526. L. Erlbaum Associates, Hillsdale (1986)
10. Rasmussen, J., Pejtersen, A.M., Goodstein, L.P.: Cognitive Systems Engineering. John Wiley & Sons, New York, NY (1994)
11. Vicente, K.J.: Cognitive work analysis: toward safe, productive and healthy computer-based work, vol. xix, p. 392. Lawrence Erlbaum Associates, Mahwah (1999)
12. Pritchett, A.R., Lee, S.M., Goldsman, D.: Hybrid-System Simulation for National Airspace System Safety Analysis. AIAA Journal of Aircraft 38(5), 835–840 (2001)
13. Davidsson, P., Logan, B., Takadama, K.: Multi-agent and multi-agent-based simulation: joint workshop MABS 2004, revised selected papers, New York, NY, USA, July 19, 2004, vol. x, p. 264. Springer, Berlin (2005)
14. Wooldridge, M.J., Jennings, N.R.: Intelligent Agents: Theory and Practice. Knowledge Engineering Review, pp. 115–152 (1995)
15. Macal, C.M., North, M.J.: Tutorial on Agent-Based Modeling and Simulation Part 2: How to Model With Agents. In: The 2006 Winter Simulation Conference (2006)
16. Russell, S.J., Norvig, P., Canny, J.: Artificial intelligence: a modern approach, 2nd edn., vol. xxviii, p. 1080. Prentice Hall, Upper Saddle River (2003)
17. Gilbert, N., Troitzsch, K.G.: Simulation for the Social Scientist. Open University Press, Buckingham (1999)
18. Goldspink, C.: Modeling Social Systems as Complex: Toward a Social Simulation Meta Model. Journal of Artificial Societies and Social Simulation 3(2) (2000)
19. Shen, W., Norrie, D.H.: Agent-Based Systems for Intelligent Manufacturing: A State-of-the-Art Survey. Knowledge and Information System, an International Journal 1(2), 129–156 (1999)
20. Agent-based Design and Simulation of Supply Chain Systems. In: Barbuceanu, M., Teigen, R., Fox, M.S. (eds.) Proceedings of the 6th Workshop on Enabling Technologies Infrastructure for Collaborative Enterprises (WET-ICE 1997), IEEE, MIT, Cambridge, MA (1997)
21. Huang, C.C.: Using Intelligent Agents to Manage Fuzzy Business Process. IEEE Transactions on Systems, Man, and Cybernetics - Part A: Systems and Humans 31(6), 508–523 (2001)
22. Ankenbrand, T., Tomassini, M.: Agent-Based Simulation of Multiple Financial Markets. In: Neural Network World, pp. 397–405 (1997)
23. Mizuta, H., Yamagata, Y.: Agent-based Simulation for Economic and Environmental Studies. In: JSAI 2001 Workshop (2001)
24. Bianco, L., et al.: Modelling and simulation in air traffic management. In: Transportation analysis, vol. vii, p. 202. Springer, Berlin (1997)
25. Lee, S.-M., Pritchett, A.R., Corker, K.: Evaluating Transformations of the Air Transportaiton System Through Agent-Based Modeling and Simulation. In: 7th USA/Europe ATM R&D Seminar, Barcelona, Spain (2007)

26. Modeling the NAS: A grand challenge for the simulation community. In: Wieland, F.P., et al. (eds.) First International Conference on Grand Challenges for Modeling and Simulation (2002)
27. Bonabeau, E.: Agent-based modeling: methods and techniques for simulating human systems. Proc. Natl. Acad. Sci. (2002)
28. Hayes, C.C.: Agents in a Nutshell - A Very Brief Introduction. IEEE Transactions on Knowledge and Data Engineering 11(1), 127–132 (1999)
29. Shoham, Y.: Agent-oriented Programming. Artificial Intelligence 60, 51–92 (1993)
30. Uhrmacher, A.M.: Concepts of Object- and Agent-Oriented Simulation. Transactions of the Society of Computer Simulation 14(2), 59–67 (1997)
31. Agent-Oriented Software Engineering. In: Jennings, N.R., Wooldridge, M.J. (eds.) Proceedings of the 9th European Workshop on Modeling Autonomous Agents in a Multi-Agent World: Multi-Agent System Engineering, MAAMAW-1999 (2000)
32. Bradshaw, J.M.: Software agents, vol. x, p. 480. AAAI Press, Menlo Park, CA (1997)
33. Anderson, J.: A Generic Distributed Simulation System for Intelligent Agent Design and Evaluation. In: The Tenth Conference on AI, Simulation and Planning, AIS-2000, Society for Computer Simulation International (2000)
34. Examining air transportation safety issues through agent-based simulation incorporating human performance models. In: Pritchett, A.R., et al. (eds.) Proceedings of the IEEE/AIAA 21st Digital Avionics Systems Conference (2002)
35. Wickens, C.D., et al.: Flight to the Future: Human Factors in Air Traffic Control, vol. xi, p. 368. National Academy Press, Washington (1997)
36. Shah, A.P., Pritchett, A.R.: Agent-Based Modeling and Simulatin of Socio-Technical Systems. In: Rouse, B., Boff, K. (eds.) Organizaitonal Simulation (2005)
37. Lee, S.M., et al.: Developing Human Performance Models Using Apex/CPM-GOMS for Agent-Based Modeling and Simulation. In: The 2004 Advanced Simulation Technologies Conference (ASTC 2004), Arlington, VA (2004)
38. Callantine, T.: Agents for Analysis and Design of Complex Systems. In: The 2001 International Conference on Systems, Man, and Cybernetics (2001)
39. Fujimoto, R.M.: Parallel and distributed simulation systems. Wiley series on parallel and distributed computing, vol. xvii, p. 300. Wiley, New York (2000)
40. Ghosh, S., Lee, T.S.: Modeling and asynchronous distributed simulation: analyzing complex systems, vol. xxxi, p. 300. IEEE Press, New York (2000)
41. Sloman, A., Poli, R.: SIM_AGENT: A toolkit for exploring agent designs. In: Tambe, M., Müller, J., Wooldridge, M.J. (eds.) IJCAI-WS 1995 and ATAL 1995. LNCS, vol. 1037, pp. 392–407. Springer, Heidelberg (1996)
42. Laird, J.E., Newell, A., Rosenbloom, P.S.: SOAR: An architecture for general intelligence. Artificial Intelligence 33, 1–64 (1987)
43. Anderson, J.R., Libiere, C.: The Atomic Components of Thought. Lawrence Erlbaum Associates, Mahwah, NJ (1998)
44. Logan, B., Theodoropoulos, G.: The Distributed Simulation of Multi-Agent systems. Proceedings of the IEEE 89(2), 174–185 (2001)
45. Uhrmacher, A.M., Gugler, K.: Distributed, Parallel Simulation of Multiple, Deliberative Agents. In: Bruce, D., Lorenzo, D., Turner, S. (eds.) Proceedings of the 14th Workshop on Parallel and Distributed Simulation (PADS 2000), pp. 101–108. IEEE Computer Society, Bologna, Italy (2000)
46. Jefferson, D.R.: Virtual Time. ACM Transactions on Programming Languages and Systems 7(3), 404–425 (1985)

47. Carothers, C.D., Perumall, K.S., Fujimoto, R.M.: Efficient Optimistic Parallel Simulations Using Reverse Computation. ACM Transactions on Modeling and Computer Simulation 9(3), 224–253 (2000)
48. Mirtich, B.: Timewarp Rigid Body Simulation. In: SIGGRAPH 2000, Association for Computing Machinery, New York (2000)
49. Barkat, M., Books24x7 Inc.: Signal detection and estimation, 2nd ed. (Text) (2005)
50. Swets, J.A., Pickett, R.M.: Evaluation of diagnostic systems: methods from signal detection theory, vol. xiv, p. 253. Academic Press, New York (1982)
51. Swets, J.A.: Measuring the Accuracy of Diagnostic Systems. Science 240, 1285–1293 (1998)
52. Bishop, C.M.: Neural networks for pattern recognition, vol. xvii, p. 482. Clarendon Press; Oxford University Press, Oxford, New York (1995)
53. Bradley, A.P.: The Use of the Area under the ROC Curve in the Evaluation of Machine Learning Algorithms. Pattern Recognition 30(6), 1145–1159 (1997)
54. Kuchar, J.K.: Methodology for Alerting-System Performance Evaluation. Journal of Guidance, Control, and Dynamics 19(2), 438–444 (1996)
55. Werbos, P.J.: The roots of backpropagation: from ordered derivatives to neural networks and political forecasting. In: Adaptive and learning systems for signal processing, communications, and control, vol. xii, p. 319. J. Wiley & Sons, New York (1994)
56. Rumelhart, D.E., McClelland, J.L., University of California San Diego. PDP Research Group: Parallel distributed processing: explorations in the microstructure of cognition. In: Computational models of cognition and perception, vol. 2, MIT Press, Cambridge, Mass (1986)
57. Picton, P.: Neural networks, vol. xii, p. 195. Palgrave, New York (2000)
58. Freeman, J.A., Skapura, D.M.: Neural networks: algorithms, applications, and programming techniques. In: Repr. with corrections. ed. Computation and neural systems series, vol. xiii, p. 401. Addison-Wesley, Reading, Mass (1992)
59. Hoffmann, N.: Simulating neural networks, p. 244. Vieweg, Wiesbaden (1994)
60. Wu, J.-K.: Neural networks and simulation methods. In: Electrical engineering and electronics, vol. 87, xiv, p. 431. M. Dekker, New York (1994)
61. Dash, M., Liu, H.: Feature Selection for Classification. Intelligent Data Analysis 1(3), 131–156 (1997)
62. Plutowski, M., Sakata, S., White, H.: Cross-validation estimates IMSE. In: Advances in Neural Information Processing System, vol. 6, pp. 391–398 (1994)
63. Ippolito, C.A., Pritchett, A.R.: Software architecture for a reconfigurable flight simulator. In: The AIAA Modeling and Simulation Technologies Conference, Denver, CO (2000)
64. Roberts, C.A., Dessouky, Y.M.: An Overview of Object-Oriented Simulation. Simulation 70(6), 359–368 (1998)
65. Andrews, J.W.: A Relative Motion Analysis of Horizontal Collision Avoidance, M.I.T. Lincoln Laboratory (1978)

# Chapter 10

# Group-Oriented Service Provisioning in Next-Generation Network

Vedran Podobnik[1], Vedran Galetic[2], Krunoslav Trzec[2], and Gordan Jezic[1]

[1] University of Zagreb, Faculty of Electrical Engineering and Computing, Croatia
[2] Ericsson Nikola Tesla, Croatia
`{vedran.podobnik,gordan.jezic}@fer.hr`,
`{vedran.galetic,krunoslav.trzec}@ericsson.com`

**Abstract.** The chapter deals with group-oriented service provisioning in next-generation network (NGN). It consists of three parts: the first bringing forth user profile creation and semantic comparison; the second explaining user profile clustering and semantic classification; and the third describing social network creation and analysis. The multi-agent system A-STORM (*Agent-based Service and Telecom Operations Management*) is presented and elaborated as part of the proof-of-concept prototype which demonstrates provisioning of group-oriented services within NGN. As a group-oriented service, the RESPIRIS (*Recommendation-based Superdistribution of Digital Goods within Implicit Social Networks*) service is implemented and provisioned by using prototype's agents. The proposed provisioning scenario is set forth, as well as provisioning process analysis presented.

**Keywords:** Intelligent Software Agents, Group-oriented Services, Semantic Clustering, Ontology-based User Profiles.

## 1 Introduction

The future of telecommunication industry is directed towards creating systems aware of user preferences, device capabilities, and communication context, and simultaneously enabling dynamic user group formation defined by similar characteristics (e.g., user preferences, user device and/or user context). Consequently, telecommunication operators (telcos) have recognized the importance of dynamic formation of user groups according to similar characteristics. The topic of this chapter is how using Semantic Web and software agent technologies can enable dynamic social networking in the environment of next-generation network (NGN).

The evolved NGN should aim at taking changing user demands into account and at creating spontaneous, adaptive services that can be delivered anytime, anywhere, to any device that its user prefers. Therefore, the realization of the full potential of convergence will make it necessary for operators to deploy dynamic, cooperative and business-aware consistent knowledge layer in the network architecture in order to enable ubiquitous personalized services. Providing such context-aware services

D. Srinivasan & L.C. Jain (Eds.): Innovations in MASs and Applications – 1, SCI 310, pp. 277–298.
springerlink.com        © Springer-Verlag Berlin Heidelberg 2010

transparently to the user is not only challenging from a network point of view, but also imposes rigorous requirements on the service provisioning.

Semantic Web technologies are rather novel but very amenable grounding for user clustering, while software agents have proven to be very suitable for user profile management and telecommunication processes enhancements. The idea of semantic reasoning has resulted in a number of languages. Among these are Resource Data Framework (RDF), RDF Schema (RDFS) and the Web Ontology Language (OWL). Information retrieval from RDF and OWL ontologies is performed by using various query languages. These languages are often loosely based on the Structured Query Language (SQL) syntax, but are performed on different data model; instead of relational database, the data being queried is represented as a graph consisting of subject-verb-object (SVO) triples. Such languages are RDF Data Query Language (RDQL) and Sesame RDF Query Language (SeRQL).

This chapter is organized as follows. In Section 2, we define group-oriented services. Section 3 describes NGN, and Section 4 presents how to create an ontology-based profile of telecommunication service user. Section 5 brings forth user profile clustering, and Section 6 elaborates semantic classification of user profiles. In Section 7 a multi-agent system enabling service provisioning in NGN is presented, as well as a proof-of-concept implementation done in Java. Section 7 proposes ideas for future research work and concludes the chapter.

## 2  Group-Oriented Services

Simultaneous development of mobile devices and telcos' infrastructure resulted in increasing complexity of mobile services [25]. The future of mobile communications is evolving from *linear services* (i.e., traditional services where the user cannot influence the predefined service provisioning procedure) towards new *non-linear services* (i.e., interactive services where the user participates in the service provisioning procedure, tailoring the service to his/her preferences, device and/or context) [4]. The non-linear services were available only for fixed network users until recently. In this chapter we study a special type of non-linear services: *group-oriented services* for *mobile users*:

- we define *group-oriented service* as a service in whose provisioning there cannot participate just one user, but a set of users with certain similarities (e.g., similar preferences, devices and/or context);
- we define *mobile users* as users possessing mobile devices (e.g., mobile phone or PDA).

The main idea behind the group-oriented services is to group mobile users into clusters taking into account users' interests, their mobile devices' characteristics and context in which they find themselves while requesting a service. To achieve this it is necessary to introduce a rather new approach in the service provisioning process: building *implicit social networks* of mobile users. Unlike explicit social networks (e.g., Facebook[1], MySpace[2] or LinkedIn[3]), implicit networks are built autonomously based

---

[1] http://www.facebook.com
[2] http://www.myspace.com
[3] http://www.linkedin.com

on similarities of user profiles, without the interference of users themselves, in order to provide useful information for telcos [25].

## 3  Next-Generation Network

Next-generation network (NGN) [3] [10] [26] [28] emerged as a fusion of enhanced IP-based network [1] and telecommunication network. The ITU-T Recommendation Y.2001 [19] provides a general definition of the NGN as a packet-based network able to provide telecommunication services and to make use of multiple-broadband QoS-enabled (*Quality of Service*) transport technologies and in which service-related functions are independent from underlying transport-related technologies. NGN enables unfettered access for users to networks and competing service providers and/or services of their choice. Additionally, the NGN concept introduces multiple convergences (see Fig. 1):

- *network convergence*, describing the integration of various wireline and wireless access technologies;
- *terminal convergence*, describing the introduction of adaptive services which can be delivered anytime, anywhere, and to any device the user prefers;
- *content convergence*, describing the ever-growing tendency of digitizing various forms of information, and;
- *business convergence*, describing the fusion of the ICT (*Information and Communication Technology*) and media industries.

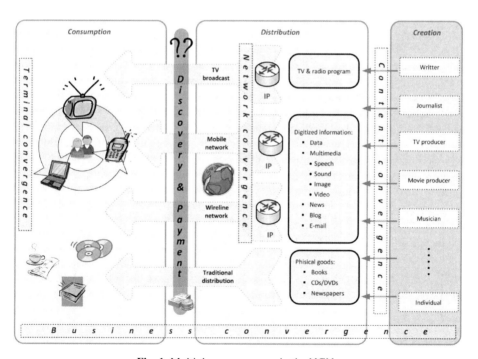

**Fig. 1.** Multiple convergences in the NGN

# 4  User Profiles

Telecom service provisioning paradigm is shifting from CRM (*Customer Relationship Management*) to CMR (*Customer-Managed Relationship*) [39]. The CMR is a relationship in which a company (i.e., telecom operator) uses a methodology, software, and perhaps Internet capabilities to encourage the customer (i.e., telecom service user) to control access to information and ordering. When a company applies the CMR paradigm its customers should be in control of the information pertaining to them such as their profile and transaction history. Moreover, a company should model its processes having in mind that customers' needs and feelings have the priority or are at least of equal weight to the company's needs and desires. Additionally, CMR paradigm allows customers to define how they communicate with the company, what services or products they will purchase, and how they will pay.

The above mentioned requisites brought up by the CRM paradigm considerably influence the design of company's ICT systems: traditional ICT solutions must be upgraded with mechanisms which enable advanced customer profile management. Namely, creation, storage and continuous update of customer profiles, as well as autonomous analysis of customer profiles and matchmaking of those profiles with service/product descriptions, should be enabled. When a company which introduces the CRM paradigm acts on the market as a telecom operator, its integrated ICT system must not support just traditional telecom operations but also enable advanced customer (i.e., telecom service user) profile management [7] [8] [9] (Fig. 2).

## 4.1  Related Work on User Profiling

The W3C is working on the CC/PP[4] (*Composite Capabilities/Preferences Profile*), an RDF-based specification which describes device capabilities and user preferences used to guide the adaptation of content presented to that device. It is structured to allow a client to describe its capabilities by reference to a standard profile, accessible to an origin server or other sender of resource data, and a smaller set of features that are an addition to or different than the standard profile. A set of CC/PP attribute names, permissible values and associated meanings constitute a CC/PP vocabulary.

OMA's (*Open Mobile Alliance*) UAProf[5] (*User Agent Profile*) specification, based on the CC/PP, is concerned with capturing classes of mobile device capabilities which include the hardware and software characteristics of the device. Such information is used for content formatting, but not for content selection purposes. The UAProf specification does not define the structure of user preferences in the profile.

---

[4] CC/PP specifications: *http://www.w3.org/Mobile/CCPP/*
[5] UAProf specification: *http://www.openmobilealliance.org/*

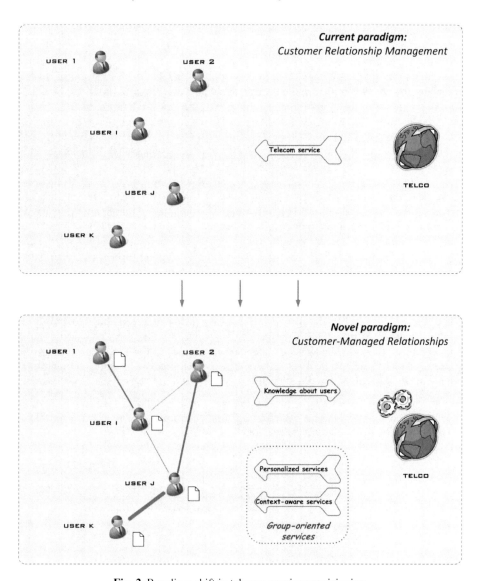

**Fig. 2.** Paradigm shift in telecom service provisioning

## 4.2 The Semantic Web

The Semantic Web [11] [37] [38] is a vision in which knowledge [2] is organized into conceptual spaces according to meaning, and keyword-based searches are replaced by semantic query answering [11]. Semantic Web languages, such as *Resource Data Framework*[6] *(RDF), RDF Schema*[7] *(RDFS)* and the *Web Ontology Language*[8] *(OWL),*

---

[6] RDF specifications: *http://www.w3.org/RDF/*
[7] RDFS specifications: *http://www.w3.org/TR/rdf-schema/*
[8] OWL specifications: *http://www.w3.org/TR/owl-features/*

can be used to maintain detailed user profiles. With the help of various query languages, based on *Structured Query Language (SQL)* syntax, it is possible to perform efficient semantic profile matchmaking once the profiles have been created according to a certain standard. Such matchmaking enables us to perform clustering according to true, semantic similarities, rather than keyword matchmaking. Semantic queries are the main means of information retrieval (IR) used in current research in this area. Inspiration for a query-based style of reasoning stems directly from the widespread propagation of RDBMS *(Relational Database Management Systems)*. Semantic query languages have a number of features in which they differ from SQL queries due to Semantic Web knowledge, which can be either *asserted* (explicitly stated) or *inferred* (implicit), being *network structured,* rather than *relational.* Also, the Semantic Web assumes an OWM *(Open World Model)* in which the failure to derive a fact does not imply the opposite [12], in contrast to *closed world* reasoning where all propositions whose verity cannot be ascertained are considered false [13].

In our implementation information is retrieved by means of RDQL *(RDF Data Query Language)* and SeRQL *(Sesame RDF Query Language)* queries. The Sesame [14] repository with OWL support [15] is utilized to store the required knowledge.

### 4.3 Proposed User Profiles

We extended the CC/PP and UAProf specifications and mapped them to the OWL ontology in order to create the telecommunication service user profile, as shown in List. 1. Opening rows (i.e., rows 1-10) and the closing row (i.e., row 43) must be defined according to XML and RDF specifications, while the remaining rows (i.e., rows 11-42) contain information about the user. We distinguish five different types of user information:

- *Rows 11-13* and the *row 42* – Every profile is an instance of a *UserDeviceProfile* (an ontology defining *UserDeviceProfile* is shown in Fig. 3. The profile is described with 20 attributes, as follows;

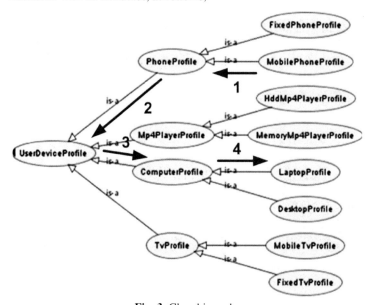

**Fig. 3.** Class hierarchy

- *Rows 14-21* – Six attributes defining *user preferences*;
- *Rows 22-28* – Five attributes defining the user device *hardware*;
- *Rows 29-33* – Three attributes defining the user device *software*;
- *Rows 34-41* – Six attributes defining the user *context*;

Attributes can be classified into one of the following types:

- *Interval*: An interval attribute is defined by a continuous linear scale divided into equal intervals (e.g., in List. 1 *hasAvailableMemory* could be qualified as an interval attribute with integer values);
- *Ordinal* (or *rank*): An ordinal attribute has multiple states that can be ordered in a meaningful sequence. The distance between two states increases when they are further apart in the sequence and the intervals between these consecutive states might be different (e.g., in List. 1 *hasPrefferedQoS* could be qualified as an ordinal attribute with values *Bronze, Silver,* and *Gold*);
- *Nominal* (or *categorical*): A nominal attribute takes on multiple states, but these states are not ordered in any way (e.g., in List. 1 *hasPrefferedLanguage* is a nominal attribute with values *English, Deutsch*, and *Hrvatski*).
- *Binary*: A binary attribute is a nominal attribute that has only two possible states (e.g., in List. 1 *hasPrefferedDeliveryType* can be *streaming* or *nonstreaming*).

```
1   <?xml version="1.0"?>
2   <rdf:RDF
3       xmlns="http://www.tel.fer.hr/astorm/InfoServiceV2.owl#"
4       xmlns:is="http://www.tel.fer.hr/astorm/InfoServiceV2.owl#"
5       xmlns:rdf="http://www.w3.org/1999/02/22-rdf-syntax-ns#"
6       xmlns:xsd="http://www.w3.org/2001/XMLSchema#"
7       xmlns:rdfs="http://www.w3.org/2000/01/rdf-schema#"
8       xmlns:owl="http://www.w3.org/2002/07/owl#"
9       xml:base="http://www.tel.fer.hr/astorm/User1.owl">
10
11  <!-- profile for User1's mobile phone Sony Ericsson K700 -->
12      <is:MobilePhoneProfile rdf:ID="SonyEricssonK700">
13
14  <!-- hardware -->
15      <is:hasAvailableMemory rdf:datatype="http://www.w3.org/2001/XMLSchema#int">18000</is:hasAvailableMemory>
16      <is:hasHorizontalScreenResolution rdf:datatype="http://www.w3.org/2001/XMLSchema#int">180</is:hasHorizontalScreenResolution>
17      <is:hasVerticalScreenResolution rdf:datatype="http://www.w3.org/2001/XMLSchema#int">230</is:hasVerticalScreenResolution>
18      <is:hasScreenBitsPerPixel rdf:datatype="http://www.w3.org/2001/XMLSchema#int">16</is:hasScreenBitsPerPixel>
19      <is:hasImei rdf:datatype="http://www.w3.org/2001/XMLSchema#string">35461002-303538-0-34</is:hasImei>
20
21  <!-- software -->
22      <is:hasOs rdf:resource="http://www.tel.fer.hr/astorm/InfoServiceV2.owl#BasicOs"/>
23      <is:hasBrowser rdf:resource="http://www.tel.fer.hr/astorm/InfoServiceV2.owl#SonyEricssonBrowser"/>
24      <is:hasJavaVersion rdf:datatype="http://www.w3.org/2001/XMLSchema#int">15</is:hasJavaVersion>
25
26  <!-- user preferences -->
27      <is:hasPreferredInformationType rdf:resource="http://www.tel.fer.hr/astorm/InfoServiceV2.owl#PlainText"/>
28      <is:hasPreferredInformationService rdf:resource="http://www.tel.fer.hr/astorm/InfoServiceV2.owl#CroatiaPoliticsInstance"/>
29      <is:hasPreferredLanguage rdf:resource="http://www.tel.fer.hr/astorm/InfoServiceV2.owl#English"/>
30      <is:hasPreferredGenre rdf:resource="http://www.tel.fer.hr/astorm/InfoServiceV2.owl#RockMusic"/>
31      <is:hasPreferredQoS rdf:resource="http://www.tel.fer.hr/astorm/InfoServiceV2.owl#Silver"/>
32      <is:hasPreferredDeliveryType rdf:resource="http://www.tel.fer.hr/astorm/InfoServiceV2.owl#NonStreaming"/>
33
34  <!-- context -->
35      <is:hasEnvironment rdf:resource="http://www.tel.fer.hr/astorm/InfoServiceV2.owl#InnerSpace"/>
36      <is:hasLocation rdf:resource="http://www.tel.fer.hr/astorm/InfoServiceV2.owl#Ina"/>
37      <is:hasCoordinatesX rdf:datatype="http://www.w3.org/2001/XMLSchema#float">50.21389</is:hasCoordinatesX>
38      <is:hasCoordinatesY rdf:datatype="http://www.w3.org/2001/XMLSchema#float">48.21389</is:hasCoordinatesY>
39      <is:atTime rdf:resource="http://www.tel.fer.hr/astorm/InfoServiceV2.owl#Night"/>
40      <is:hasSocialActivity rdf:resource="http://www.tel.fer.hr/astorm/InfoServiceV2.owl#WritingPresentation"/>
41
42      </is:MobilePhoneProfile>
43  </rdf:RDF>
```

**List. 1.** An example of a user profile

# 5   Semantic Clustering

Clustering is a process that results in partitioning a set of objects, which are described by a set of attributes, into clusters. Clustering algorithms rely on distance measures that are in relation to the similarity between objects that need to be grouped. Consequently, objects in resulting clusters are more similar to one another than to those in other clusters.

## 5.1 Common Clustering Methods

Common clustering methods include *partition-based* and *hierarchical* approach, and *Kohonen neural network*, each of which uses a specific distance measure.

The partition-based approach includes *k-means* and *partitioning around medoids* (PAM). The idea is to partition a set of objects into multiple non-overlapping clusters. A partition-based technique creates an initial partition depending on a specific number of clusters and then attempts to improve the partition iteratively by moving objects between or among clusters.

The hierarchical clustering approach starts by building a binary clustering hierarchy (i.e., a tree). Each leaf node represents an object to be clustered. Hierarchical clustering methods can be further classified into *agglomerative* or *divisive* clustering, depending on whether the clustering hierarchy is formed in a bottom-up or top-down fashion. The hierarchical agglomerative clustering (HAC) algorithm uses a bottom-up strategy, starting with as many clusters as there are objects. Based on an inter-cluster similarity measure of choice, the two most similar clusters then merge to form a new cluster. This process is continued until either a hierarchy emerges, where a single cluster remains at the top of the hierarchy containing all the target objects, or the process reaches a specified termination condition (e.g., inter-cluster similarity less than a specified threshold). In contrast, the hierarchical divisive clustering algorithm employs a top-down strategy, starting with all the objects in a single cluster. The cluster is then divided into its two most distinct clusters. The division process is repeated until either each object is part of a distinct cluster or the process reaches a specified condition (e.g., intra-cluster similarity of each cluster greater than a specified threshold).

A Kohonen neural network[9] is also known as a self-organizing map. The network is an unsupervised two-layer neural network. Each input node corresponds to a coordinate axis in the input attribute vector space. Each output node corresponds to a node in a two-dimensional grid. The network is fully connected. During the training phase, all objects to be clustered are fed into the network repeatedly in order to adjust the connection weights in such a way that the distribution of the output nodes represents that of the input objects. The input vector space distribution serves as the criterion. Since we used the partition-based approach, the following sections provide a more detailed preview of these algorithms.

### 5.1.1 Partitioning around Medoids
The PAM [16] algorithm partitions an initial set of $n$ objects into $k$ clusters by first finding a representative object for each cluster. A representative, that is often called a *medoid*, is the most centrally located object in a cluster. Once $k$ medoids have been selected, each nonselected object is classified into the cluster of the medoid closest to the object, based on the distance. The algorithm then repeatedly tries improving the partition by substituting a non-selected object for medoid whenever such substitutions add to the clustering quality.

---

[9] Kohonen neural network: http://mlab.taik.fi/~timo/som/thesis-som.html

### 5.1.2  K-Means Algorithm

K-means algorithm starts by initializing (randomly or by some heuristic) the set of $k$ means (i.e., *controids*). Afterwards, it firstly constructs an initial partition by associating each of $n$ objects with the closest of $k$ centroids: as a result $k$ clusters are generated. Secondly, the k-means algorithm recalculates centroids of every cluster based on objects currently associated with that particular cluster: new centroid is calculated as the mean of all objects in the cluster. The algorithm repeats alternate application of these two steps until convergence, which is obtained when the points no longer switch clusters, centroids are no longer changed or a certain criteria is reached. Our implementation performs centroid recalculation for a given number of iterations so the optimal clustering cannot always be reached. In our case study we repeat the procedure in ten attempts, so we introduce a reliability measure. After all attempts have been performed reliability is determined for each obtained partition by dividing the number of attempts resulting in that particular partition with the total number of attempts. Throughout the testing, reliability for this algorithm has been mostly above 80%.

### 5.2  K-Means Constructive Clustering Based on Semantic Matchmaking

In our case, we compare user profiles and partition users into groups. Such partitioning enables the telecommunication operator to enhance the provisioning of group-oriented services. In our implementation we do not use standard distance measures to compare profiles, but rather take advantage of a novel approach based on semantic matchmaking [5] [6]. Table 1 shows the comparison of two user profiles. Each attribute in the profile is asserted individually, while the final result is the arithmetic mean of individual attribute scores. The semantic matchmaking procedure is carried out as follows:

- *Position within the class hierarchy*: Each profile is an instance of a certain class from the ontology. Fig. 3 shows how class hierarchy position is transformed into a real number that represents the similarity between two classes, or objects. Greater distance between two classes implies less similarity between classes' instances: we can see that the *MobilePhoneProfile* and the *LaptopProfile* classes are four "steps" away from each other in the hierarchy. The similarity score is calculated by division of 1 by the number steps (in this case 4, so the similarity score equals 0,25);
- *Common attribute types*: When comparing binary and nominal attributes the result is either *0* (if the values are not equal), or *1* (if the values are identical). When comparing ordinal attributes the result is a number between *0* and *1*, depending on the rank of each value. A comparison result of two attributes is a ratio between the smaller and the greater attribute between the two: e.g., when comparing *Silver* and *Gold* levels of *QoS* the similarity score is 0,5;
- *Attributes with object values*: Some attributes' values contain references to other class instances. They can also be compared using the already mentioned approach using the class hierarchy position.

**Table 1.** Profile comparison results

| Attribute | Type | Value (profile A) | Value (profile B) | Score |
|---|---|---|---|---|
| ID | abstract | Mobile 1 | Laptop 1 | none |
| Class | class | MobilePhoneProfile | LaptopProfile | 0,250 |
| **User preferences** | | | | |
| InformationType | instance | PlainText | Avi | 0,250 |
| InformationService | instance | CroatiaPoliticsInstance | MoviesInstance | 0,142 |
| Language | instance | English | Hrvatski | 0,500 |
| Genre | instance | RockMusic | ThrillerMovie | 0,250 |
| QoS | instance | Silver | Gold | 0,500 |
| DeliveryType | instance | NonStreaming | Streaming | 0,500 |
| **Hardware** | | | | |
| AvailableMemory | integer | 18000 | 1000000 | 0,018 |
| HorizontalResolution | integer | 180 | 1600 | 0,113 |
| VerticalResolution | integer | 230 | 1050 | 0,219 |
| BitsPerPixel | integer | 16 | 32 | 0,500 |
| **Software** | | | | |
| Os | instance | BasicOs | WindowsVista | 0,500 |
| Browser | instance | SonyEricssonBrowser | MozillaFirefox | 0,500 |
| JavaVersion | integer | 15 | 16 | 0,94 |
| **Context** | | | | |
| Environment | instance | InnerSpace | InnerSpace | 1,000 |
| Location | instance | Ina | TrgBanaJelacica | 0,250 |
| CoordinatesX | float | 50,21459 | 50,21779 | 0,990 |
| CoordinatesY | float | 48,21344 | 48,74144 | 0,990 |
| Time | instance | Night | Night | 1,000 |
| SocialActivity | instance | WritingPresentation | CoffeDrinking | 0,250 |
| | | | *Profile similarity* | *0,483* |

For clustering, we use *k-means* algorithm extended with the *constructive clustering analysis* method that determines the quality of the partition [16]. The main task of *constructive clustering* analysis method is finding the optimal number of clusters for a given set of objects. This method uses the *silhouette measure* to determine the optimal number of clusters. To perform *constructive clustering* it is necessary to use one of the non-constructive analysis algorithms (e.g., *k-means*) to perform clustering for a given number of clusters. The *silhouette measure* is calculated for each object in the given set taking into account the resulting partition. It is calculated as follows:

$$s(i) = \begin{cases} 1 - \frac{a(i)}{b(i)} & if\ a(i) < b(i) \\ 0 & if\ a(i) = b(i),\ or\ the\ object\ is\ the\ only\ object\ in\ the\ cluster\ \mathcal{A} \\ \frac{a(i)}{b(i)} - 1 & if\ b(i) < a(i) \end{cases} \quad (1)$$

where $s(i)$ is the *silhouette measure* of the object $o_i$, cluster $\mathcal{A}$ is the cluster containing the object $o_i$, $a(i)$ is the average distance (the calculation of distance between two objects (i.e., user profiles) is defined in Table 1) of object $o_i$ to all other objects in cluster $\mathcal{A}$, and $b(i)$ is the average distance of the object $o_i$ to all other objects in cluster $\mathcal{B}$, which is the *neighbour* of object $o_i$. The *neighbour* is the cluster closest to the observed object, different from cluster $\mathcal{A}$. Usually it is the second best choice for $o_i$. The next step is calculating the *average silhouette width* for the object set, calculated as an average value of *silhouette measures* of all objects for the given partition. When we obtain the *average silhouette width* for each number of clusters $k$ between 2 and $n-1$ (where $n$ denotes the total number of objects to be clustered) the highest one is called the *silhouette coefficient*, and the $k$ for which the *silhouette coefficient* is achieved is the optimal number of clusters. List. 2 presents the calculation of *silhouette coefficient* when clustering is done upon 20 user profiles. Firstly, *k-means* algorithm is used for clustering user profiles into 2 to 19 clusters (iterations $k = 2$ to $k = 19$, respectively) and then the *average silhouette width* for every iteration is calculated. One notices that the highest *average silhouette width* is achieved when $k = 3$, so the optimal number of clusters is 3.

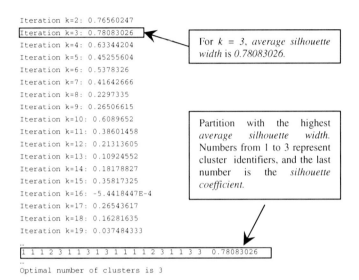

```
Iteration k=2:  0.76560247
Iteration k=3:  0.78083026
Iteration k=4:  0.63344204
Iteration k=5:  0.45255604
Iteration k=6:  0.5378326
Iteration k=7:  0.41642666
Iteration k=8:  0.2297335
Iteration k=9:  0.26506615
Iteration k=10: 0.6089652
Iteration k=11: 0.38601458
Iteration k=12: 0.21313605
Iteration k=13: 0.10924552
Iteration k=14: 0.18178827
Iteration k=15: 0.35817325
Iteration k=16: -5.4418447E-4
Iteration k=17: 0.26543617
Iteration k=18: 0.16281635
Iteration k=19: 0.037484333
...
1 1 1 2 3 1 1 3 1 3 1 1 1 1 2 3 1 1 3 3   0.78083026
...
Optimal number of clusters is 3
```

For $k = 3$, *average silhouette width* is *0.78083026.*

*Partition with the highest average silhouette width.* Numbers from 1 to 3 represent cluster identifiers, and the last number is the *silhouette coefficient.*

**List. 2.** An example of k-means constructive clustering based on semantic matchmaking

## 6 Semantic Classification

A test[10] shows that semantic clustering algorithm is of $O(n^2)$ time complexity; in order to semantically cluster 5 profiles it takes about 2,5 s, for 10 profiles 10 s, whereas for 20 profiles the time required for semantic clustering reaches up to 40 s (Fig. 4). The number of a telco's service users can reach up to several millions. After filtering them according to various factors (e.g., location), there can still be hundreds or even thousands of users' profiles to be clustered, which would require time of the order of magnitude up to $10^7$ s, which is equivalent to several months' period. This example clearly shows that performing semantic clustering of all profiles whenever a new one is added would by all means be highly irrational. In order to improve the system's performances and cost-effectiveness, a machine learning-based classification method is used.

Each learning method can be categorized as either a *lazy method* or an *eager method*. While the main task of eager methods is to define the general, explicit description of a target function valid for the whole feature space, lazy methods

---

[10] The computer with following characteristics was used for the test: processor *Intel Core 2 Duo @ 1.73 GHz*, memory *2 GB DDR2*.

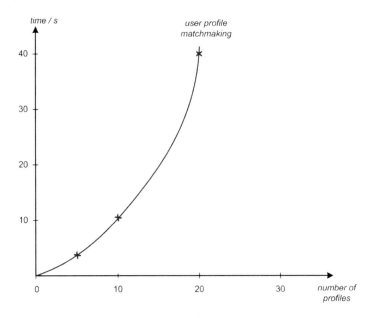

**Fig. 4.** Duration of the semantic clustering procedure depending on the number of profiles

operate in the opposite way – the classification assessment is deferred until a new instance is given that need be classified [40]. Lazy methods are also called *instance-based learning methods* as they construct hypotheses directly from the training instances themselves [41]. Therefore, for every new instance a new approximation of the target function is assessed. This approximation is performed only locally, i.e., within a small range around the new instance. Even though lazy methods are rather cost-ineffective as each new instance necessitates processing in order to generate the target function approximation, such approximation-based approach is highly beneficial for rather complex target functions. One example of lazy instance-based methods is *k-Nearest Neighbour* (*k-NN*) method.

### 6.1 *k-NN* Method Basics

*k-NN* is one of the most thoroughly analyzed algorithms in machine learning, due in part to its age and in part to its simplicity [40]. It is a classification algorithm grounded on the preposition that the similarity of instances' target function values corresponds to their relative nearness. This algorithm assumes all instances correspond to vectors in the n-dimensional Euclidean space, where their attributes are represented by their coordinates. Even though a target function can theoretically also take continuous values, in this chapter we will concentrate only on those taking discrete values. If $V$ is the set of all values present within the observed instance's $k$ nearest neighbours, then the new instance $x_q$ is assigned the classification value calculated by the following formula:

$$\hat{f}(x_q) \leftarrow \text{argmax}_{v \in V} \sum_{i=1}^{k} \delta(v, f(x_i)) \tag{2}$$

where $\delta$ is a function called Kronecker delta and is defined as:

$$\delta(a, b) = \begin{cases} 1, a = b \\ 0, a \neq b \end{cases} . \tag{3}$$

The following example depicts the essence of $k$-$NN$ method. Fig. 5 shows a new instance $x_q$ in a two-dimensional space where there are other already categorized instances. The instance $x_q$ should also be categorized as either "+" or "-", which are in this case the only possible values of the target function. The new instance will be assigned the value held by the majority of $k$ instances that are nearest to it. The relative nearness is assessed by calculating the standard Euclidean distance. Generally, Euclidean distance between two instances $x_i$ and $x_j$ is defined as follows:

$$d(x_i, x_j) \overset{\text{def}}{=} \sqrt{\sum_{r=1}^{n}[a_r(x_i) - a_r(x_j)]^2} \tag{4}$$

where $a_r(x)$ denotes the value of the $r$th attribute of instance $x$. In the observed example $n=2$, as instances are represented as vectors in a two-dimensional space.

Furthermore, special attention should be paid to the fact that value assignment depends greatly on the predefined $k$. In particular, if $k=1$, then $x_q$ will take the same value as its closes neighbour, namely, "+". However, for $k=5$, $x_q$ will be assigned the classification value of its five closest neighbours, which is in this case "-", as shown in Fig. 5. Therefore, defining the appropriate $k$ can be crucial and of rather delicate nature.

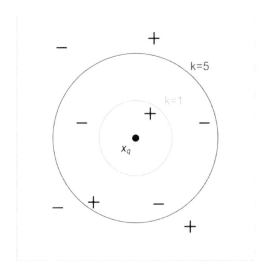

**Fig. 5.** The $k$-$NN$ classification illustration

Moreover, it is logical to suppose that nearer instances influence the classification assessment more than the further ones. In that sense, the expression (Formula 1) can be modified in the following way:

$$\hat{f}(x_q) \leftarrow \text{argmax}_{v \in V} \sum_{i=1}^{k} w_i \delta(v, f(x_i)) \tag{5}$$

where $w_i$ is the weight inversely proportional to the distance between instances $x_q$ and $x_i$, i.e.,

$$w_i = \frac{1}{d(x_q, x_i)^2} \tag{6}$$

## 6.2  Case Study: Classification of User Profiles by Using *k-NN* Method

The benefits of using *k-NN* in the proposed manner will be demonstrated through the following case study. It consists of two parts:

1.  *k-NN* verification  – comparing the outcome of clustering the existing *n* profiles and classifying the *n+1*[st] with the outcome of clustering of all *n+1* profiles;
2.  processing time assessment with regard to different *k*.

Each of the *n* profiles, as described in Section 4, comprises of *l* attributes. Their values determine where the instances, represented as vectors, will be placed in the Euclidean space. We devised and tested four different approaches of classifying the new profile. These approaches can be categorized according to whether *k-NN* is performed in *l*- or 2-dimensional space.

a)  Running *k-NN* in *l*-dimensional space – instances need not undergo the multidimensional scaling method. There are two possible means of categorizing the new instance:
    1.  Bypassing the construction of similarity matrix and simply running *k-NN* regarding all *n+1* *l*-dimensional vectors;
    2.  Constructing the similarity matrix and running *k-NN* regarding only *k* profiles that are most similar to the new one;
b)  Running *k-NN* in 2-dimensional space – in order to have the profiles represented as 2-dimensional vectors at disposition, performing multidimensional scaling procedure is inevitable. The two corresponding ways of categorization are as follows:
    1.  Letting all *n+1* profiles undergo multidimensional scaling and then running *k-NN*;
    2.  Having constructed the similarity matrix, performing multidimensional scaling over *k* profiles that are most similar to the observed one.

Time required for classification of a new profile among 19 already existing ones depends on predefined *k*, but by no means exceeds 100 ms. Moreover, methods (a.1), (a.2) and (b.2) are proven much more time-efficient than method (b.1). Namely, the classification using methods (a.1), (a.2) and (b.2) requires about 16 ms for $k$ being less than 10, whereas method (b.1) requires little less than 100 ms for the same $k$.

It is a reasonable objection that during the classification of a number of new examples, due to $k$-$NN$'s heuristic nature, clusters may not anymore comprise instances that are the most similar to each other, as they ought to. Therefore, semantic clustering can be performed every time when a predefined number of new profiles has been reached.

# 7  Agent-Based Telecommunication Service Provisioning

The Agent-based Service and Telecom Operations Management (A-STORM) multi-agent system is part of the proof-of-concept prototype that deals with agent-based service provisioning in the NGN. This section briefly addresses software agents as a technology used to implement the prototype and the NGN as its environment. The description of the prototype architecture follows and is accompanied by a case study that illustrates a group-oriented telecommunication service.

## 7.1  Agents

The dynamic and distributed nature of services/resources in the NGN requires telecom stakeholders not only to respond to requests, but also to intelligently anticipate and adapt to their environment. Software agents are the computing paradigm, which is very convenient for the creation of programs able to conform to the requirements set forth. A software agent is a program which autonomously acts on behalf of its principal, while carrying out complex information and communication tasks that have been delegated to it. From the owner's point of view, agents improve his/her efficiency by reducing the time required to execute personal and/or business tasks. A system composed of several software agents, collectively capable of reaching goals that are difficult to achieve by an individual agent or a monolithic system, is called a multi-agent system (MAS) [29].

An agent [17] [31] [32] [33] [34] must possess some intelligence grounded on its *knowledge base*, *reasoning mechanisms* and *learning capabilities*. The intelligence of an agent is a prerequisite for all its other characteristics. Depending on the assignment of a particular agent, there are differences in types of information contained in its knowledge base. However, generally this information can be divided into two parts – the *owner's profile* and the agent's *knowledge about its environment*. It is very important to notice that the agent's knowledge base does not contain static information. Adversely, the agent continuously updates its owner's profile according to its owner's latest needs. This allows the agent to efficiently represent its owner, thus realizing the *calm technology* concept. Calm technology is that which serves us, but does not demand our focus or attention [35]. Furthermore, the agent also updates knowledge regarding its environment with the latest events from its ambience and the current state of observed parameters intrinsic to its surroundings, thus realizing *context-awareness*. Context-awareness describes the ability of an agent to provide results that depend on changing context information [36]. In our model, we differentiate *situation context* (e.g., user location and environment temperature) from *capability context* (e.g., features of a user's terminal). An agent executes tasks *autonomously* without any interventions from its owner, making it an invisible servant. An agent must be *reactive*, so it can properly and timely respond to impacts from its environment. An agent not only reacts to excitations from its environment, but also takes initiatives coherent to its tasks. A well-defined objective is an inevitable prerequisite for *proactivity*. An efficient software agent collaborates with other agents from its surroundings: i.e. it is *cooperative*. If an agent is capable of migrating between heterogeneous network nodes, it is classified as a *mobile* software agent [30]. An agent has a lifetime throughout which the persistency of its identity and its states should be retained. Thus, it is characterized by *temporal continuity*.

## 7.2  Multi-agent System for Service Provisioning in NGN

The A-STORM multi-agent system was implemented using the Java Agent DEvelopment (JADE[11]) framework. JADE is a software framework used for developing agent-based applications in compliance with the Foundation for Intelligent Physical Agents (FIPA[12]) specifications.

Today we are witnessing the fusion of the Internet and mobile networks into a single, but extremely prominent and globally ubiquitous converged network [18] [19]. One of the fundamental principals in the converged network is the separation of services from transport [20]. This separation represents the horizontal relationship aspect of the network where the transport stratum and the service stratum can be distinguished. The transport stratum encompasses the processes that enable three types of connectivity, namely, user-to-user, user-to-service platform and service platform-to-service platform connectivity. On the other hand, the service stratum comprises processes that enable (advanced) telecom service provisioning assuming that the earlier stated types of connectivity already exists. From the vertical relationship aspect each stratum can have more layers where a data (or user) plane, a control plane and a management plane in each layer can be distinguished.

In our prototype intelligent agents are introduced into the control plane and the management plane of the service stratum. Those agents are in charge of gathering and processing context information that is required for different service provisioning tasks. In particular, the prototype has been developed in order to explore possibilities for implementation of ontology-based user profiling/clustering, context-aware service personalization and rule-based software deployment in the converged network. Fig. 6 shows the architecture of A-STORM multi-agent system used to implement the proof-of-concept prototype.

Business-driven provisioning agents (i.e., the Business Manager Agent, the Provisioning Manager Agent) perform provisioning operations according to business strategies defined in a rule-based system. Service deployment agents (i.e., the Deployment Coordinator Agent, Remote Maintenance Shell agents) provide end-to-end solutions for efficient service deployment by enabling software deployment and maintenance at remote systems. Context management agents (i.e., the Charging Manager Agent, Group Manager Agent, Session Manager Agent, Preference Manager

---

[11] http://jade.tilab.com
[12] http://www.fipa.org

Agent) gather context information from network nodes and devices (e.g., trigger events in SIP/PSTN call model, balance status, device location) and enable user personalization through the execution of context-dependent personal rules. A more detailed description of these agents and their functionalities can be found in [21] [22].

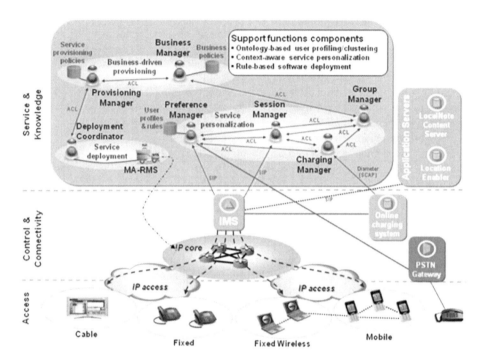

**Fig. 6.** A-STORM proof-of-concept prototype architecture

### 7.3 Proof-of-Concept Group-Oriented Service: Recommendation-Based Superdistribution of Digital Goods within Implicit Social Networks (RESPIRIS)

Superdistribution [42] [43], the concept the RESPIRIS (*Recommendation-based Superdistribution of Digital Goods within Implicit Social Networks*) service is based on, is the combined distribution and market scheme for digital goods involving buyers (i.e., end-users) in the distribution process in such a way that they redistribute a good to other legitimate buyers. The principle of superdistribution is tightly connected with the viral marketing phenomenon [23] [44]. Superdistribution can be explicit (end-users exchange digital goods) or implicit (end-users exchange recommendations for digital goods, while digital goods are distributed by businesses that act as providers (e.g., telecom operator)). Industry is continuously comprehensively experimenting with the concept of superdistribution. Existing solutions have even been cast in form of a standard for the mobile domain by the Open Mobile Alliance (OMA). Lots of telecom operators provision services based on that standard, while mobile device manufacturers produce devices that support it. However, all of the existing solutions represent explicit superdistribution scheme. This explicitness can be identified from two perspectives: not only is exchange of digital goods done directly between end-users (implying rather complex scenarios as Digital Rights Management (DRM) issues must be considered), but also the superdistribution groups are defined explicitly by providers or end-users. The innovativeness of the RESPIRIS service lies within the fact that it represents the implicit superdistribution scheme. This implicitness is twofold. Firstly, superdistribution groups are generated dynamically and implicitly by telecoms based on mobile user profiles. Namely, each subscriber of the proposed service is represented by a corresponding profile wherein his/her preferences are described (e.g., what sort of digital good he/she is interested in, of which genre, etc.). User profiles are used for semantic clustering of users into superdistribution groups. The groups users are allocated into are based on users' preferences' similarities and built autonomously, hence the term *implicit social network*. Secondly, end-users exchange recommendations for digital goods, while digital goods are distributed by telecom operators. Therefore, such solution is much simpler from the perspective of DRM and does not demand that user devices have hardware/software support for OMA superdistribution standards.

The demonstration of the RESPIRIS scenario, situated into A-STORM multi-agent system, follows. The participating A-STORM agents are described in the previous section, while the presented sequence of actions is accompanied by a sequence diagram in Fig. 7. If a particular user, e.g., $user_i$, is interested in the RESPIRIS service then his/her Preference Manager Agent $PrefMA_i$ sends a registration request to the Provisioning Manager Agent (ProvMA) (1). (Registration cancellation is also performed in an analogous manner.) The ProvMA provides the Group Manager Agent (GMA) with a list of $m$ user profiles whose users subscribed to the RESPIRIS service (2). The GMA allocates these profiles into $k$ groups according to users' preferences. This is achieved by performing semantic clustering algorithms described in Section 5.2. The GMA provides the ProvMA with a list of assembled groups and associated profiles (3). In a particular group $G_x$ consisting of $n$ ($n < m$) profiles, the ProvMA informs each of the $n$ PrefMAs of all other PrefMAs within that particular group (4). If the $user_i$ wants to purchase some content, the corresponding $PrefMA_i$ first needs to ascertain that all requirements imposed by the particular content provisioning service are met. Therefore, it consults the ProvMA (5a), which performs matching of the user profile with the service profile [9] and determines whether additional software is necessary for the chosen content to render properly on the user's device. If so, the ProvMA sends the list of required software to the Deployment Coordinator Agent (DCA) (5b), which forwards it, along with deployment procedure specification, to the Remote Maintenance Shell (RMS) agent [24] (5c). The RMS agent deploys and installs the software to the user's mobile device. The $PrefMA_i$ can then download the content so it sends a message to the Business Manager Agent (BMA) requesting to purchase the content and the rights to resell it to $x$ ($x<n$) other users whose profiles belong to its group (6). The BMA sends the requested content to the $PrefMA_i$ and the rights to resell it (7). Being authorised to resell the content, the $PrefMA_i$ has the liberty to sporadically distribute recommendations to any subset of PrefMAs belonging to its group, e.g. $PrefMA_s..PrefMA_t$ (8). Should any of PrefMAs that have received the

recommendation, say, the PrefMA$_j$, decide to purchase the same content, it informs the BMA about the purchase, along with information on who it received the recommendation from (9).

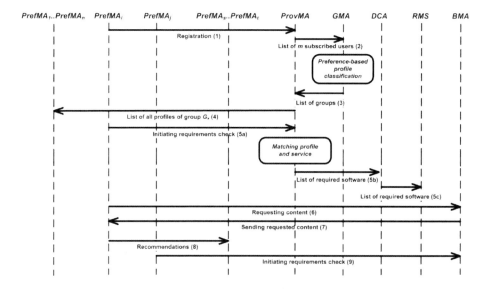

**Fig. 7.** The RESPIRIS scenario

This approach in digital goods distribution ensures that all involved parties benefit from successful transactions between end-users. Namely, let alone the fact that providing digital goods does not imply almost any production expenses, the service provider need not perform resource-consuming tasks such as advertisement. The advertisement becomes redundant as users occasionally receive recommendations. End-users benefit from this solution as it facilitates choosing from a large number of generally available digital goods, as they are provided with the ability to concentrate on a smaller number of recommended goods. What makes these goods potentially interesting is the fact that recommendations are sent by users from the group assembled according to the similarity in preferences of users whose profiles it encompasses. If a user opts for purchasing a recommended good, not only does the content provider gain income, but the reseller also profits, in accordance with its contract with the provider. In fact, the reseller gains credit for every successful transaction it induces, manifesting by means of reducing the cost of a future purchase. For instance, the user A purchases some digital content from the content provider for a price of $X$ money units and recommends it to the user B. The user B is offered to buy the recommended content at a lower price (e.g., $0.9 \times X$ money units). If he/she decides to do so, the content provider also credits the user A by reducing the price of his/her next purchase (e.g., for, $0.2 \times X$ money units).

# 8   Conclusion and Future Work

In this chapter, a possible solution for group-oriented service provisioning in Next-Generation Network (NGN) is proposed. A group-oriented service is defined as a telecommunication service that is provisioned to a group of users with certain similarities (e.g., similar preferences, devices and/or context). The main idea behind group-oriented services is to group users into clusters taking into account users' interests, characteristics of their devices and the context in which they find themselves while requesting a service. User profile creation and semantic comparison are described as well as user profile clustering and semantic classification. Common clustering methods and various algorithms are presented and compared. Classification of user profiles by using *k-NN* method is taken into consideration. As a proof-of-concept, the multi-agent system A-STORM (*Agent-based Service and Telecom Operations Management*) enabling service provisioning in NGN is presented and elaborated. Namely, an example of group-oriented services RESPIRIS (*Recommendation-based Superdistribution of Digital Goods within Implicit Social Networks*), is implemented and a provisioning scenario is set forth.

Future work will include an integrated risk and performance analysis from the telco viewpoint addressing typical superdistribution schemes (i.e., business strategies). In particular, stability robustness of the business strategies will be considered using the evolutionary game-theoretic analysis of the superdistribution process. Consequently, our future research will be aimed at designing agents' behaviors that enable evolutionary stable viral marketing schemes in order to provide sufficient incentives and motivation for the telecom providers at large to apply the proposed multi-agent system.

**Acknowledgements.** This work was carried out within research projects 036-0362027-1639 "Content Delivery and Mobility of Users and Services in New Generation Networks", supported by the Ministry of Science, Education and Sports of the Republic of Croatia, and "Agent-based Service & Telecom Operations Management", supported by Ericsson Nikola Tesla, Croatia.

# References

1. Podobnik, V., Lovrek, I.: Multi-Agent System for Automation of B2C Processes in the Future Internet. In: 27th IEEE Conference on Computer Communications (INFOCOM) Workshops, pp. 1–4. IEEE Press, Phoenix (2008)
2. Ackoff, R.L.: From Data to Wisdom. Journal of Applied System Analysis 16, 3–9 (1989)
3. Knightson, K., Morita, N., Towle, T.: NGN architecture: generic principles, functional architecture, and implementation. IEEE Communications Magazine 43(10), 49–56 (2006)
4. Feijoo, C., et al.: The emergence of IP interactive multimedia services and the evolution of the traditional audiovisual public service regulatory approach. Telematics and Informatics 24(4), 272–284 (2007)
5. Grimnes, G.A., Edwards, P., Preece, A.: Instance based clustering of semantic web resources. In: Bechhofer, S., Hauswirth, M., Hoffmann, J., Koubarakis, M. (eds.) ESWC 2008. LNCS, vol. 5021, pp. 303–317. Springer, Heidelberg (2008)

6. Dietze, S., Gugliotta, A., Domingue, J.: Conceptual Situation Spaces for Semantic Situation-Driven Processes. In: Bechhofer, S., Hauswirth, M., Hoffmann, J., Koubarakis, M. (eds.) ESWC 2008. LNCS, vol. 5021, pp. 599–613. Springer, Heidelberg (2008)
7. Bonnin, J., Lassoued, I., Hamouda, Z.B.: Automatic multi-interface management through profile handling. Mobile Networks and Applications 14(1), 4–17 (2009)
8. Panayiotou, C., Samaras, G.: mPERSONA: personalized portals for the wireless user: An agent approach. Mobile Networks and Applications 9(6), 663–677 (2004)
9. Frkovic, F., Podobnik, V., Trzec, K., Jezic, G.: Agent-Based User Personalization Using Context-Aware Semantic Reasoning. In: Lovrek, I., Howlett, R.J., Jain, L.C. (eds.) KES 2008, Part I. LNCS (LNAI), vol. 5177, pp. 166–173. Springer, Heidelberg (2008)
10. Podobnik, V., Petric, A., Trzec, K., Jezic, G.: Software Agents in New Generation Networks: Towards the Automation of Telecom Processes. In: Jain, L.C., Nguyen, N.T. (eds.) Knowledge Processing and Decision Making in Agent-Based Systems, pp. 71–99. Springer, Heidelberg (2009)
11. Antoniou, G., van Harmelen, F.: Semantic Web Primer. MIT Press, Cambridge (2004)
12. Walton, C.: Agency and the Semantic Web. Oxford University Press, New York (2007)
13. Grimm, S., Motik, B.: Closed World Reasoning in the Semantic Web through Epistemic Operators. In: 1st OWL: Experiences and Directions Workshop (OWLED), Galway (2005)
14. Broekstra, J., Kampman, A., van Harmelen, F.: Sesame: A Generic Architecture for Storing and Querying RDF and RDF Schema. In: Horrocks, I., Hendler, J. (eds.) ISWC 2002. LNCS, vol. 2342, pp. 54–68. Springer, Heidelberg (2002)
15. Kiryakov, A., Ognyanov, D., Manov, D.: OWLIM – A Pragmatic Semantic Repository for OWL. In: Dean, M., Guo, Y., Jun, W., Kaschek, R., Krishnaswamy, S., Pan, Z., Sheng, Q.Z. (eds.) WISE 2005 Workshops. LNCS, vol. 3807, pp. 182–192. Springer, Heidelberg (2005)
16. Wei, C.P., Hu, P.J.-H., Kung, L., Tan, J.: E-Health Intelligence: A Multiple-Level Approach for E-Health Data Mining. In: Tan, J. (ed.) E-Health Care Information Systems: An Introduction for Students and Professionals, pp. 330–351. Wiley, San Francisco (2005)
17. Nwana, H.S.: Software Agents: An Overview. Knowledge and Engineering Review 11(3), 205–244 (1996)
18. Berners-Lee, T., Fischetti, M.: Weaving the Web. Harper San Francisco, New York (1999)
19. ITU-T Recommendation Y.2001: General Overview of NGN (2004)
20. ITU-T Recommendation Y.2011: General principles and general reference model for Next Generation Networks (2004)
21. Petric, A., et al.: An Agent Based System for Business-driven Service Provisioning. In: AAAI 2007 Workshop on Configuration, pp. 25–30. AAAI Press, Vancouver (2007)
22. Petric, A., et al.: Agent-Based Support for Context-Aware Provisioning of IMS-Enabled Ubiquitous Services. In: Kowalczyk, R., Vo, Q.B., Maamar, Z., Huhns, M. (eds.) SOCASE 2009. LNCS, vol. 5907, pp. 71–82. Springer, Heidelberg (2009)
23. Cattelan, R.G., He, S., Kirovski, D.: Prototyping a novel platform for free-trade of digital content. In: 12th Brazilian Symposium on Multimedia and the Web (WebMedia), pp. 79–88. ACM, Natal (2006)
24. Kusek, M., et al.: Mobile agent based software operation and maintenance. In: 7th International Conference on Telecommunications (ConTEL), pp. 601–608. IEEE, Zagreb (2003)
25. Basuga, M., et al.: The MAgNet: Agent-based Middleware Enabling Social Networking for Mobile Users. In: 10th International Conference on Telecommunications (ConTEL), pp. 89–96. IEEE, Zagreb (2009)

26. Vrdoljak, L., Bojic, I., Podobnik, V., Kusek, M.: The AMiGO-Mob: Agent-based Middleware for Group-oriented Mobile Service Provisioning. In: 10th International Conference on Telecommunications (ConTEL), pp. 97–104. IEEE, Zagreb (2009)
27. Skorin-Kapov, L., Mosmondor, M., Dobrijevic, O., Matijasevic, M.: Application-Level QoS Negotiation and Signaling for Advanced Multimedia Services in IMS. IEEE Communications 45(7), 108–117 (2007)
28. Podobnik, V., Trzec, K., Jezic, G.: Context-Aware Service Provisioning in Next-Generation Networks: An Agent Approach. International Journal of Information Technology and Web Engineering 2(4), 41–62 (2007)
29. Jennings, N., Sycara, K., Wooldridge, M.: A Roadmap of Agent Research and Development. Journal of Autonomous Agents and Multi-Agent Systems 1(1), 7–36 (1998)
30. Bradshaw, J.M.: Software Agents. MIT Press, Cambridge (1997)
31. Chorafas, D.N.: Agent Technology Handbook. McGraw-Hill, New York (1998)
32. Podobnik, V., Petric, A., Jezic, G.: An Agent-Based Solution for Dynamic Supply Chain Management. Journal of Universal Computer Science 14(7), 1080–1104 (2008)
33. Kusek, M., Lovrek, I., Sinkovic, V.: Agent Team Coordination in the Mobile Agent Network. In: Khosla, R., Howlett, R.J., Jain, L.C. (eds.) KES 2005. LNCS (LNAI), vol. 3681, pp. 240–246. Springer, Heidelberg (2005)
34. Trzec, K., Lovrek, I.: Field-Based Coordination of Mobile Intelligent Agents: An Evolutionary Game Theoretic Analysis. In: Apolloni, B., Howlett, R.J., Jain, L. (eds.) KES 2007, Part I. LNCS (LNAI), vol. 4692, pp. 198–205. Springer, Heidelberg (2007)
35. Weiser, M., Brown, J.S.: The Coming Age of Calm Technology. In: Dening, P.J., Metcalfe, R.M., Burke, J. (eds.) Beyond Calculation: The Next Fifty Years of Computing, Springer, New York (1997)
36. Bellavista, P., Corradi, A., Montanari, R., Tonin, A.: Context-Aware Semantic Discovery for Next Generation Mobile Systems. IEEE Communications 44(9), 62–71 (2006)
37. Leuf, B.: The Semantic Web: Crafting Infrastructure for Agency. Wiley, New York (2006)
38. Berners-Lee, T., Hendler, J., Lassila, O.: The Semantic Web. Scientific American 284(5), 34–43 (2001)
39. Moutinho, L.: Futurecast in Consumer (Mis)behaviour. In: 10th International Conference on Telecommunications (ConTEL), p. 5. IEEE, Zagreb (2009)
40. Mitchell, T.: Machine Learning. McGraw-Hill, New York (1997)
41. Russell, S., Norvig, P.: Artificial Intelligence: A Modern Approach. Prentice Hall, New Jersey (2002)
42. Kupper, A., Ahrens, S., Hess, T., Freese, B.: Superdistribution of digital content - overview, opportunities and challenges. In: ITI 5th International Conference on Information and Communications Technology (ICICT), pp. 173–179. IEEE, Cairo (2007)
43. Schmidt, A.U.: On the Superdistribution of Digital Goods. Journal of Universal Computer Science 15(2), 401–425 (2009)
44. Leskovec, J., Adamic, L.A., Huberman, B.A.: The dynamics of viral marketing. ACM Transactions on the Web 1(1), art. no. 5 (2007)

# Author Index